Optimizing Optimization

Quantitative Finance Series

Aims and Objectives

- Books based on the work of financial market practitioners and academics
- Presenting cutting-edge research to the professional/practitioner market
- Combining intellectual rigor and practical application
- Covering the interaction between mathematical theory and financial practice
- To improve portfolio performance, risk management and trading book performance
- Covering quantitative techniques

Market

Brokers/Traders; Actuaries; Consultants; Asset Managers; Fund Managers; Regulators; Central Bankers; Treasury Officials; Technical Analysis; and Academics for Masters in Finance and MBA market.

Series Titles

Economics for Financial Markets
Performance Measurement in Finance
Real R&D Options
Advance Trading Rules, Second Edition
Advances in Portfolio Construction and Implementation
Computational Finance
Linear Factor Models in Finance
Initial Public Offering
Funds of Hedge Funds
Venture Capital in Europe
Forecasting Volatility in the Financial Markets, Third Edition
International Mergers and Acquisitions Activity Since 1990
Corporate Governance and Regulatory Impact on Mergers and Acquisitions
Forecasting Expected Returns in the Financial Markets
The Analytics of Risk Model Validation
Computational Finance Using C++ and C#
Collectible Investments for the High Net Worth Investor

Series Editor

Dr. Stephen Satchell

Dr. Satchell is the Reader in Financial Econometrics at Trinity College, Cambridge and Visiting Professor at Birkbeck College, City University Business School and University of Technology, Sydney. He edits three journals: *Journal of Asset Management, Journal of Derivatives and Hedge Funds* and *Journal of Risk Model Validation.*

Optimizing Optimization
The Next Generation of Optimization Applications and Theory

Stephen Satchell

AMSTERDAM • BOSTON • HEIDELBERG • LONDON • NEW YORK
OXFORD • PARIS • SAN DIEGO • SAN FRANCISCO • SINGAPORE
SYDNEY • TOKYO
Academic Press is an imprint of Elsevier

Academic Press is an imprint of Elsevier
30 Corporate Drive, Suite 400, Burlington, MA 01803, USA
525 B Street, Suite 1900, San Diego, California 92101-4495, USA
84 Theobald's Road, London WC1X 8RR, UK

Library of Congress Cataloging-in-Publication Data

British Library Cataloguing-in-Publication Data
A catalogue record for this book is available from the British Library.

ISBN: 978-0-12-374952-9

For information on all Academic Press publications
visit our Web site at www.elsevierdirect.com

Printed in the United States of America

08 09 10 9 8 7 6 5 4 3 2 1

Contents

List of Contributors

Almira Biglova holds a PhD in mathematical and empirical finance and is currently a Risk Controller at DZ Bank International in Luxembourg. She held previous positions as a research assistant in the Department of Econometrics, Statistics, and Mathematical Finance at the University of Karlsruhe in Germany, an Assistant Professor in the Department of Mathematics and Cybernetics on the Faculty of Informatics at Technical University of Ufa in Russia, and a financial specialist in the Department of Financial Management at Moscow Credit Bank in Russia. She specializes in mathematical modeling and numerical methods. Dr. Biglova has published more than 30 research papers in finance.

Sebastian Ceria is CEO of Axioma, Inc. Before founding Axioma, Ceria was an Associate Professor of Decision, Risk, and Operations at Columbia Business School from 1993 to 1998. He was honored with the prestigious Career Award for Operations Research from the National Science Foundation, and recognized as the "best core teacher" by his students and the school. Ceria completed his PhD in operations research at Carnegie Mellon University's Graduate School of Industrial Administration.

Giuliano De Rossi is a quantitative analyst in the Equities Quantitative Research Group at UBS, which he joined in 2006. Between 2004 and 2006, he was a Fellow and Lecturer in economics at Christ's College, Cambridge University, and a part-time consultant to London-based hedge funds. His academic research focused on signal extraction and its application to financial time series data. Giuliano holds an MSc in economics from the London School of Economics and a PhD in economics from Cambridge University.

Frank J. Fabozzi is Professor in the Practice of Finance and Becton Fellow at the Yale School of Management. He is an Affiliated Professor at the University of Karlsruhe (Germany) Institut für Statistik, Ökonometrie und Mathematische Finanzwirtschaft (Institute of Statistics, Econometrics and Mathematical Finance). He earned a doctorate in economics from the City University of New York in 1972. In 2002, he was inducted into the Fixed Income Analysts Society's Hall of Fame and is the 2007 recipient of the C. Stewart Sheppard Award given by the CFA Institute. He has served as the editor of the *Journal of Portfolio Management* since 1985.

Hal Forsey earned a PhD in mathematics at the University of Southern California and a masters in operations research from U.C. Berkeley. He is Professor of Mathematics Emeritus at San Francisco State University and has worked closely with Dr. Frank Sortino for over 25 years. Dr. Forsey wrote the source code for all of the models developed at the Pension Research Institute, including the software on the CD in the book *Managing Downside Risk*. Hal co-authored many articles with Dr. Sortino and his wide consulting experience in applied mathematics has been invaluable to the SIA executive team.

Manfred Gilli is Professor emeritus at the Department of Econometrics at the University of Geneva, Switzerland, where he taught numerical methods in economics and finance. His main research interests include the numerical solution of large and sparse systems of equations, parallel computing, heuristic optimization techniques, and numerical methods for pricing financial instruments. He is a member of the advisory board of *Computational Statistics and Data Analysis* and member of the editorial boards of *Computational Economics* and the Springer book series on *Advances in Computational Economics* and *Advances in Computational Management Science*.

A. D. Hall holds a PhD in econometrics (London School of Economics, 1976). He has taught econometrics at the Australian National University and the University of California, San Diego (both recognized internationally as leading institutions for this discipline) and finance at the School of Business, Bond University and the University of Technology, Sydney. He has publications in a number of the leading international journals in economics and econometrics, including *The Review of Economics and Statistics*, *The Review of Economic Studies*, *The International Economic Review*, *The Journal of Econometrics*, *Econometric Theory*, *Econometric Reviews*, *The Journal of Business and Economic Statistics*, *Biometrika*, and *The Journal of Time Series Analysis*. His research interests cover all aspects of financial econometrics, with a special interest in modeling the term structure of interest rates. Dr. Hall is currently the Head of the School of Finance and Economics.

John Knight is Professor of Economics at the University of Western Ontario. His research interests are in theoretical and financial econometrics, areas in which he has published extensively as well as supervised numerous doctoral dissertations.

Fiona Kolbert is a member of the Research Group at SunGard APT where she has worked since 1998. She is the lead researcher and developer of their optimizer product.

Mark Kritzman is President and CEO of Windham Capital Management, LLC, and a Senior Partner of State Street Associates. He teaches financial engineering at MIT's Sloan School and serves on the boards of the Institute for Quantitative

Research in Finance and The Investment Fund for Foundations. Mr. Kritzman has written numerous articles and is the author of six books, including *Puzzles of Finance* and *The Portable Financial Analyst*. He has an MBA with distinction from New York University and a CFA designation.

Richard Louth is a PhD student at the University of Cambridge, a Teaching Fellow at the Faculty of Economics, and a Preceptor Corpus Christi College. He holds a first-class honors degree and an MPhil with distinction from said university. His research interests include portfolio optimization, dependence modeling, and forecasting methods.

Sergio Ortobelli Lozza is Associate Professor at "Lorenzo Mascheroni" Department of University of Bergamo (Italy), Laurea in Mathematics at University of Milan, Italy (1994), PhD in mathematical finance, University of Bergamo, Italy (1999), and Researcher (1999–2002) at the University of Calabria, Italy. His research activity is primarily focused on the application of different distributional approaches to portfolio theory, risk management, and option theory.

Francois Margot is an Associate Professor at the Tepper School of Business of Carnegie Mellon University and Senior Associate Researcher at Axioma. He has written numerous papers on mixed-integer linear and nonlinear optimization. He has a Mathematical Engineer Diploma and a PhD from the Ecole Polytechnique Federale of Lausanne (Switzerland), and prior to joining Axioma on leave from Carnegie Mellon University, he has held academic positions at the University of British Columbia, Michigan Technological University, and University of Kentucky, and was an Academic Visitor at the IBM T.J. Watson Research Center. He lives in Pittsburgh and Atlanta.

Tim Matthews has almost 20 years, experience working in the risk, performance, and quantitative analytics arenas, primarily within the European investment management community. His roles have included senior positions in business development, client service, and research and product development at firms including QUANTEC, Thomson Financial, and most recently BITA Risk Solutions. Recent roles include Director of Business Development at Thomson Financial responsible for the distribution and service of their Quantitative Analytics (TQA) platforms successfully establishing a significant footprint in the EMEA region. Tim holds masters degrees in applied statistics and stochastic modeling from Birkbeck, University of London, and in Electronic Engineering from Southampton University. He is an Associate (ASIP) of the CFA Society of the UK.

Svetlozar Rachev was a co-founder and President of BRAVO Risk Management Group, originator of the Cognity methodology, which was acquired by FinAnalytica, where he serves as Chief Scientist. Rachev holds Chair-Professorship in Statistics, Econometrics, and Mathematical Finance at University of Karlsruhe, and is the author of 12 books and over 300 published articles on finance, econometrics, statistics, and actuarial science. At University of California at Santa Barbara, he founded the PhD program in mathematical and empirical finance. Rachev holds PhD (1979) and Doctor of Science (1986) degrees from Moscow University and Russian Academy of Sciences. Rachev's scientific work lies at the core of Cognity's newer and more accurate methodologies in risk management and portfolio analysis.

Anthony Renshaw is Axioma Director of Applied Research. He has written numerous articles and white papers detailing real-world results of various portfolio construction strategies. He has an AB in applied mathematics from Harvard and a PhD in mechanical engineering from U.C. Berkeley. Prior to joining Axioma, he worked on Space Station Freedom, designed washing machines and X-ray tubes at General Electric's Corporate Research and Development Center, and taught mechanical engineering as an Associate Professor at Columbia University. He holds seven US patents and has over 30 refereed publications. He lives in New York City and Hawaii.

Daryl Roxburgh is Head of BITA Risk Solutions, the London- and New York-based portfolio construction and risk solutions provider. He specializes in portfolio construction and risk analysis solutions for the quantitative, institutional, and private banking markets. He is an avid antique car collector and has advised on this topic.

Stephen Satchell is a Fellow of Trinity College, Cambridge and Visiting Professor at Birkbeck College, City University Business School and University of Technology, Sydney. He provides consulting for a range of financial institutions in the broad area of quantitative finance. He has edited or authored over 20 books in finance. Dr. Satchell is the Editor of three journals: *Journal of Asset Management, Journal of Derivatives and Hedge Funds* and *Journal of Risk Model Validation*.

Anureet Saxena is an Associate Researcher at Axioma Inc. His research concerns mixed-integer linear and nonlinear programming. He holds a PhD and masters in industrial administration from Tepper School of Business, Carnegie Mellon University, and BTech in computer science and engineering from Indian Institute of Technology, Bombay. He was awarded the 2008 Gerald L. Thompson award for best dissertation in management science. He is the numero uno winner of the Egon Balas award for best student paper, gold medalist at a national level physics competition, and a national talent scholar. He has authored papers in leading mathematical programming journals and has delivered more than a dozen invited talks at various international conferences.

Bernd Scherer is Managing Director at Morgan Stanley Investment Management and Visiting Professor at Birkbeck College. Bernd published in the *Journal of Economics and Statistics*, *Journal of Money Banking and Finance*, *Financial Analysts Journal*, *Journal of Portfolio Management*, *Journal of Investment Management*, *Risk*, *Journal of Applied Corporate Finance*, *Journal of Asset Management*, etc. He received masters degrees in economics from the University of Augsburg and the University of London and a PhD from the University of Giessen.

Katja Scherer is a Quantitative Research Analyst at BITA Risk, where she is focusing on portfolio optimization strategies and quantitative consulting projects. Previously, she was a Client Relationship Manager at Deutsche Asset Management, Frankfurt. Mrs. Scherer holds a BA degree from the University of Hagen and an MSc degree in economics from the University of Witten/Herdecke, both Germany. She also received a Finance Certificate in the Computational Finance Program at Carnegie Mellon University, New York.

Enrico Schumann is a Research Fellow with the Marie Curie RTN "Computational Optimization Methods in Statistics, Econometrics and Finance" (COMISEF) and a PhD candidate at the University of Geneva, Switzerland. He holds a BA in economics and law and an MSc in economics from the University of Erfurt, Germany, and has worked for several years as a Risk Manager with a public sector bank in Germany. His academic and professional interests include financial optimization, in particular portfolio selection, and computationally intensive methods in statistics and econometrics.

Frank A. Sortino, Chairman and Chief Investment Officer, is Professor of Finance emeritus at San Francisco State University. He founded the Pension Research Institute in 1981 as a nonprofit organization, focusing on problems facing fiduciaries. When Dr. Sortino retired from teaching in 1997, the University authorized PRI's privatization as a for-profit think tank.

PRI, founded in 1981, has conducted research projects with firms such as Shell Oil Pension Funds, the Netherlands; Fortis, the Netherlands; Manulife, Toronto, Canada; Twentieth Century Funds, City & County of San Francisco Retirement System, Marin County Retirement System, and The California State Teachers Retirement System. The results of this research have been published in many leading financial journals and magazines.

Prior to teaching, Dr. Sortino was in the investment business for more than a decade and was a partner of a New York Stock Exchange firm and Senior VP of an investment advisory firm with over $1 billion in assets. He is known internationally for his published research on measuring and managing investment risk and the widely used Sortino ratio. Dr. Sortino serves on the Board of Directors of the Foundation for Fiduciary Studies and writes a quarterly analysis of mutual funds for *Pensions & Investments* magazine. Dr. Sortino earned an MBA from U.C. Berkeley and a PhD in finance from the University of Oregon.

Laurence Wormald is Head of Research at SunGard APT where he has worked since July 2008. Previously he had served as Chief Risk Officer for a business unit of Deutsche Bank in London, and before that as Research Director for several other providers of risk analytics to institutional investors. Laurence has also worked over the last 15 years as a consultant to the Bank for International Settlements, the European Central Bank, and the Bank of England. He is currently Chairman of the City Associates Board of the Centre for Computational Finance and Economic Agents (CCFEA) at Essex University.

Section One

Practitioners and Products

1 Robust portfolio optimization using second-order cone programming

Fiona Kolbert and Laurence Wormald

Executive Summary

Optimization maintains its importance within portfolio management, despite many criticisms of the Markowitz approach, because modern algorithmic approaches are able to provide solutions to much more wide-ranging optimization problems than the classical mean–variance case. By setting up problems with more general constraints and more flexible objective functions, investors can model investment realities in a way that was not available to the first generation of users of risk models.

In this chapter, we review the use of second-order cone programming to handle a number of economically important optimization problems involving:

- Alpha uncertainty
- Constraints on systematic and specific risks
- Fund of funds with multiple active risk constraints
- Constraints on risk using more than one risk model
- Combining different risk measures

1.1 Introduction

Despite an almost-continuous criticism of mathematical optimization as a method of constructing investment portfolios since it was first proposed, there are an ever-increasing number of practitioners of this method using it to manage more and more assets. Given the fact that the problems associated with the Markowitz approach are so well known and so widely acknowledged, why is it that portfolio optimization remains popular with well-informed investment professionals?

The answer lies in the fact that modern algorithmic approaches are able to provide solutions to much more wide-ranging optimization problems than the classical mean–variance case. By setting up problems with more general constraints and more flexible objective functions, investors can model investment realities in a way that was not available to the first generation of users of risk models.

In particular, the methods of cone programming allow efficient solutions to problems that involve more than one quadratic constraint, more than one

Doi:10.1016/B978-0-12-374952-9.00001-4.

quadratic term within the utility function, and more than one benchmark. In this way, investors can go about finding solutions that are robust against the failure of a number of simplifying assumptions that had previously been seen as fatally compromising the mean–variance optimization approach.

In this chapter, we consider a number of economically important optimization problems that can be solved efficiently by means of second-order cone programming (SOCP) techniques. In each case, we demonstrate by means of fully worked examples the intuitive improvement to the investor that can be obtained by making use of SOCP, and in doing so we hope to focus the discussion of the value of portfolio optimization where it should be on the proper definition of utility and the quality of the underlying alpha and risk models.

1.2 Alpha uncertainty

The standard mean–variance portfolio optimization approach assumes that the alphas are known and given by some vector $\boldsymbol{\alpha}$. The problem with this is that generally the alpha predictions are not known with certainty—an investor can estimate alphas but clearly cannot be certain that their predictions will be correct. However, when the alpha predictions are subsequently used in an optimization, the optimizer will treat the alphas as being certain and may choose a solution that places unjustified emphasis on those assets that have particularly large alpha predictions.

Attempts to compensate for this in the standard quadratic programming approach include just reducing alphas that look too large to give more conservative estimates and imposing constraints such as maximum asset weight and sector weight constraints to try and prevent any individual alpha estimate having too large an impact. However, none of these methods directly address the issue and these approaches can lead to suboptimal results. A better way of dealing with the problem is to use SOCP to include uncertainty information in the optimization process.

If the alphas are assumed to follow a normal distribution with mean α^* and known covariance matrix of estimation errors Ω, then we can define an elliptical confidence region around the mean estimated alphas as:

$$(\alpha - \alpha^*)^{\mathrm{T}}\Omega^{-1}(\alpha - \alpha^*) \le k^2$$

There are then several ways of setting up the robust optimization problem; the one we consider is to maximize the worst-case return for the given confidence region, subject to a constraint on the mean portfolio return, α_{p}. If $\mathbf{w}$ is the vector of portfolio weights, the problem is:

$$\text{Maximize } (\text{Min}(\mathbf{w}^{\mathrm{T}}\boldsymbol{\alpha}) - \text{portfolio variance})$$

subject to

$$(\alpha - \alpha^*)^T \Omega^{-1} (\alpha - \alpha^*) \leq k^2$$

$$\alpha^{*T} w = \alpha_p$$

$$e^T w = 1$$

$$w \geq 0$$

This can be written as an SOCP problem by introducing an extra variable, α_u (for more details on the derivation, see Scherer (2007)):

$$\text{Maximize } (w^T \alpha^* - k\alpha_u - \text{portfolio variance})$$

subject to

$$w^T \Omega w \leq \alpha_u^2$$

$$\alpha^{*T} w = \alpha_p$$

$$e^T w = 1$$

$$w \geq 0$$

Figure 1.1 shows the standard mean–variance frontier and the frontier generated including the alpha uncertainty term ("Alpha Uncertainty Frontier"). The example has a 500-asset universe and no benchmark and the mean portfolio alpha is constrained to various values between the mean portfolio alpha found for the minimum variance portfolio (assuming no alpha uncertainty) and 0.9. The size of the confidence region around the mean estimated alphas (i.e., the value of k) is increased as the constraint on the mean portfolio alpha is increased. The covariance matrix of estimation errors Ω is assumed to be the individual volatilities of the assets calculated using a SunGard APT risk model. The portfolio variance is also calculated using a SunGard APT risk model.

Some extensions to this, e.g., the use of a benchmark and active portfolio return, are straightforward.

The key questions to making practical use of alpha uncertainty are the specification of the covariance matrix of estimation errors Ω and the size of the confidence region around the mean estimated alphas (the value of k). This will depend on the alpha generation process used by the practitioner and, as for the alpha generation process, it is suggested that backtesting be used to aid in the choice of appropriate covariance matrices Ω and confidence region sizes k. From a practical point of view, for reasonably sized problems, it is helpful if the covariance matrix Ω is either diagonal or a factor model is used.

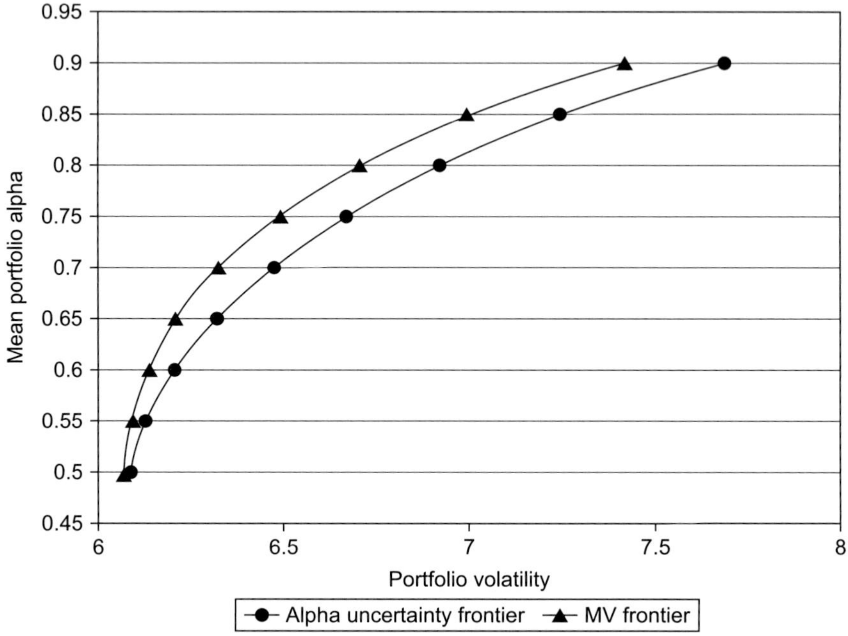

Figure 1.1 Alpha uncertainty efficient frontiers.

1.3 Constraints on systematic and specific risk

In most factor-based risk models, the risk of a portfolio can be split into a part coming from systematic sources and a part specific to the individual assets within the portfolio (the residual risk). In some cases, portfolio managers are willing to take on extra risk or sacrifice alpha in order to ensure that the systematic or specific risk is below a certain level.

A heuristic way of achieving a constraint on systematic risk in a standard quadratic programming problem format is to linearly constrain the portfolio factor loadings. This works well in the case where no systematic risk is the requirement, e.g., in some hedge funds that want to be market neutral, but is problematic in other cases because there is the question of how to split the systematic risk restrictions between the different factors. In a prespecified factor model, it may be possible to have some idea about how to constrain the risk on individual named factors, but it is generally not possible to know how to do this in a statistical factor model. This means that in most cases, it is necessary to use SOCP to impose a constraint on either the systematic or specific risk.

In the SunGard APT risk model, the portfolio variance can be written as:

$$w^T B^T B w + w^T \sum w$$

where

$w = n \times 1$ vector of portfolio weights
$B = c \times n$ matrix of component (factor) loadings
$\Sigma = n \times n$ diagonal matrix of specific (residual) variances

The systematic risk of the portfolio is then given by:

$$\text{Systematic risk of portfolio} = \sqrt{(w^T B^T B w)}$$

and the specific risk of the portfolio by:

$$\text{Specific risk of portfolio} = \sqrt{(w^T \Sigma w)}$$

The portfolio optimization problem with a constraint on the systematic risk (σ_{sys}) is then given by the SOCP problem:

$$\text{Minimize } (w^T B^T B w + w^T \Sigma w)$$

subject to

$$w^T B^T B w \leq \sigma_{sys}^2$$

$$\alpha^{*T} w = \alpha_p$$

$$e^T w = 1$$

$$w \geq 0$$

where

$\alpha^* = n \times 1$ vector of estimated asset alphas
α_p = portfolio return

One point to note on the implementation is that the $B^T B$ matrix is never calculated directly (this would be an $n \times n$ matrix, so could become very large when used in a realistic-sized problem). Instead, extra variables b_i are introduced, one per factor, and constrained to be equal to the portfolio factor loading:

$$b_i = (Bw)_i, i = 1 \cdots c$$

This then gives the following formulation for the above problem of constraining the systematic risk:

$$\text{Minimize} (b^T b + w^T \Sigma w)$$

subject to

$$b^T b \le \sigma_{sys}{}^2$$

$$\alpha^{*T} w = \alpha_p$$

$$e^T w = 1$$

$$b = Bw$$

$$w \ge 0$$

Similarly, the problem with a constraint on the specific risk (σ_{spe}) is given by:

$$\text{Minimize}(b^T b + w^T \textstyle\sum w)$$

subject to

$$w^T \textstyle\sum w \le \sigma_{spe}{}^2$$

$$\alpha^{*T} w = \alpha_p$$

$$e^T w = 1$$

$$b = Bw$$

$$w \ge 0$$

Figure 1.2 shows the standard mean–variance frontier and the frontiers generated with constraints on the specific risk of 2% and 3%, and on the systematic risk of 5%. The example has a 500-asset universe and no benchmark and the portfolio alpha is constrained to various values between the portfolio alpha found for the minimum variance portfolio and 0.9. (For the 5% constraint on the systematic risk, it was not possible to find a feasible solution with a portfolio alpha of 0.9.) Figure 1.3 shows the systematic portfolio volatilities and Figure 1.4 shows the specific portfolio volatilities for the same set of optimizations.

Constraints on systematic or specific volatility can be combined with the alpha uncertainty described in the previous section. The resulting frontiers can be seen in Figures 1.5–1.7 (the specific 3% constraint frontier is not shown because this coincides with the Alpha Uncertainty Frontier for all but the first point).

The shape of the specific risk frontier for the alpha uncertainty frontier (see Figure 1.7) is unusual. This is due to a combination of increasing the emphasis on the alpha uncertainty as the constraint on the mean portfolio alpha

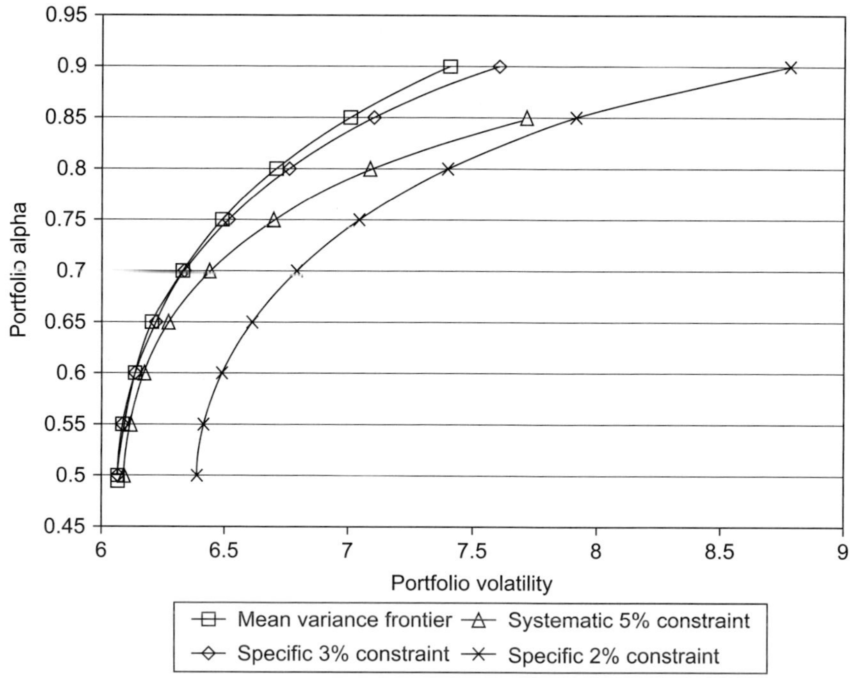

Figure 1.2 Portfolio volatility with constraints on systematic and specific risk

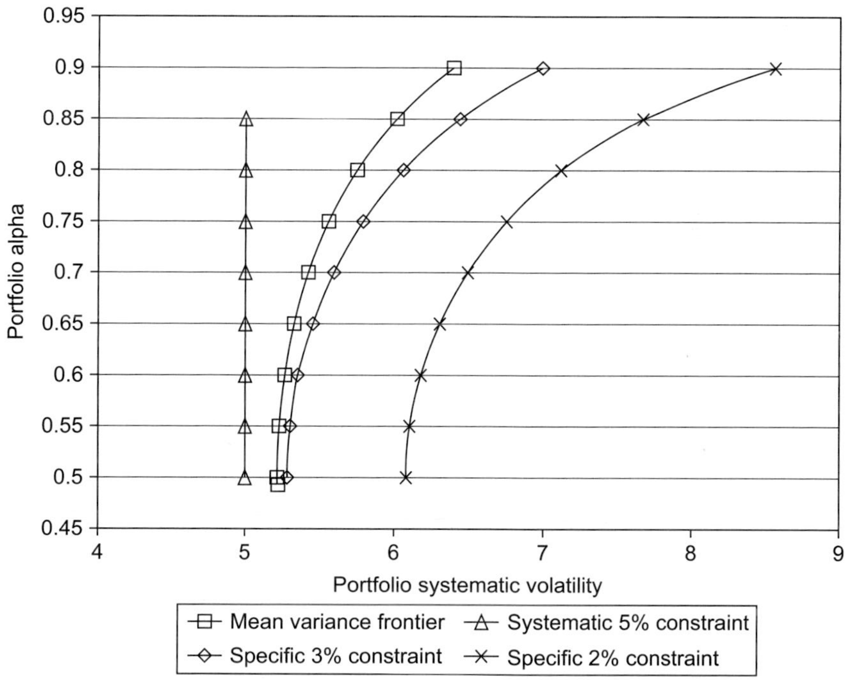

Figure 1.3 Portfolio systematic volatility with constraints on systematic and specific risk.

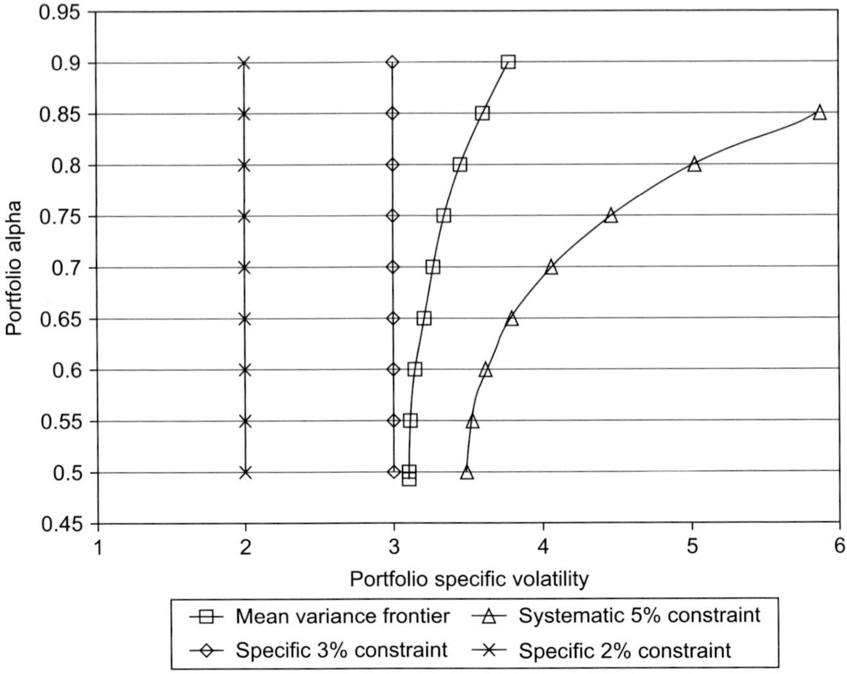

Figure 1.4 Portfolio specific volatility with constraints on systematic and specific risk.

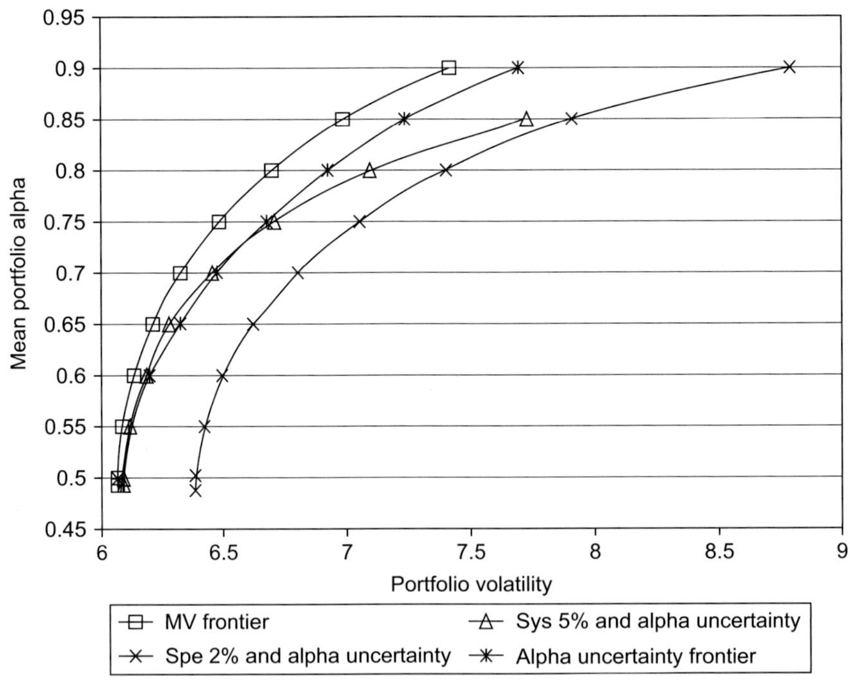

Figure 1.5 Portfolio volatility with alpha uncertainty and constraints on systematic and specific risk.

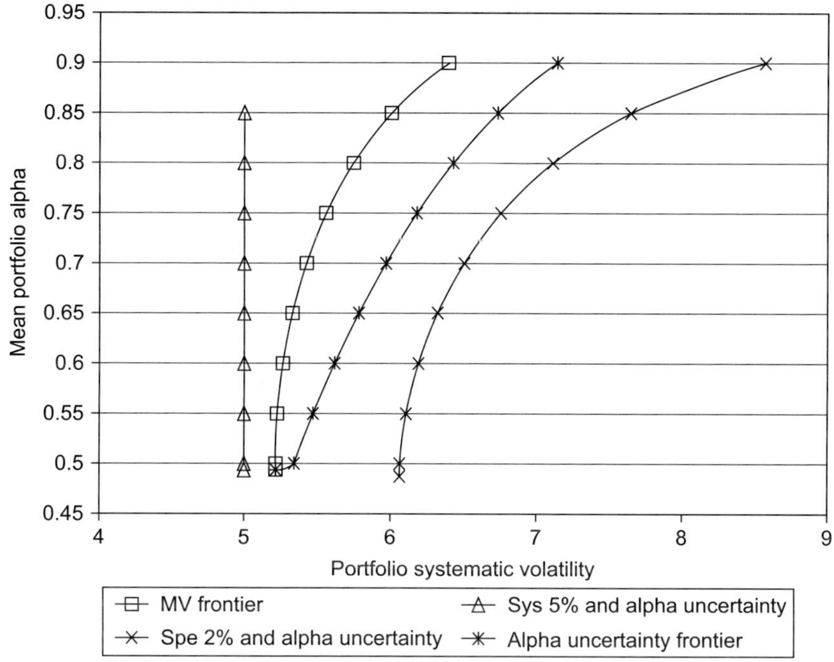

Figure 1.6 Portfolio systematic volatility with alpha uncertainty and constraints on systematic and specific risk.

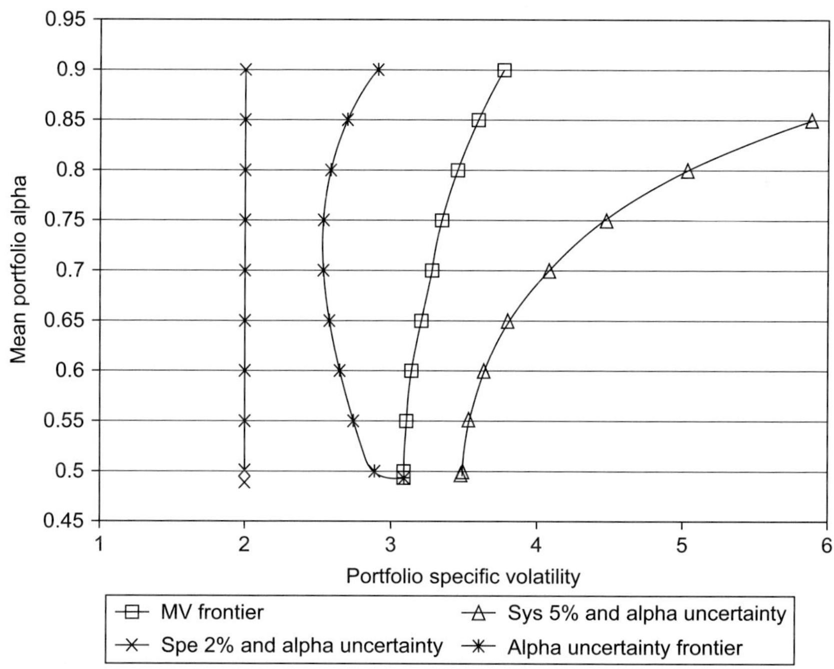

Figure 1.7 Portfolio specific volatility with alpha uncertainty and constraints on systematic and specific risk.

increases, and the choice of covariance matrix of estimation errors. In a typical mean–variance optimization, as the portfolio alpha increases, the specific risk would be expected to increase as the portfolio would tend to be concentrated in fewer assets that have high alphas. However, in the above alpha uncertainty example, because the emphasis increases on the alpha uncertainty term, and the covariance matrix of estimation errors is a matrix of individual asset volatilities, this tends to lead to a more diversified portfolio than in the pure mean–variance case. It should be noted that with a different choice of covariance matrix of estimation errors, or if the emphasis on the alpha uncertainty is kept constant, a more typical specific risk frontier may be seen.

Whilst the factors in the SunGard APT model are independent, it is straightforward to extend the above formulation to more general factor models, and to optimizing with a benchmark and constraints on active systematic and active specific risk.

1.4 Constraints on risk using more than one model

With the very volatile markets that have been seen recently, it is becoming increasingly common for managers to be interested in using more than one model to measure the risk of their portfolio.

In the SunGard APT case, the standard models produced are medium-term models with an investment horizon of between 3 weeks and 6 months. However, SunGard APT also produces short-term models with an investment horizon of less than 3 weeks. Some practitioners like to look at the risk figures from both types of model. Most commercial optimizers designed for portfolio optimization do not provide any way for them to combine the two models in one optimization so they might, for example, optimize using the medium-term model and then check that the risk prediction using the short-term model is acceptable. Ideally, they would like to combine both risk models in the optimization, for example, by using the medium-term model risk as the objective and then imposing a constraint on the short-term model risk. This constraint on the short-term model risk requires SOCP.

Other examples of possible combinations of risk models that may be used by practitioners are:

- SunGard APT Country and SunGard APT Region Models
- Risk models from two different vendors, or a risk model from a vendor alongside one produced internally
- Different types of risk model, e.g., a statistical factor model, one such as those produced by SunGard APT, and a prespecified factor model

One way of using both risk models in the optimization is to include them both in the objective function:

$$\text{Minimize } [x_1((\text{w} - \text{b})^T \text{B}_1{}^T \text{B}_1 (\text{w} - \text{b}) + (\text{w} - \text{b})^T \textstyle\sum_1 (\text{w} - \text{b})) \\ + x_2((\text{w} - \text{b})^T \text{B}_2{}^T \text{B}_2 (\text{w} - \text{b}) + (\text{w} - \text{b})^T \textstyle\sum_2 (\text{w} - \text{b}))]$$

subject to

$$\alpha^{*T} w = \alpha_p$$

$$e^T w = 1$$

$$w \leq w_{max}$$

$$w \geq 0$$

where

w	=	$n \times 1$ vector of portfolio weights
b	=	$n \times 1$ vector of benchmark weights
B_i	=	$c \times n$ matrix of component (factor) loadings for risk model i
Σ_i	=	$n \times n$ diagonal matrix of specific (residual) variances for risk model i
x_i	=	weight of risk model i in objective function ($x_i \geq 0$)
α^*	=	$n \times 1$ vector of estimated asset alphas
α_p	=	portfolio return
w_{max}	=	$n \times 1$ vector of maximum asset weights in the portfolio

This is a standard quadratic programming problem and does not include any second-order cone constraints but does require the user to make a decision about the relative weight (x_i) of the two risk terms in the objective function. This relative weighting may be less natural for the user than just imposing a tracking error constraint on the risk from one of the models. Figure 1.8 shows frontiers with tracking error measured using a SunGard APT medium-term model (United States August 2008) for portfolios created as follows:

- Optimizing using the medium-term model only
- Optimizing using the short-term model only
- Optimizing including the risk from both models in the objective function, with equal weighting on the two models

The same universe and benchmark has been used in all cases and they each contain 500 assets, and the portfolio alpha is constrained to values between 0.01 and 0.07.

Figure 1.9 shows the frontiers for the same set of optimizations with tracking errors measured using a SunGard APT short-term model (United States August 2008).

It can be seen from Figures 1.8 and 1.9 that optimizing using just one model results in relatively high tracking errors in the other model, but including terms from both risk models in the objective function results in frontiers for both models that are close to those generated when just optimizing with the individual model.

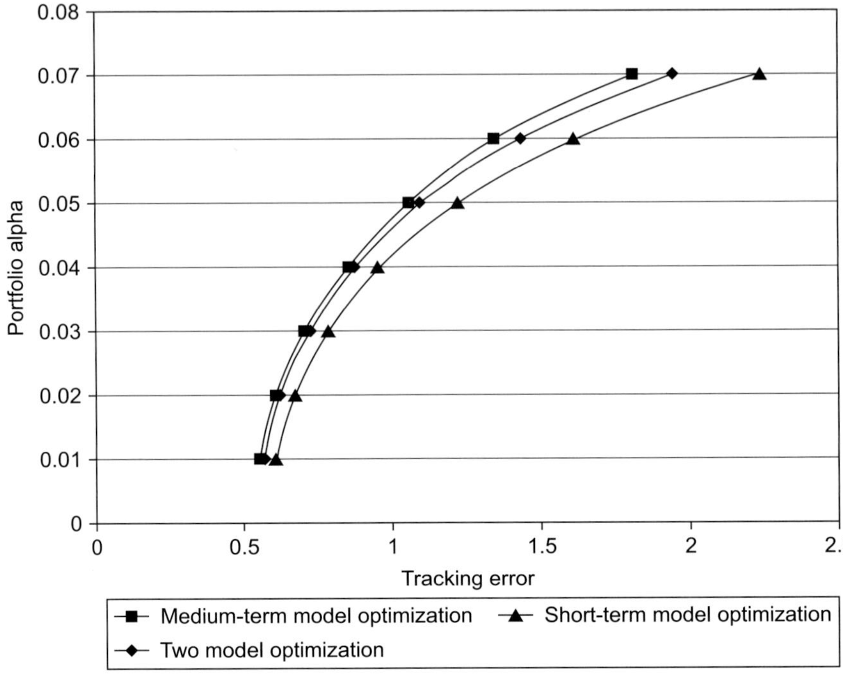

Figure 1.8 Tracking error measured using the SunGard APT medium-term model.

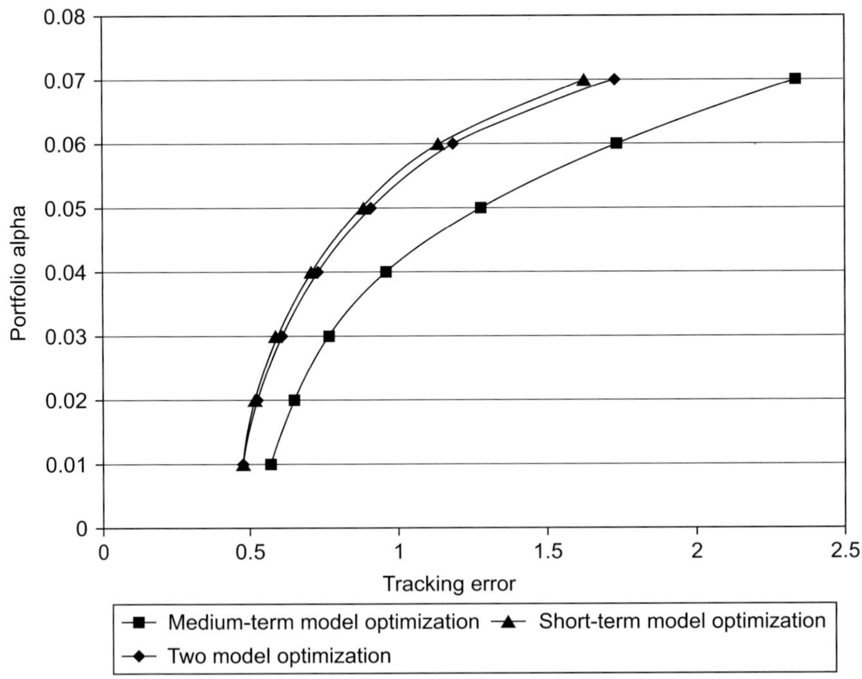

Figure 1.9 Tracking error measured using the SunGard APT short-term model.

Using SOCP, it is possible to include both risk models in the optimization by including the risk term from one in the objective function and constraining on the risk term from the other model:

$$\text{Minimize}[(w - b)^T B_1{}^T B_1 \ (w - b) + (w - b)^T \Sigma_1 (w - b)]$$

subject to

$$(w - b)^T B_2{}^T B_2 (w - b) + (w - b)^T \Sigma_2 (w - b) \le \sigma_{a2}{}^2$$

$$\alpha^{*T} \ w = \alpha_p$$

$$e^T w = 1$$

$$w \le w_{max}$$

$$w \ge 0$$

where σ_{a2} = maximum tracking error from the second risk model.

Figure 1.10 shows the effect of constraining on the risk from the short-term model, with an objective of minimizing the risk from the medium-term model, with a constraint on the portfolio alpha of 0.07. The tracking errors from just optimizing using one model without any constraint on the other model, and optimizing including the risk from both models in the objective function, are also shown for comparison.

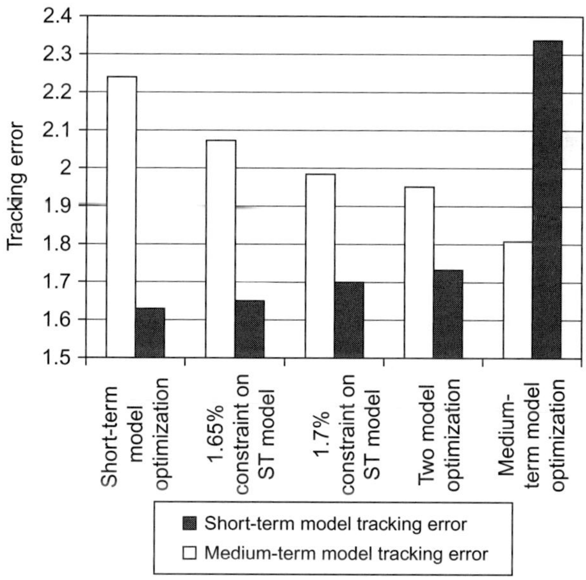

Figure 1.10 Tracking errors with constraints on the SunGard APT short-term model tracking error.

Whilst the discussion here has concerned using two SunGard APT risk models, it should be noted that it is trivial to extend the above to any number of risk models, and to more general risk factor models.

1.5 Combining different risk measures

In some cases, it may be desirable to optimize using one risk measure for the objective and to constrain on some other risk measures. For example, the objective might be to minimize tracking error against a benchmark whilst constraining the portfolio volatility. Another example could be where a pension fund manager or an institutional asset manager has an objective of minimizing tracking error against a market index, but also needs to constrain the tracking error against some internal model portfolio.

This can be achieved in a standard quadratic programming problem format by including both risk measures in the objective function and varying the relative emphasis on them until a solution satisfying the risk constraint is found. The main disadvantage of this is that it is time consuming to find a solution and is difficult to extend to the case where there is to be a constraint on more than one additional risk measure. A quicker, more general approach is to use SOCP to implement constraints on the risk measures.

The first case, minimizing tracking error, whilst constraining portfolio volatility, results in the following SOCP problem when using the SunGard APT risk model:

$$\text{Minimize}[(\mathrm{w}-\mathrm{b})^{\mathrm{T}}\mathrm{B}^{\mathrm{T}}\mathrm{B}(\mathrm{w}-\mathrm{b})+(\mathrm{w}-\mathrm{b})^{\mathrm{T}}\textstyle\sum(\mathrm{w}-\mathrm{b})]$$

subject to

$$\alpha^{*\mathrm{T}}\,\mathrm{w}=\alpha_{\mathrm{p}}$$

$$\mathrm{w}^{\mathrm{T}}\mathrm{B}^{\mathrm{T}}\mathrm{B}\mathrm{w}+\mathrm{w}^{\mathrm{T}}\textstyle\sum\mathrm{w}\leq\sigma^2$$

$$\mathrm{e}^{\mathrm{T}}\mathrm{w}=1$$

$$\mathrm{w}\leq\mathrm{w}_{\mathrm{max}}$$

$$\mathrm{w}\geq 0$$

where

$\mathrm{w} = n \times 1$ vector of portfolio weights

$\mathrm{b} = n \times 1$ vector of benchmark weights

$\mathrm{B} = c \times n$ matrix of component (factor) loadings

$\Sigma = n \times n$ diagonal matrix of specific (residual) variances

σ = maximum portfolio volatility

α^* = $n \times 1$ vector of estimated asset alphas

α_p = Portfolio return

w_{max} = $n \times 1$ vector of maximum asset weights in the portfolio

An example is given below where an optimization is first run without any constraint on the portfolio volatility, but with a constraint on the portfolio alpha. The optimization is then rerun several times with varying constraints on the portfolio volatility, and the same constraint on the portfolio alpha. The universe and benchmark both contain 500 assets. The resulting portfolio volatilities and tracking errors can be seen in Figure 1.11.

The second case, minimizing tracking error against one benchmark, whilst constraining tracking error against some other benchmark, results in the following SOCP problem when using the SunGard APT risk model:

$$\text{Minimize}[(w - b_1)^T B^T B(w - b_1) + (w - b_1)^T \Sigma(w - b_1)]$$

subject to

$$\alpha^{*T} w = \alpha_p$$

$$(w - b_2)^T B^T B(w - b_2) + (w - b_2)^T \Sigma(w - b_2) \leq \sigma_{a2}^2$$

$$e^T w = 1$$

$$w \leq w_{max}$$

$$\mathbf{w} \geq 0$$

where

b_1 = $n \times 1$ vector of weights for benchmark used in objective function

b_2 = $n \times 1$ vector of weights for benchmark used in constraint

σ_{a2} = maximum tracking error against second benchmark

An example of this case is given below where an optimization is first run without any constraint on the tracking error against the internal model portfolio, but with a constraint on the portfolio alpha, minimizing the tracking error against a market index. The optimization is then rerun several times with varying constraints on the tracking error against the internal model portfolio, and the same constraint on the portfolio alpha. The universe and benchmark both contain 500 assets. The resulting tracking errors against both the market index and the internal model portfolio can be seen in Figure 1.12.

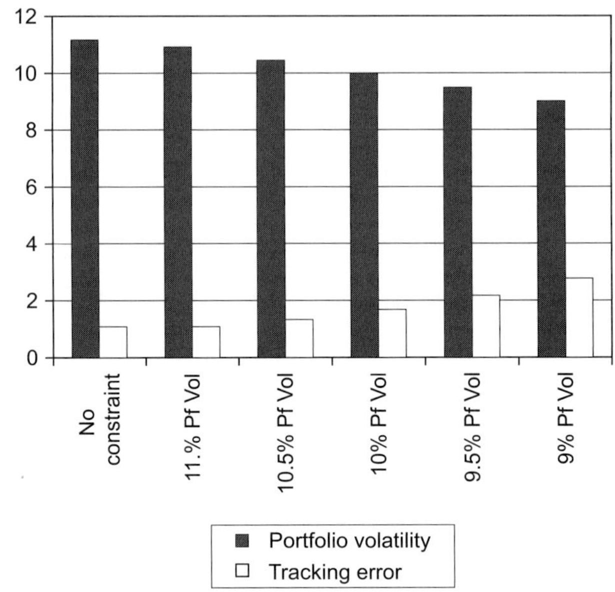

Figure 1.11 Risk with portfolio volatility constrained.

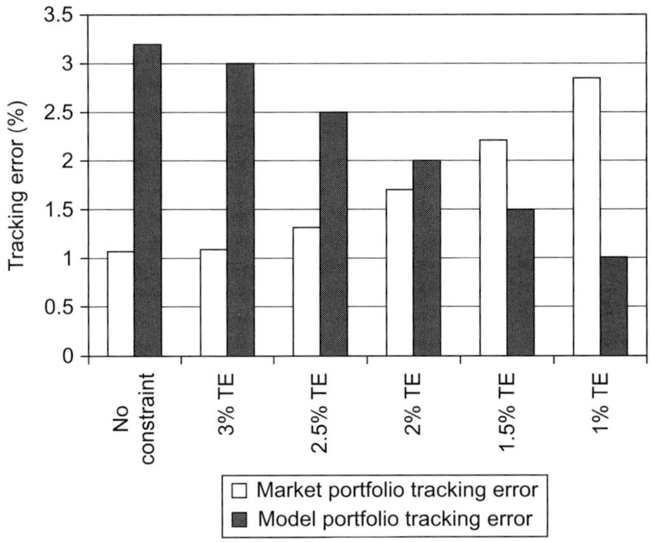

Figure 1.12 Risk with tracking error constrained against a model portfolio.

1.6 Fund of funds

An organization might want to control the risk of all their funds against one benchmark, but give fund managers different mandates with different benchmarks and risk restrictions. If the managers each individually optimize their own fund against their own benchmark, then it can be difficult to control the overall risk for the organization. From the overall management point of view, it would be better if the funds could be optimized together, taking into account the overall benchmark. One way to do this is to use SOCP to impose the tracking error constraints on the individual funds, and optimize with an objective of minimizing the tracking error of the combined funds against the overall benchmark, with constraints on the minimum alpha for each of the funds. Using the SunGard APT risk model, this results in the following SOCP problem:

$$\text{Minimize } (w_c - b_c)^T B^T B(w_c - b_c) + (w_c - b_c)^T \Sigma(w_c - b_c)$$

subject to

$$w_c = \Sigma_i f_i w_i, \Sigma_i f_i = 1, f_i \geq 0$$

$$(w_i - b_i)^T B^T B(w_i - b_i) + (w_i - b_i)^T \Sigma(w_i - b_i) \leq \sigma_{ai}{}^2, i = 1 \cdots m$$

$$e^T w_i = 1, i = 1 \cdots m$$

$$w_i \geq 0, w_i \leq \max_i, i = 1 \cdots m$$

$$\alpha_i^{*T} w_i \geq \alpha_{pi}, i = 1 \cdots m$$

where

m = number of funds

w_i = $n \times 1$ vector of portfolio weights for fund i

b_i = $n \times 1$ vector of benchmark weights for fund i

w_c = $n \times 1$ vector of weights for overall (combined) portfolio

f_i = weight of fund i in overall (combined) portfolio

b_c = $n \times 1$ vector of overall benchmark weights

B = $c \times n$ matrix of component (factor) loadings

Σ = $n \times n$ diagonal matrix of specific (residual) variances

σ_{ai} = maximum tracking error for fund i

$\max_i$ = $n \times 1$ vector of maximum weights for fund i

α^*_i = $n \times 1$ vector of assets alphas for fund i

α_{pi} = minimum portfolio alpha for fund i

In the example given below, we have two funds, and the target alpha for both funds is 5%. The funds are equally weighted to give the overall portfolio. Figure 1.13 shows the tracking error of the combined portfolio and each of the funds against their respective benchmarks where the funds have been optimized individually.

In this case, the tracking error against the overall benchmark is much larger than the tracking errors for the individual funds against their own benchmarks.

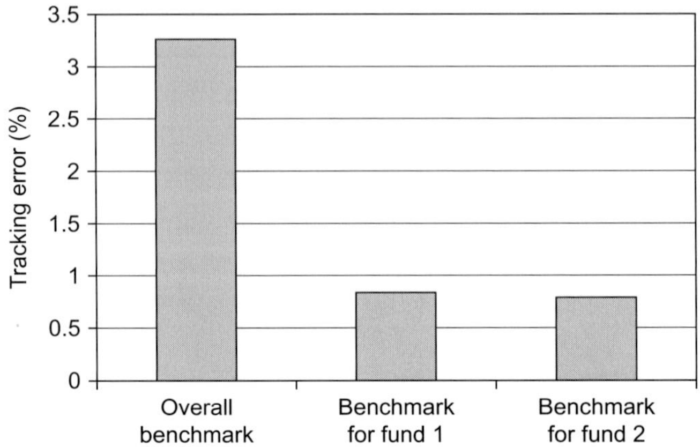

Figure 1.13 Tracking errors when optimizing funds individually.

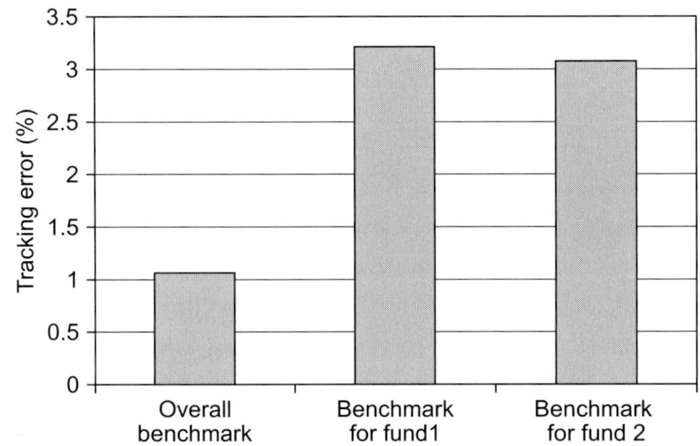

Figure 1.14 Tracking errors when optimizing funds together.

This sort of situation would arise when the overall benchmark and the individual fund benchmarks are very different, e.g., in the case where the overall benchmark is a market index and the individual funds are a sector fund and a value fund. It is unlikely to occur when both the overall and individual fund benchmarks are very similar, for instance, when they are all market indexes.

Figure 1.14 shows the tracking errors when the combined fund is optimized with the objective of minimizing tracking error against the combined benchmark, subject to the constraints on alpha for each of the funds, but without the constraints on the individual fund tracking errors.

Figure 1.15 shows the results of optimizing including the SOCP constraints on the tracking errors for the individual funds.

From the organization's perspective, using SOCP to constrain the individual fund tracking errors whilst minimizing the overall fund tracking error should achieve

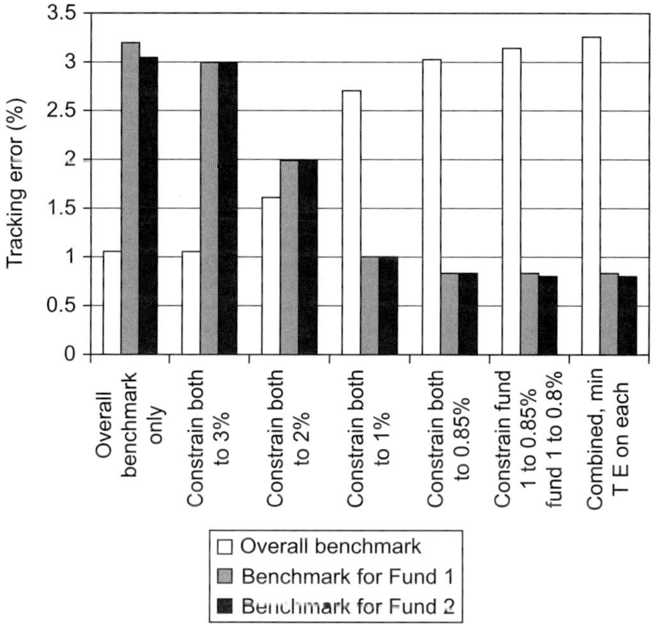

Figure 1.15 Tracking errors with constraints on risk for each fund.

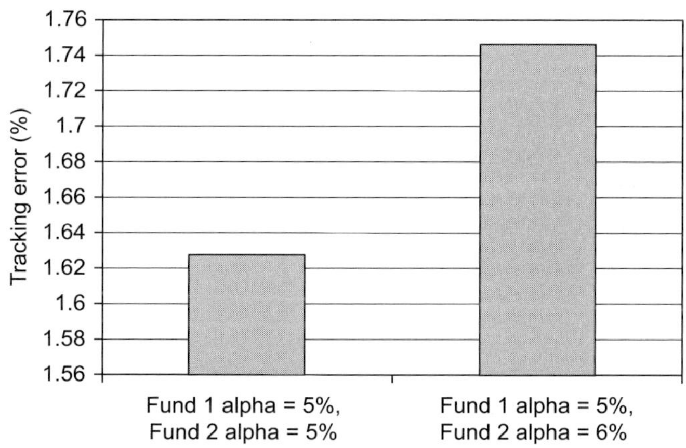

Figure 1.16 Tracking error with different alpha constraints on Fund 2.

their goal. However, there is a question as to whether this is a fair method of optimization from the point of view of the individual managers. Suppose that instead of both managers in the above example having a minimum portfolio alpha requirement of 5%, one of the managers decides to target a minimum portfolio alpha of 6%. If they are still both constrained to have a maximum individual tracking error against their own benchmark of 2%, it can be seen from Figure 1.16 that the tracking error for the overall fund against the overall benchmark will increase.

The organization might decide that this new tracking error against the overall benchmark is too high and, to solve this problem, will impose lower tracking error restrictions on the individual funds. This could be considered to be unfairly penalizing the first fund manager as the reason the overall tracking error is now too high is because of the decision by the second manager to increase their minimum portfolio alpha constraint. It is tricky to manage this issue and it may be that the organization will need to consider the risk and return characteristics of the individual portfolios generated by separate optimizations on each of the funds both before setting individual tracking error constraints, and after the combined optimization has been run to check that they appear fair.

1.7 Conclusion

SOCP provides powerful additional solution methods that extend the scope of portfolio optimization beyond the simple mean–variance utility function with linear and mixed integer constraints. By considering a number of economically important example problems, we have shown how SOCP approaches allow the investor to deal with some of the complexities of real-world investment problems. A great advantage in having efficient methods available to generate these solutions is that the investor's intuition can be tested and extended as the underlying utility or the investment constraints are varied.

Ultimately, it is not the method of solving an optimization problem that is critical—rather it is the ability to comprehend and set out clearly the economic justification for framing an investment decision in terms of a trade-off of risk, reward and cost with a particular form of the utility function and a special set of constraints. There are many aspects of risky markets behavior that have not been considered here, notably relating to downside and pure tail risk measures, but we hope that an appreciation of the solution techniques discussed in this chapter will lead to a more convincing justification for the entire enterprise of portfolio optimization, as the necessary rethinking of real-world utilities and constraints is undertaken.

References

Alizadeh, F., & Goldfarb, D. (2003). Second-order cone programming. *Mathematical Programming*, *95*, 3–51.

Fabozzi, F. J., Focardi, S. M., & Kolm, P. N. (2006). *Financial modelling of the equity market*. Hoboken: John Wiley & Sons.

Lobo, M., Vandenberghe, L., Boyd, S., & Lebret, H. (1998). Applications of second-order cone programming. *Linear algebra and its applications*, *284*, 193–228.

Scherer, B. (2007). Can robust portfolio optimisation help to build better portfolios? *Journal of Asset Management*, *7*, 374–387.

2 Novel approaches to portfolio construction: multiple risk models and multisolution generation

Sebastian Ceria, Francois Margot, Anthony Renshaw and Anureet Saxena

Executive Summary

This chapter highlights several novel methods for portfolio optimization. One approach involves using more than one risk model and a systematic calibration procedure is described for incorporating more than one risk model in a portfolio construction strategy. The addition of a second risk model can lead to better overall performance than one risk model alone provided that the strategy is calibrated so that both risk models affect the optimal portfolio solution. In addition, the resulting portfolio is no more conservative than the portfolio obtained with one risk model alone.

The second approach addresses the issue of generating multiple interesting solutions to the portfolio optimization problem. It borrows the concept of "elasticity" from Economics, and adapts it within the framework of portfolio optimization to evaluate the relative significance of various constraints in the strategy. By examining heatmaps of portfolio characteristics derived by perturbing constraints with commensurable elasticities, it offers insights into trade-offs associated with modifying constraint bounds. Not only do these techniques assist in enhancing our understanding of the terrain of optimal portfolios, they also offer the unique opportunity to visualize trade-offs associated with mathematically intractable metrics such as transfer coefficient. A carefully designed case study elucidates the practical utility of these techniques in generating multiple interesting solutions to portfolio optimization.

These methods are representative of Axioma's new approach to portfolio construction that creates real-world value in optimized portfolios.

2.1 Introduction

Optimization as an applied discipline has a history spanning more than six decades. Although highly successful in a variety of real-world applications such as supply chain management, manufacturing, scheduling, etc., the penetration

Doi:10.1016/B978-0-12-374952-9.00002-6.

of optimization in the financial world has been less than remarkable. While the practice of using optimization-based techniques in constructing portfolios has gained wide acceptance in the past decade, it has still not reached its full potential. The community of portfolio managers (PMs) continues to perceive optimizers as black box tools that only partly capture the complexities of real-world portfolios. We believe that there are two factors that have hindered the growth of financial optimization.

First, unlike other application areas, uncertainty has a focal position in the world of financial optimization. Risk management, the practice of managing and controlling the inherent stochastic nature of portfolio returns, is an indispensable component of any quantitative model. It entails complex interaction between statistics and optimization, and is often encumbered with problems ranging from parameter misestimation to overfitted models. With the emergence of global markets, fading international trading boundaries, and current economic turbulence, the risk model business itself has undergone a paradigm shift. A contemporary PM trading in stocks on NYSE to Nikkei needs access to a robust risk model that can provide multiple views of risk, and can quickly adapt to the ever-changing financial landscape. These factors make software design for financial optimization extremely challenging.

Second, constructing an optimal portfolio is only one aspect of the practice of portfolio management. An optimized portfolio often undergoes a series of refinements and alterations each of which addresses a practical concern that was not captured by the underlying mathematical model. As appealing as it may seem, a single optimal portfolio hardly provides any insight to complement the tremendous amount of expertise it takes to mutate the output of a black box solver into a practically tradable portfolio. For instance, it offers little insight into the manner in which the constraints present in the strategy affect the choice of the optimal portfolio. While some of the constraints in a strategy, such as budget constraint, are mandatory, several others just represent tentative guidelines that the PM is expected to follow. A single optimal portfolio restricts the PM from exploring portfolios that violate some of these tentative constraints but have other extremely desirable characteristics such as expected return, transfer coefficient, implied beta, etc. To overcome these shortcomings, software providers need to break from the image of being black box optimizers and design flexible products that assist the PM in portfolio design from inception through trade execution.

As a leading provider of portfolio optimization tools, Axioma has undertaken several initiatives to address these concerns. Our product suite employs ideas from robust optimization to tackle parameter misestimation problems, thereby giving rise to portfolios that are less sensitive to estimation errors. Our "Robust" risk models are updated on a daily basis to reflect the latest changes in the financial markets across the globe. Our modeling environment allows PMs to incorporate multiple risk models, thereby giving them a better risk assessment. This chapter offers a snapshot of the latest research developments at Axioma that further enhance our capacity to address challenging problems in financial optimization. The rest of the chapter is organized as follows.

In Section 2.2, we present a systematic calibration procedure for incorporating more than one risk model in a portfolio construction strategy. The addition of a second risk model can lead to better overall performance than one risk model alone provided that the strategy is calibrated so that both risk models affect the optimal portfolio solution. Our computational results illustrate that there is a substantial, synergistic benefit in using multiple risk models in a rebalancing.

In Section 2.3, we address the issue of generating multiple interesting solutions to the portfolio optimization problem. We borrow the concept of "elasticity" from Economics, and adapt it within the framework of portfolio optimization to evaluate the relative significance of various constraints in the strategy. We show that examining heatmaps of portfolio characteristics derived by perturbing constraints with commensurable elasticities can offer crucial insights into trade-offs associated with modifying constraint bounds. Not only do these techniques assist in enhancing our understanding of the terrain of optimal portfolios, they also offer the unique opportunity to visualize trade-offs associated with mathematically intractable metrics such as transfer coefficient. The section concludes with a carefully designed case study to highlight the practical utility of these techniques in generating multiple interesting solutions to portfolio optimization.

2.2 Portfolio construction using multiple risk models

The question addressed in this section is how best to incorporate a second risk model into an existing portfolio construction strategy that already utilizes a primary risk model. The primary and secondary risk models could be fundamental factor risk models, statistical factor risk models, dense, asset–asset covariance matrices computed from the historical time series of asset returns, or any combination of these. The second risk model could also be simply a diagonal specific variance matrix, in which case the second risk prediction may be difficult to compare with the primary risk prediction. How do we determine if the second risk model is beneficial, superfluous, or deleterious? Should the second risk model be incorporated into the portfolio construction strategy at all, and if so, how should the strategy parameters be calibrated (or recalibrated, in the case of the existing strategy parameters) to best take advantage of the second risk model?

Several authors have argued that one of the contributors to the poor performance of quantitative hedge funds in August 2007 was that many quantitative managers use the same commercially available risk models (Ceria, 2007; Foley, 2008). The use of such a small number of similar risk models may have led to these managers making similar trades when deleveraging their portfolios during this period. Using more than one risk model may be helpful in ameliorating this problem since the second risk model diversifies portfolio positions.

When the two risk models are comparable, simple "averaging" approaches are possible. The PM could explicitly average the risk predictions of both models and construct a portfolio whose risk target was limited by the average model prediction.

If the portfolio's tracking error limit is defined explicitly by the primary risk model, then averaging methods are not attractive. In these cases, the second risk model must be incorporated into the portfolio construction strategy by either adding it to the objective term (either as risk or variance) or by adding a new risk constraint using the second model. The obvious question that arises is calibration. When the second risk model is added to the objective, a risk aversion constant must be selected; when it is added as a risk constraint, a risk limit must be chosen.

Here, we incorporate the second risk model into the portfolio construction strategy as a second, independent risk limit constraint and propose a calibration strategy that ensures that both risk model constraints (primary and secondary) are binding for the optimized portfolio. If the risk limits are not calibrated in this fashion, then it is possible (and often likely) that only one risk model constraint, the more conservative one, will be binding for the optimal portfolio solution. The other risk model will be superfluous. Although one can imagine scenarios where this might be desirable—say, over the course of a backtest where the most conservative solution is desired—the calibration procedure described here is superior in at least three respects. First, the procedure enables a PM to ensure his or her intended outcome regardless of whether that intention is to have one, both, or neither risk models binding. Second, as illustrated in this chapter, in many cases there is substantial, synergistic benefit when both risk models are simultaneously binding. And third, we explicitly avoid overly conservative solutions. In fact, throughout this chapter, we only consider portfolios that are just as conservative (i.e., have identical risk) as the portfolios obtained using one risk model alone. More conservative solutions are not considered.

The second risk constraint interacts with both the primary risk constraint and with the other constraints in the portfolio construction strategy. As shown by the results reported here, properly calibrating a portfolio construction strategy with multiple risk models often leads to superior portfolio performance over using any one single risk model alone.

Here, we performed calibrations for three different, secondary risk model constraints from October 31, 2005 to October 31, 2006. We then test the performance of the calibrated portfolio construction strategies out-of-sample from October 31, 2006 to October 31, 2007. The primary risk model in each case is Axioma's Japanese daily, fundamental factor model. For the first example, the second risk model is Axioma's Japanese daily, statistical factor model that is used to constrain active risk. Hence, in this example, there are two different, but comparable risk models in the portfolio construction strategy. In the second example, the Axioma Japanese statistical factor model is used as the second risk model but to constrain total portfolio risk instead of active risk. In the third example, the second risk model is the active specific risk prediction from the Japanese fundamental model. All three examples lead to superior portfolio performance both in- and out-of-sample tests. The results of the calibration procedure illustrate the parameter regions where both risk models are binding, and give guidance on how to adjust the other portfolio parameters in the portfolio construction strategy.

Example 1. Constraining active risk with fundamental and statistical risk models

In our first example, we calibrate a simple portfolio construction strategy using Axioma's two Japanese risk models. The primary risk model is the fundamental factor model. The second risk model is the statistical factor model. We take the largest 1000 assets in the TOPIX exchange as our universe and benchmark. This is a broad-market benchmark that is similar to (but not identical to) the TOPIX 1000 index. We rebalance the portfolio monthly from October 31, 2005 to October 31, 2006. This consists of only 12 rebalancings, which is small but nevertheless illustrates the second risk model calibration procedure.

The portfolio construction strategy maximizes expected, active return. Our naïve expected return estimates are the product of the assets' factor exposures in Axioma's Japanese fundamental factor risk model and the factor returns over the 20 days prior to rebalancing. These expected return estimates are used for illustration only and are not expected to be particularly accurate.

At each rebalancing, we impose the following constraints, parameterized using the three variables *X*, TO, and *Y*:

- Long-only holdings
- Maximum tracking error[1] of 4% (the primary risk model constraint)
- Active asset holding bounds of ±*X*%
- Maximum, monthly, round-trip portfolio turnover of TO%
- Maximum active statistical risk model risk of *Y*% (the second risk model constraint)

The portfolio starts from an all-cash position, so the turnover constraint is not applied in the first rebalancing.

With *X* and TO fixed, at each rebalancing, we determine the range of *Y* over which both risk constraints are binding. We then compute various statistics such as the number of assets held and the realized portfolio return for the optimal portfolios within this range.

Figure 2.1 shows results for TO = 30%. The horizontal axis gives the values for the maximum asset bound constraint (*X*), and the vertical axis gives the level of active risk (*Y*) from the statistical factor model. The average number of assets held is shown by the black-and-white contour plot. The average number of assets held varies from 112 to 175. The white regions in the figure indicate regions in which at least one risk model is not binding. The white region above the contour plot represents solutions in which the statistical factor model's risk constraint is sufficiently large that it is not binding and therefore superfluous. The optimal solutions in this region are identical to those on the upper edge of the contour plot. The white region below the contour plot is the region in

[1] Tracking error is the predicted active risk with respect to the market-cap weighted universe.

which the statistical factor model's risk constraint is sufficiently tight that the tracking error constraint from the primary, fundamental factor model is no longer binding. In this region, the primary tracking error constraint is irrelevant, and the portfolios are more conservative than indicated.

Even though both risk models are comparable, the shaded region lies entirely below 4% of active risk. If we had naively created a portfolio construction strategy in which both risk models were limited to 4%, then the second risk model would have been superfluous (nonbinding) in all optimized solutions.

Although the results may look like an efficient frontier, they are not. The entire shaded region corresponds to optimal portfolios, not just those on its boundaries. The results are the averages over the 12 rebalancings, so the boundaries for any particular rebalancing may be somewhat different than those shown in the figure. We have used a 12-month calibration period to limit these differences. The maximum tracking error of 4% is shown by the dashed line.

The shaded contours in Figure 2.1 form two distinct regions. For a max asset holdings bound (X) of less than 3%, the tighter asset bounds increase the number of assets held. In addition, tighter statistical factor model risk constraints also increase the number of assets held. For a max asset holdings bound (X) greater than 3%, more assets are held by either tightening the statistical

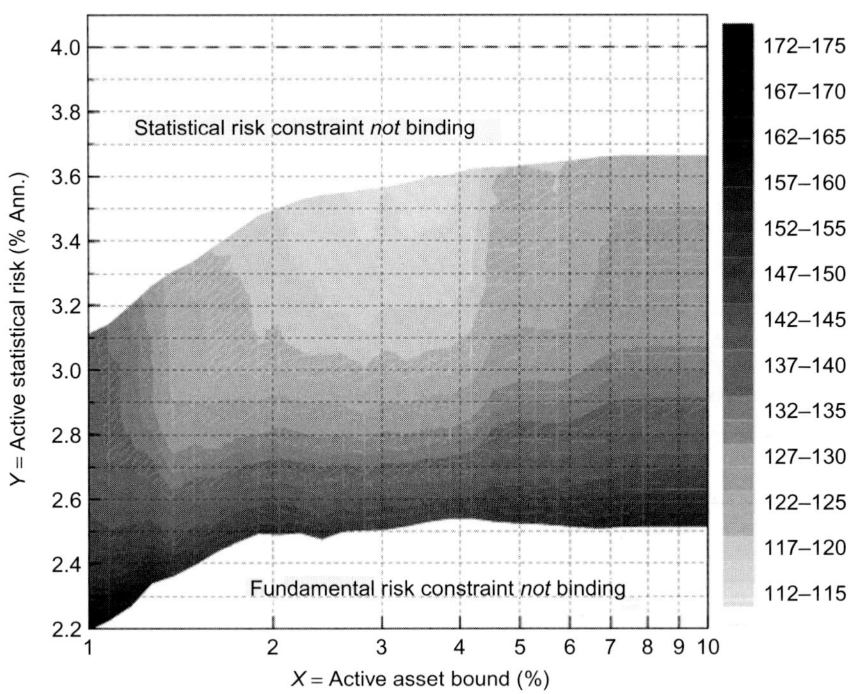

Figure 2.1 The narrow, shaded contour gives the average number of asset held when both risk models are binding. TO = 30%.

factor model risk constraint or by raising the max asset bound (X) constraint. The most diversified portfolios are obtained by doing both simultaneously.

Figures 2.2 and 2.3 show contours of the cumulative (in this case, annual), active return and the single, worst, monthly active return over the course of the 12 rebalancings. Neither of these results includes transaction costs or market impact.

Adding the Japanese statistical factor model risk constraint generally increases the portfolio return and improves the worst monthly return. The strategy parameters of $X = 1.0\%$ and $Y = 2.3\%$ produce a large cumulative return and the smallest worst monthly return. This solution performs better than any of the solutions without the secondary statistical factor model risk constraint (the white region above the shaded regions). These parameter values are the suggested strategy calibration for TO = 30%.

Figures 2.4–2.6 below show the same calibration results for TO = 15%, 30%, and 60%. The figures show that TO has a profound effect on the portfolios obtained: tighter TO constraints narrow the region over which both risk models are simultaneously binding. As a practical matter, for strategies similar to TO = 15%, it may be difficult to keep a strategy in the dual-binding region unless it is calibrated explicitly. (The shaded regions shown here are averages over all 12 rebalancings.)

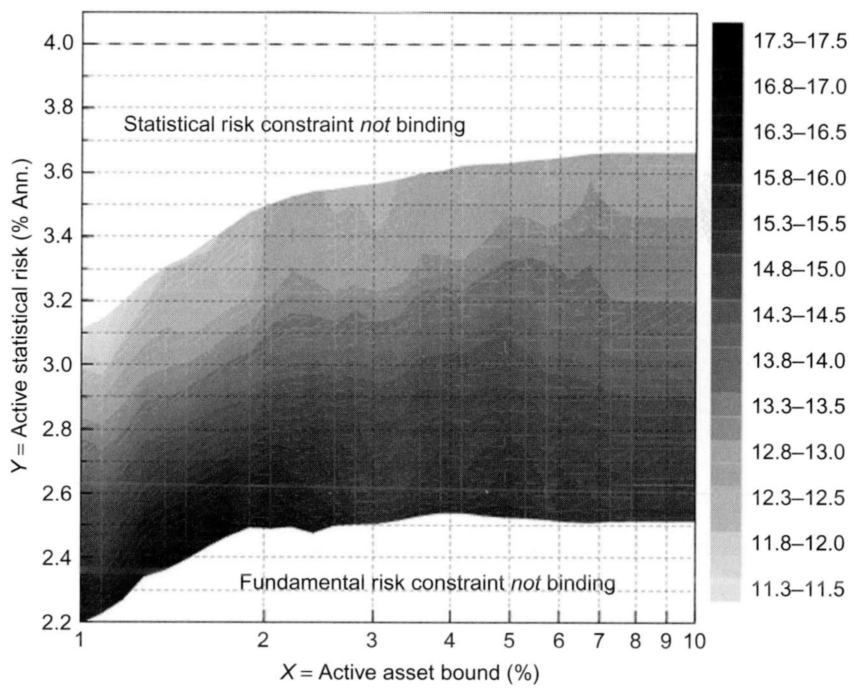

Figure 2.2 Cumulative, active return (%) when both risk models are binding. TO = 30%.

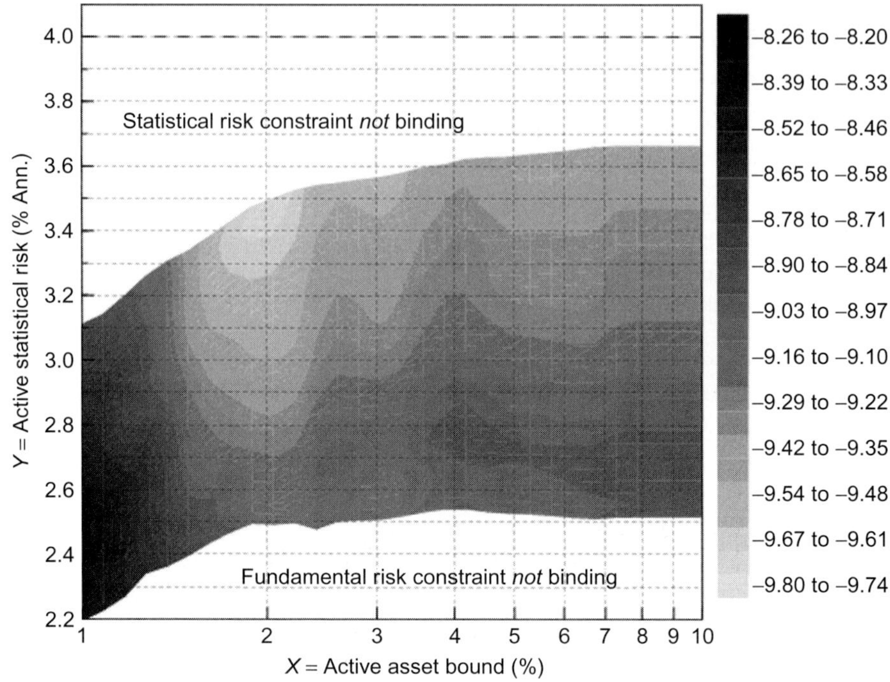

Figure 2.3 The worst, monthly, active return (%) when both risk models are binding. TO = 30%.

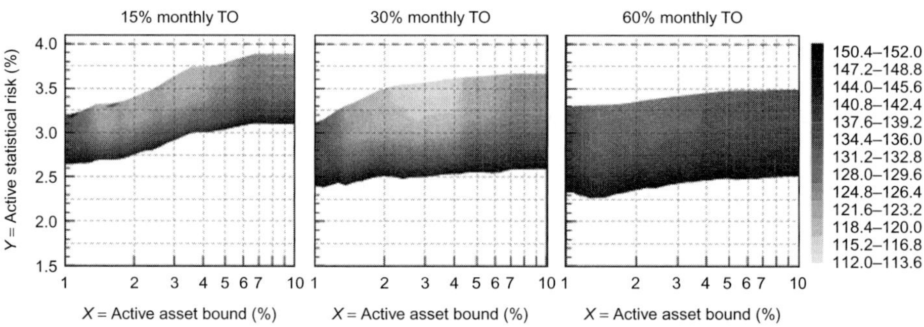

Figure 2.4 The number of asset held versus asset bound and the statistical risk model risk constraint for TO = 15%, 30%, and 60%.

The case with TO = 30% is shown here in the same way as previously shown in Figures 2.1–2.3, and is shown in order to facilitate comparison with the other TO cases.

In all three TO cases, portfolio performance is improved by adding the second risk model constraint and making it as tight as possible without making the primary risk constraint nonbinding.

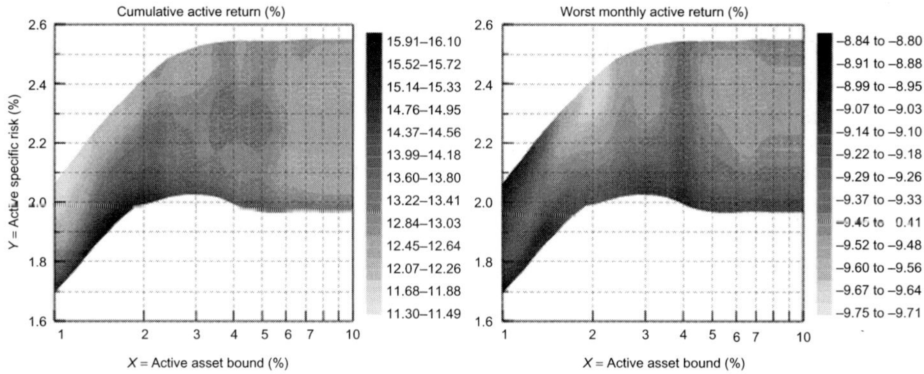

Figure 2.8 The cumulative and worst-case monthly active returns versus asset bound and active specific risk limit for TO = 30%.

Figure 2.8 shows contour plots of the cumulative and worst monthly active returns as functions of the asset bound X and the specific risk model risk constraint level Y for TO = 30%.

The results are similar to those reported in the previous examples. The best solution occurs when $X = 1.1\%$ and $Y = 1.8\%$.

2.2.1 Out-of-sample results

We performed an out-of-sample backtest from October 31, 2006 to October 31, 2007. We considered six different strategies, three without the second risk model and three with the second risk model. The cases correspond to the optimal parameters determined by the results above. These different cases are summarized in the top of Table 2.1. TO = 30% and the primary tracking error from the fundamental factor model is 4%. When running the backtest, we start from October 31, 2005 and run 24 monthly rebalancings so that the turnover for the first, out-of-sample month is meaningful. The results reported are only for the 12 out-of-sample rebalancings.

Table 2.1 shows the performance statistics for these six cases. In all cases, the addition of a properly calibrated second risk model constraint leads to superior portfolio performance as measured by either the strategy's annual active return or its information ratio. In general, the risk metrics—volatility or worst monthly return—changed only slightly (in either direction) while the return increases substantially. Of course, there are only 12 out-of-sample returns, so the performance statistics have relatively large standard errors. Nevertheless, the performance statistics have moved in a favorable direction when the second risk model was added to the strategy and all the parameters were calibrated adequately.

Table 2.1 Out-of-sample results from October 31, 2006 to October 31, 2007

	Second risk model					
	Statistical act. risk		Statistical tot. risk		Specific risk	
Asset bound	1.0%	1.0%	1.2%	1.2%	1.1%	1.1%
Second risk bound	None	2.3%	None	14.5%	None	1.8%
Average assets held	128.4	149.33	128.4	126.17	127.3	169.33
Annual active return	3.71%	4.23%	3.58%	8.34%	3.528%	5.218%
Worst monthly return	−1.68%	−2.00%	−1.79%	−1.98%	−1.866%	−1.888%
Monthly volume	1.43%	1.29%	1.55%	1.59%	1.464%	1.500%
Information ratio	0.748	0.949	0.668	1.518	0.696	1.004
Transfer coefficient	41.6%	42.9%	42.0%	41.5%	42.1%	42.6%

For the three different, second risk model constraints examples considered, the largest out-of-sample improvement over the use of a single risk model occurred in the second case, when the second risk constraint, using the statistical factor model, was used to limit the total risk of the portfolio.

2.2.2 Discussion and conclusions

In our examples, we have added the second risk model as a risk constraint. It could also be incorporated into the objective function of the portfolio construction strategy, as could be done with the primary risk model constraint as well. When a risk model is added to the objective function, it *always* affects the solution. There are no solutions corresponding to the nonbinding risk constraint case. This can be advantageous in that it is impossible to choose calibration parameters that render a risk model superfluous. In fact, many PMs mistakenly believe that the only way to obtain solutions on the efficient frontier is to include the risk term(s) in the objective function.[2]

There are, however, disadvantages to including risk in the objective function. First and foremost, the actual risk predicted is unknown. It may or may not be close to the targeted tracking error. In addition, although the risk terms always affect the solution, it can be difficult to determine by how much they affect the solution or when the other portfolio construction constraints are dominating

[2] This is likely a legacy of the successful marketing of quadratic optimization programs that could *only* handle risk in the objective function. SOCP solvers do not suffer this handicap.

the optimization portfolio solution. In the examples above, the predicted risk could be directly compared to their constraint limits. When risk is in the objective function, however, the only way to determine its significance is to solve the problem twice—once with the risk in the objective function and a second time with the risk aversion parameter altered (or omitted entirely)—and then compare the resulting portfolios. The PM would then have to estimate whether or not the differences in the portfolios were meaningful.

Second, when risk is included in the objective function, the proper calibration values can vary significantly, depending on the expected returns used, and often makes little intuitive sense. If a PM is given a 4% tracking error target, he or she must first translate this 4% target into a corresponding risk aversion value. This risk aversion value may vary considerably from one portfolio construction strategy to another even if all the strategies target 4% tracking error. Next, if a second risk model is added to the objective function, the PM will have to calibrate *both* risk aversion constants once again. The risk aversion found using one risk model is unlikely to be the same when two risk models are present as the objective maximizes the weighted sum of the variances.

On balance, even though the same solutions can be found incorporating risk into the objective function, the lack of intuition and the inability to use previously obtained risk aversion values appear to be practical drawbacks for incorporating risk into the objective function.

Using more than one risk model in a portfolio construction strategy allows a PM to exploit the fact that different risk models measure and capture risk differently. Having both a fundamental and statistical risk model simultaneously in the strategy ensures that the optimized portfolio reflects both points of view. The benefit is derived by the differences captured by both risk models, not which risk model is "better." If a PM believes one risk model is "better" than another, then he or she can simply use the "better" risk model as the primary risk model. When properly calibrated, the final two-risk model portfolio is not any more conservative than the one-risk model portfolio.

The best strategy parameters involve the interaction of all the constraints in the portfolio construction strategy. The results presented here illustrate that interaction of four constraints—tracking error, asset holdings bounds (X), second risk model's risk constraint (Y), and turnover (TO). When calibrating a second risk model, it may be necessary to alter (even loosen) previously calibrated constraint values in order to obtain the best results.

The results presented in this section illustrate that the addition of a properly calibrated second risk model constraint can lead to superior portfolio performance measured either by the strategy's annual active return or its information ratio.

2.3 Multisolution generation

In this section, we discuss a general framework for generating multiple interesting solutions to portfolio optimization problems.

Portfolio optimization is as much about gaining insights into various components—objectives and constraints—of the strategy as it is about constructing the final set of asset holdings. Consequently, our multisolution generator is geared toward generating a set of portfolios that not only provides the PM a large universe of portfolios to select from, but also gives him or her additional information to gain insights into the mechanics of portfolio construction. In our attempt to design such a generator, we identified the following three questions that every PM strives to answer.

1. What is the impact of modifying a constraint bound on the optimal portfolio?
2. What are the trade-offs in jointly varying pairs of constraint bounds?
3. Is there a good way of detecting pairs of constraints whose joint variation has non-trivial impact on the overall model?

What is the impact of modifying a constraint bound on the optimal portfolio?

For the sake of illustration, consider a strategy aimed at minimizing tax liability subject to a set of constraints that includes among others a constraint that the expected return cannot be less than a certain predetermined level, say $ER^* = 2.14\%$. A PM might be interested in understanding the marginal impact of modifying ER* on the minimum attainable tax liability. Figure 2.9 shows a typical tax liability expected return frontier.

As it is evident from the figure, a significant decrease in tax liability can be obtained by reducing ER* by a small amount. It is exactly this kind of insight that we endeavor to capture through our multisolution generator and that cannot be obtained from a single optimal portfolio.

What are the trade-offs in jointly varying pairs of constraint bounds?

Consider a PM trying to rebalance his or her portfolio under a 5% tracking error constraint and a 20% turnover constraint, and let us suppose that the

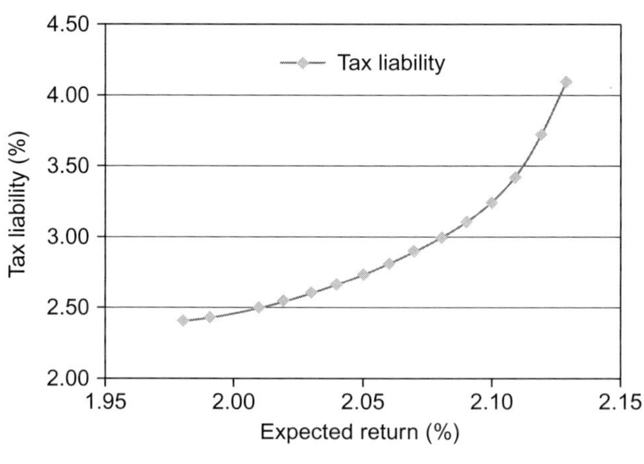

Figure 2.9 Tax liability—expected return frontier.

portfolio optimizer returns a solution with expected return of 3%. Suppose there exists an alternative solution with expected return of 3%, tracking error of 4.5%, and turnover of 22%. While this solution is infeasible to the PM's original strategy, it is still of interest to him or her; the alternative solution offers the same level of expected return at a lower tracking error although it entails a small increase in the turnover.

Traditional portfolio optimization techniques restrict the vision of a PM to only those portfolios that strictly satisfy all the constraints in his or her strategy. We believe that marginally infeasible solutions such as the one presented above can play a crucial role in guiding the decision-making process of a PM. For the sake of illustration, consider the heatmap representation of optimal expected returns for various combinations of tracking error and turnover values as depicted in Figure 2.10. Such a representation is insightful and can help a PM discover opportunity pockets, i.e., attractive combinations of tracking error and turnover values that cannot be discovered by tools geared toward generating just one solution. In principle, all of these trade-offs could theoretically be considered in an objective with appropriate weights to capture the economic benefit. However, choosing the correct weights is difficult and subject to trial and error, limiting the overall usefulness of such an approach.

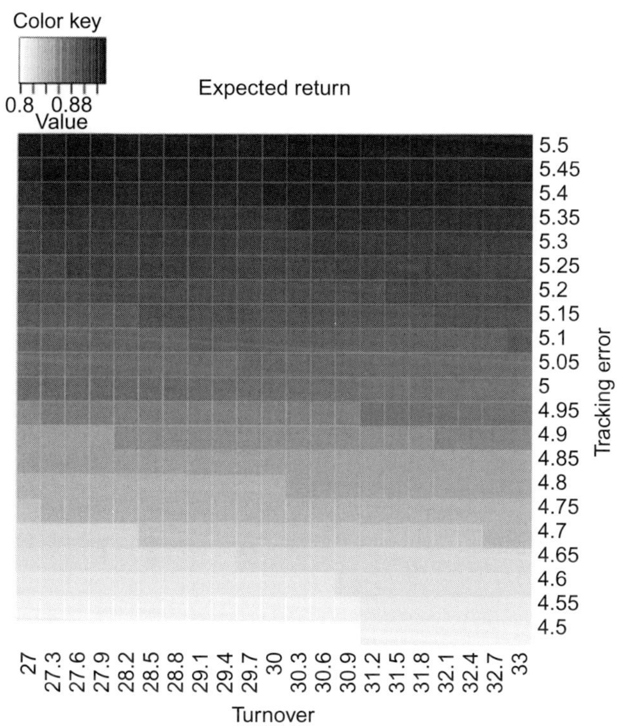

Figure 2.10 Turnover—tracking error—expected return heatmap.

A financial model provided as input to a strategy can only capture a portion of the information available to a PM. Often, a PM has access to information bits or "market hunches" that are difficult to model but nevertheless can be useful in making an informed decision. Heatmap representations such as the one shown in Figure 2.10 can be used to partially overcome this shortcoming. The example that follows elaborates on this observation.

Consider a PM rebalancing his or her portfolio in view of an expected economic downturn. Suppose the fund managed by the PM has traditionally held long positions in entertainment and food products industries, and the strategy used by the PM has tentative limits on the amount of exposure to each one of these industries. As the economy turns sour, the demand for entertainment and food products is likely to go down. For instance, people might want to save money by cutting down on visits to entertainment theaters. Similarly, they might reduce emphasis on environmentally conscious and ethically produced food items (such as organic food) and replace them by regular food products. However, as is well known in Economics, the demand for *luxury* goods such as entertainment products is significantly more elastic with respect to variations in average income than the demand for *essential* goods such as food products. Despite being a well-known fact, the PM has very limited means of meaningfully incorporating this economic wisdom within his or her strategy. Ideally, the PM would like to understand the impact on the optimal expected return that results as he or she reduces his or her exposure to the more elastic entertainment industry, and increases his or her exposure to the relatively inelastic food products industry. Heatmaps such as the one shown in Figure 2.10 can be tremendously helpful in depicting these kinds of trilateral dependencies involving expected return and exposures to a pair of industries.

Is there a good way of detecting pairs of constraints whose joint variation has nontrivial impact on the overall model?

A strategy with k constraints has $k(k-1)/2$ pairs of constraints that can be examined simultaneously to perform trade-off analysis discussed above. Admittedly, only a small fraction of these pairs of constraints are likely to be of practical interest, and identifying such pairs can be a difficult task. One could argue that a PM can use his or her judgment to select pairs of constraints and study the associated heatmaps. Such an approach has an immediate shortcoming, namely, that it is strongly rooted in the PM's preconceived view of the portfolio construction process, and hence never surprises him or her. In other words, such an approach deters the PM from examining combinations of constraints that are unlikely to yield interesting outcomes despite being performance bottlenecks of the strategy.

Furthermore, sometimes heatmaps obtained by jointly varying a pair of constraints are completely determined by variations in only one of the constraints. For instance, consider the heatmap shown in Figure 2.11 obtained by jointly varying the tracking error and sector bound constraints. Clearly, in this case,

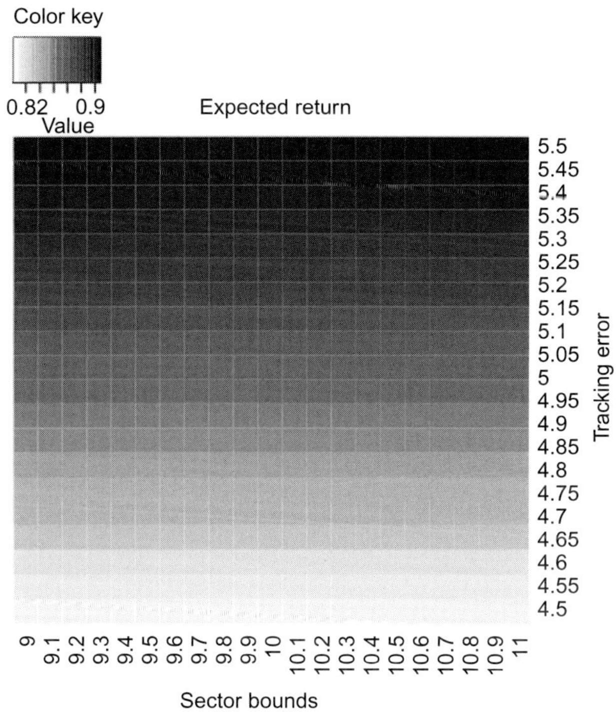

Figure 2.11 Sector bounds—tracking error—expected return heatmap.

the variation of the optimal expected return is completely determined by the tracking error and is insensitive to the variation in the asset bound constraint. Ideally, a PM would like to have at his or her disposal an automatic tool that can discard such uninteresting pairs of constraints and help him or her focus on pairs that directly impact the key performance determinants.

Next, we introduce the concept of "Constraint Elasticity" and discuss how it can be used to address the questions raised in this section.

2.3.1 Constraint elasticity

Constraint elasticity measures the responsiveness of the optimal objective value to perturbations in the constraint bounds. More precisely, the elasticity of the model with respect to a constraint in the strategy is defined to be the percentage change in the optimal objective value that arises due to 1% change in the constraint bounds.[3]

[3] Naturally, the concept of constraint elasticity gives useful results only for constraints with nonzero bounds. Fortunately, most constraints of interest such as tracking error constraint, turnover constraint, etc. satisfy this condition.

Similar to various concepts in Economics such as price elasticity of supply or income elasticity of demand, the concept of constraint elasticity is unit independent. In other words, by working with percentage changes, and not "unit" changes, in constraint bounds it allows us to assess the impact of changing a heterogeneous set of constraints such as tracking error constraints, turnover constraint, industry, asset, sector, style bound constraints, etc. Recall that these constraints can have widely varying bounds that can differ by orders of magnitude. Consequently, comparing models obtained by perturbing these constraints by a single unit hardly gives any insights into their marginal impact on the overall model.

Next, we discuss an example to illustrate the concept of constraint elasticity and the manner in which it can be used to answer the three questions raised above. Consider the following strategy, referred to as "ActiveLongOnly" strategy in the remainder of this chapter.

Maximize expected return

s.t

- Budget constraint
- Tracking error w.r.t a benchmark (maximum 5%)
- Turnover constraint (maximum 30%)
- Asset, industry, sector, style bound constraints
- Threshold holding and trading constraints
- Average daily volume holding and trading constraint
- Beta constraint

Table 2.2 displays the elasticities of various constraints in the ActiveLongOnly strategy.

The first column of the table reports the constraint name, while the second column reports the elasticity of the respective constraint. Thus, according to the following table, 1% change in the tracking error bound from 5% to 5.05% will produce a 0.759% change in the optimal expected return. Similarly, the elasticities of the remaining constraints quantitatively assess the impact of changing the constraint bound on the optimal objective value.

Table 2.2 Constraint elasticities (ActiveLongOnly strategy)

Constraint	Elasticity
Tracking error	0.75902
Maximum turnover	0.04825
Asset bounds	0.01277
Sector bounds	0.00291
Threshold trade	0.00003
Beta	0.00000
Style bounds	0.00000
Industry bounds	0.00000
Threshold holding	0.00000

Next, let us try to understand the impact of jointly varying the upper bounds on tracking error and turnover. Figure 2.10 gives a heatmap representation of the optimal expected return subject to these perturbations. The information displayed in Figure 2.10 has two interesting ramifications. First, it gives a graphical display of the optimal expected return terrain for various combinations of tracking error and turnover. Such a terrain can be very helpful in understanding trade-offs and making an informed decision. Furthermore, if the specific values of tracking error and turnover used in the original strategy were tentative, then the heatmap can be used as a guiding tool in choosing more desirable values for these characteristics.

Second, consider the level curves of the heatmap displayed in Figure 2.10. These level curves plot combinations of tracking error and turnover values that give rise to the same optimal expected return. Note that the level curves of Figure 2.10 are arcs of concentric ellipses centered on the top right corner of the graph. Alternatively, this means that as we decrease the tracking error, we lose a certain portion of optimal expected return that we can recover by slightly increasing the turnover. Since the elasticities of the tracking error constraint is roughly 15 times higher than that of the turnover constraint, the arcs of ellipses representing the level curves in Figure 2.10 are elongated along the turnover axis and compressed along the tracking error axis. Note that as the ratio of these elasticities increases so does the distortion of ellipses representing the level curves. In the extreme case when the ratio of elasticities gets significantly high, the elliptic level curves transform into flat lines representing complete dominance of one constraint in determining the optimal expected return. For instance, consider the heatmap shown in Figure 2.11 obtained by jointly perturbing the sector bounds constraint and the tracking error constraint. Recall that the elasticity of the tracking error constraint (0.759) is almost 260 times higher than that of the sector bounds constraint (0.0029). Consequently, the level curves in Figure 2.11 are flat lines that are insensitive to variations in the sector bounds constraint. Clearly, heatmaps such as the one presented in Figure 2.11 offer very little additional insight to a PM.

To summarize, constraint elasticities can be used as yardsticks to assess the relative importance of various constraints. Furthermore, jointly perturbing constraints with comparable elasticities elucidates interesting aspects of the optimal portfolio terrain. Perturbing constraints with vastly different elasticities, on the other hand, offers little insights. Thus, constraint elasticities can also be used to identify pairs of constraints that when perturbed jointly will possibly lead to insightful outcomes.

For certain kinds of strategies that give rise to *convex* optimization models, constraint elasticities can be computed using the constraint attribution utility in our product suite. For strategies that do not satisfy this criterion, we use more sophisticated techniques that go beyond the scope of this discussion.

2.3.2 Intractable metrics

A PM typically evaluates a portfolio using a wide variety of metrics. While some of these metrics, such as expected return, tax characteristics, market impact, transaction cost, etc., are easy to model and can be incorporated as objectives in a strategy, there are others such as transfer coefficient, implied beta, ratio of specific risk to tracking error, etc. that are not directly amenable to existing optimization techniques. This latter class of metrics, referred to as *intractable metrics* in the sequel, nonetheless convey important information about a portfolio and can influence a PM's eventual choice. The example that follows illustrates how heatmaps can be used to gain insights into these intractable metrics.

Consider the "ActiveLongOnly" strategy discussed above. Figure 2.12 plots the transfer coefficient of optimal portfolios that result from combinations of tracking error and turnover values depicted in Figure 2.10. Similar to level curves of Figure 2.10, the level curves in Figure 2.12 give insight into the trend of transfer coefficients of these optimal portfolios, which can be very helpful in portfolio construction. For instance, a PM might be indifferent to portfolios whose expected return is more than a certain predetermined level, and among all such portfolios he or she might be interested in choosing one with maximum transfer coefficient. This can be easily accomplished by selecting an appropriate level curve from Figure 2.10 and overlaying it over the heatmap in Figure 2.12.

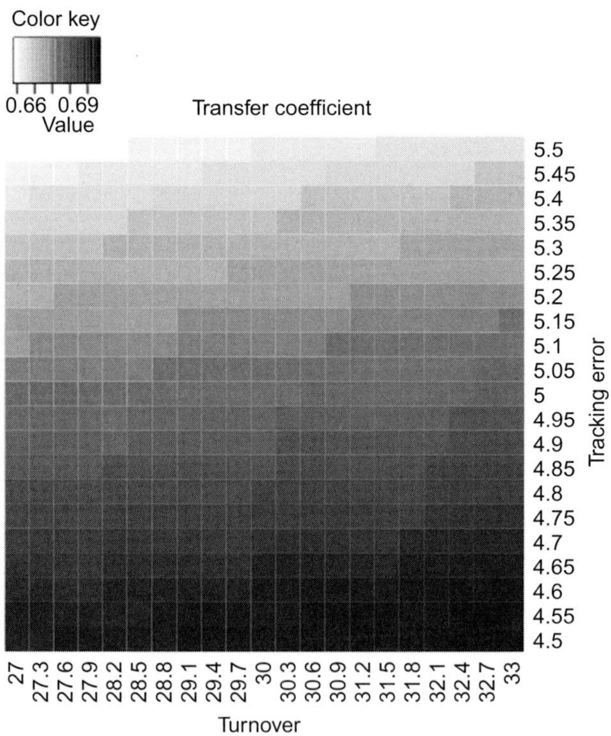

Figure 2.12 Turnover—tracking error—transfer coefficient heatmap.

Interestingly, such an exercise not only helps the PM discover portfolios with better transfer coefficient but also gives him or her specific guidelines, namely the combination of tracking error and turnover values, about how to modify his or her strategy to attain them.

Furthermore, once the PM has demarcated the universe of desirable portfolios by overlaying the heatmaps of expected return and transfer coefficients, he or she can move on to examine heatmaps of other metrics such as implied beta (Figure 2.13), ratio of specific and tracking error (Figure 2.14), etc., thus zooming into the universe of portfolios that are of interest to him or her.

Case study

We conclude this section by discussing an illustrative example. Consider a PM trying to rebalance his or her portfolio using the following strategy.

Maximize expected return

s.t

- Budget constraint
- Tracking error constraint (at most 2%)
- Turnover constraint (at most 10%)
- Asset, industry, sector, style bounds constraint
- Active beta constraint (at most 1%)
- Threshold holding and trading constraint
- Average daily volume holding and trading constraint

Table 2.3 gives the key characteristics of the starting portfolio. Given the souring economic conditions, the PM is inclined toward pursuing investment options with as little tracking error as possible, even if that entails violating some of the other constraints by a small amount. Furthermore, portfolios with higher transfer coefficient are deemed extremely desirable. Since the account under consideration is taxable, the PM is particularly interested in tax-efficient investment options although he or she admits that minimizing tax liability is to be used only as a secondary objective with maximizing expected return being the primary one.

Table 2.4 gives the key characteristics of the optimal portfolio, referred to as P0 in the sequel, obtained by rebalancing the starting portfolio using the strategy mentioned above. Note that the tracking error bound of 2% is not attained, whereas the entire turnover budget of 10% is utilized. P0 has an expected return of 0.2794% and a transfer coefficient of 0.6391. Next, we discuss a series of steps that the PM can take to improve P0. The flowchart shown in Figure 2.15 gives a brief summary of what follows.

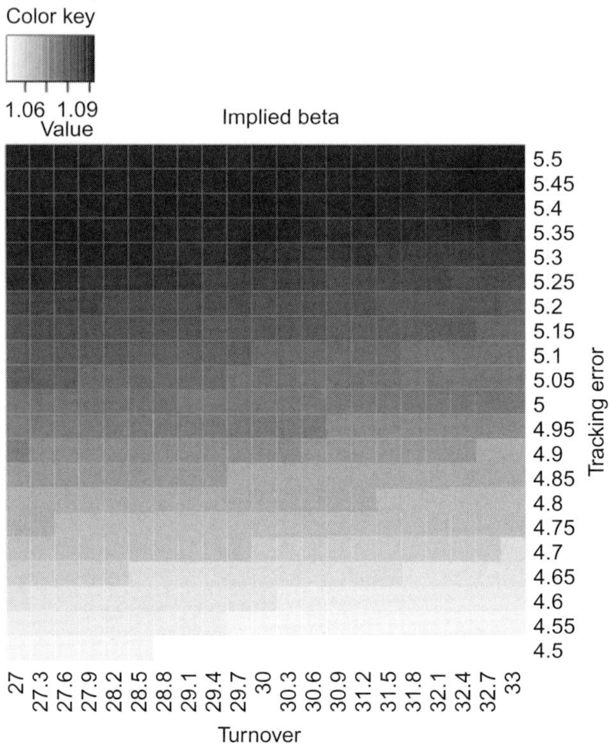

Figure 2.13 Turnover—tracking error—implied beta heatmap.

Table 2.5 displays the elasticities of various constraints in the strategy. Note that the tracking error and turnover constraints have significantly higher elasticities than the remaining constraints. Figure 2.16 plots the optimal expected return for various combinations of tracking error and turnover values, while Figure 2.17 plots the transfer coefficients of the respective portfolios. Since both of these constraints have similar elasticities, the level curves of Figure 2.16 are equally responsive to perturbations in the tracking error and turnover bounds. In contrast, the level curves of Figure 2.10 are much more responsive to variation in the highly elastic tracking error constraint than to variations in the relatively inelastic turnover constraint. Figure 2.17 suggests that portfolios with higher transfer coefficients can be obtained by decreasing the tracking error bound and increasing the turnover limits. By overlaying the expected return heatmap (Figure 2.16) over the transfer coefficient heatmap (Figure 2.17), we determined that increasing the turnover to 10.40% while decreasing the tracking error to 1.84% should yield portfolios with higher transfer coefficient and comparable expected return. Table 2.6 gives the key characteristics of the resulting optimal portfolio, referred to as P1 in the sequel. While both P0 and P1 have similar expected returns, P1 has two additional desirable characteristics,

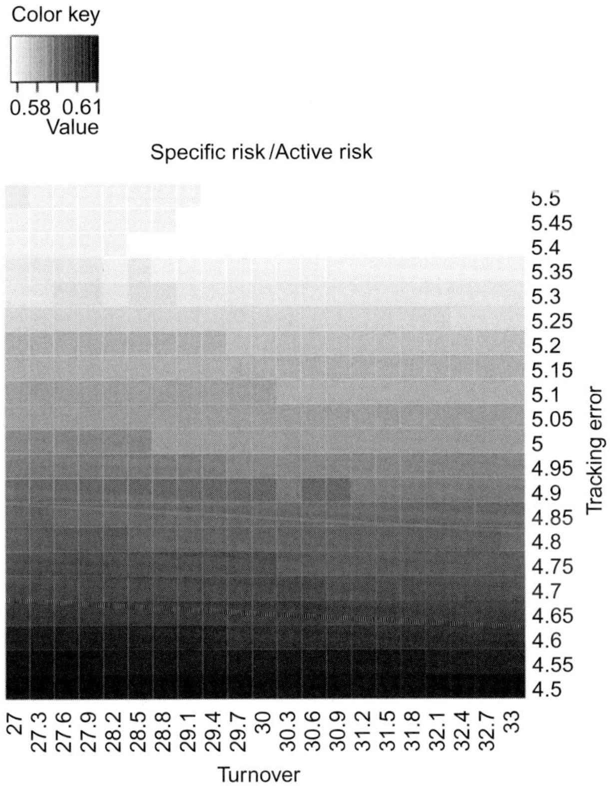

Figure 2.14 Turnover—tracking error—ratio of specific and active risk heatmap.

Table 2.3 Starting portfolio

Summary statistic	Value
Expected return	0.2488%
Transfer coefficient	0.6047
Implied beta	1.0148
Tracking error	1.87%
Tax liability	\$5,104.72
Turnover	–
Realized short-term gains	\$11,687.70
Realized short-term losses	\$13,594.12
Net realized short-term gains/losses	(\$1,906.41)
Realized long-term gains	\$41,000.23
Realized long-term losses	\$5,062.38
Net realized long-term gains/losses	\$35,937.85

Table 2.4 Portfolio P0

Summary statistic	Value
Expected return	0.2794%
Transfer coefficient	0.6391
Implied beta	1.0107
Tracking error	1.98%
Tax liability	$5,427.28
Turnover	10.00%
Realized short-term gains	$13,409.61
Realized short-term losses	$18,897.15
Net realized short-term gains/losses	($5,487.54)
Realized long-term gains	$47,579.42
Realized long-term losses	$5,910.04
Net realized long-term gains/losses	$41,669.39

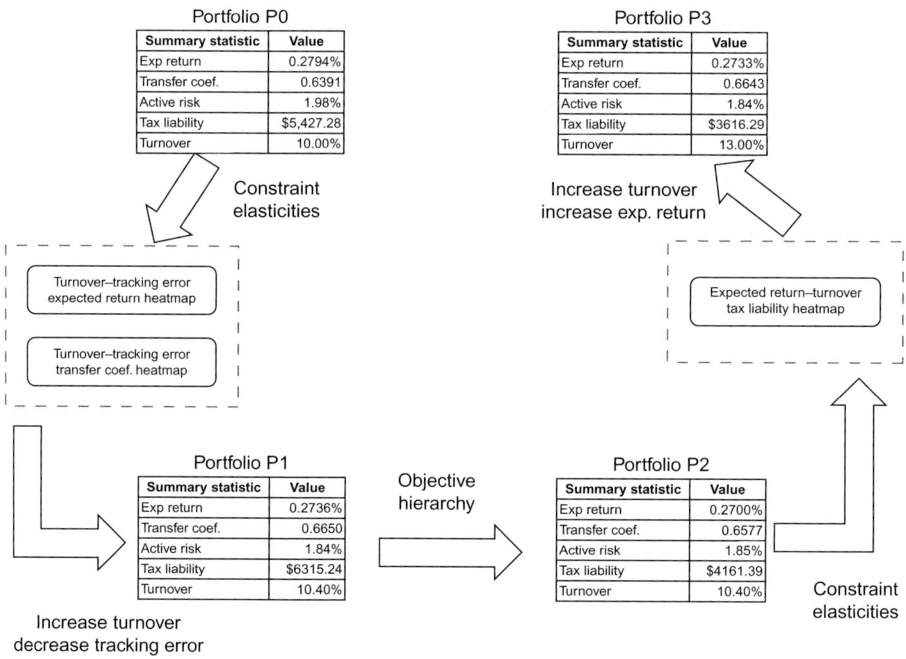

Figure 2.15 Flowchart.

namely, higher transfer coefficient and lower tracking error. P1, however, has an undesirably high tax liability.

In order to generate portfolios with lower tax liability, we modified the objective function of the above strategy to minimize tax liability. Since we are

Table 2.5 Constraint elasticities

Constraint	Elasticity
Tracking error	0.1687
Turnover	0.1563
Industry bounds	0.0978
Style bounds	0.0784
Sector bounds	0.0323
Active beta	0.0256
Asset bounds	0.0162
Threshold holding	0.0013
Threshold trade	0.0000

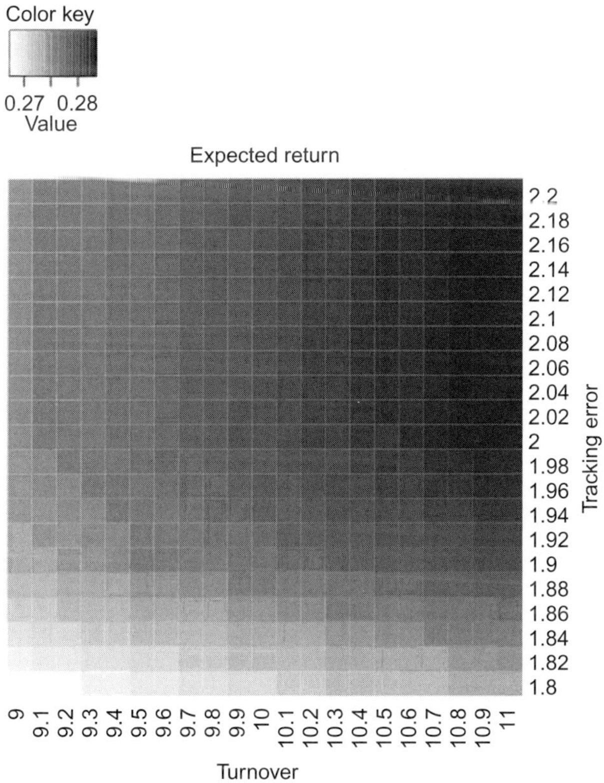

Figure 2.16 Turnover—tracking error—expected return heatmap.

interested in portfolios whose expected return is similar to that of P0 and P1, we added a constraint to the strategy limiting the expected return to be at least 0.27%. Table 2.7 gives the characteristics of the resulting optimal portfolio, referred to as P2 in the sequel. Clearly, P2 has a lower tax liability than P0

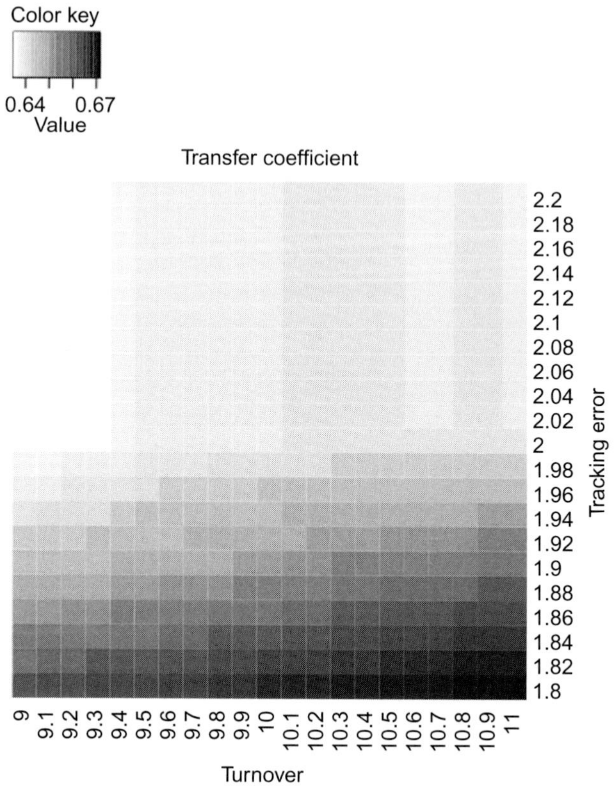

Figure 2.17 Turnover—tracking error—transfer coefficient heatmap.

Table 2.6 Portfolio P1

Summary statistic	Value
Expected return	0.2736%
Transfer coefficient	0.6650
Implied beta	1.0106
Tracking error	1.84%
Tax liability	$6,315.24
Turnover	10.40%
Realized short-term gains	$11,687.70
Realized short-term losses	$13,594.12
Net realized short-term gains/losses	($1,906.41)
Realized long-term gains	$54,855.18
Realized long-term losses	$10,847.18
Net realized long-term gains/losses	$44,008.01

Table 2.7 Portfolio P2

Summary statistic	Value
Expected return	0.2700%
Transfer coefficient	0.6577
Implied beta	1.0099
Tracking error	1.84%
Tax liability	$4,161.39
Turnover	10.40%
Realized short-term gains	$11,687.70
Realized short-term losses	$13,594.12
Net realized short-term gains/losses	($1,906.41)
Realized long-term gains	$43,812.42
Realized long-term losses	$14,163.38
Net realized long-term gains/losses	$29,649.04

Table 2.8 Constraint elasticities (minimize tax liability strategy)

Constraint	Elasticity
Tracking error	16.9601
Turnover	2.5490
Expected return constraint	1.9965
Industry bounds	0.6024
Style bounds	0.3178
Active beta	0.3089
Sector bounds	0.2160
Asset bounds	0.1505
Threshold holding	0.0163
Threshold trade	0.0121

or P1. However, P2 also has a lower transfer coefficient and lowest expected return among the three portfolios P0, P1, and P2. This suggests that it might be worthwhile exploring portfolios that are similar to P2 but have better transfer coefficient and expected return.

Table 2.8 displays the elasticities of various constraints in the strategy that was used to generate P2. Note that the tracking error constraint has a very high elasticity, suggesting that portfolios with lower tax liability can be obtained by increasing the tracking error bound. Since the PM is interested in low-risk portfolios, we forgo this option and examine the trade-offs between the next two most responsive constraints, namely the turnover constraint and the expected return constraint. Figure 2.18 plots the minimum tax liability for various combinations of turnover and expected return values. Evidently, portfolios with higher expected return and lower tax liability can be obtained by increasing the turnover value. Consequently, we increased turnover value to

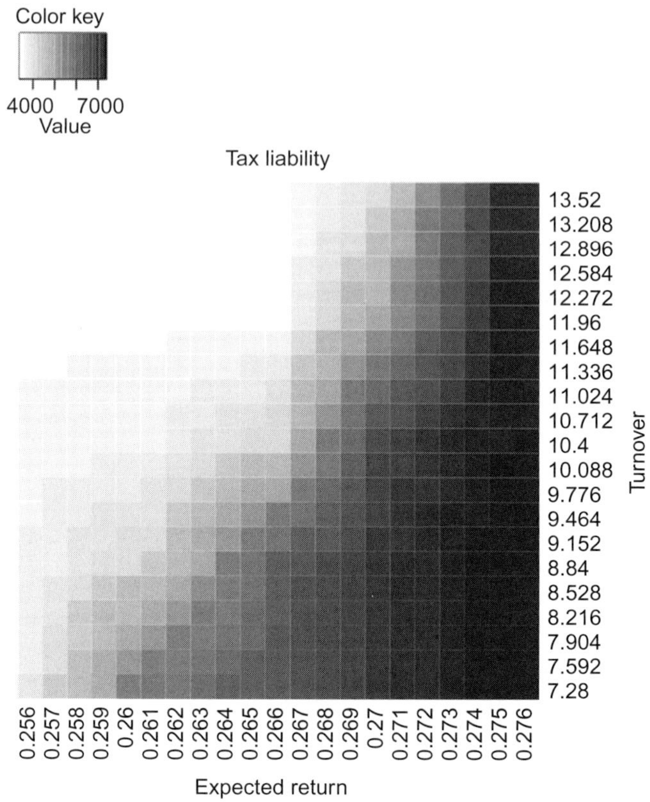

Figure 2.18 Expected return—turnover—tax liability heatmap.

13% and the lowest allowable expected return to 0.2733%, and reoptimized the resulting model to obtain the P3 portfolio whose key characteristics are reported in Table 2.9. Note that P3 has significantly lower tax liability even though its expected return, tracking error, and transfer coefficient are comparable to that of P1. Furthermore, P3 has higher realized long-term losses and lower realized long-term gains as compared to P1. In other words, P3 uses the additional 2.6% turnover to harvest losses that in turn are used to offset gains thereby reducing the overall tax liability.

This entire exercise can be viewed as a systematic attempt to improve the optimal portfolio P0 derived from the original strategy. At various stages of this exercise, we used the concept of elasticity to select pairs of constraints whose joint perturbations was most likely to impact the optimal objective value. Heatmaps associated with such pairs of constraints were used to gain insights about the optimal portfolio terrain, and also to determine the most appropriate perturbation values. The final portfolio P3 has better transfer coefficient, tracking error, and tax characteristics than P0 despite having almost the same expected return.

Table 2.9 Portfolio P3

Summary statistic	Value
Expected return	0.2733%
Transfer coefficient	0.6643
Implied beta	1.0094
Tracking error	1.84%
Tax liability	\$3,616.29
Turnover	13.00%
Realized short-term gains	\$11,687.70
Realized short-term losses	\$13,594.12
Net realized short-term gains/losses	(\$1,906.41)
Realized long-term gains	\$42,114.52
Realized long-term losses	\$16,099.53
Net realized long-term gains/losses	\$26,014.99

Indeed, one can do all of these steps manually without the aid of tools and concepts developed in this section. The resulting enterprise, however, will be tedious, vulnerable to trial and error, and most importantly deprive the PM from exploiting the latest developments in computing technology. By system atically automating various steps in this procedure and augmenting them with specifically tailored proprietary algorithms, we shoulder the computational burden of a PM thereby allowing him or her to focus exclusively on his or her central goal—*portfolio design and analysis*.

2.4 Conclusions

This chapter has presented several new methods for approaching portfolio construction. The high-level conclusions are the following:

1. The addition of a second risk model can lead to better performance than using one risk model alone when the second risk model constraint is calibrated so that both risk model constraints are simultaneously binding. In many cases, a good calibration tightens the second risk model constraint until the primary risk model constraint just remains binding. Calibration of the other portfolio construction parameters is essential. In particular, since the region over which both risk models are binding can be narrow, guessing what value to use (for X and Y in our examples) can lead to only one risk model being binding for the optimal solution, even when both risk models are comparable. The other constraints in the portfolio construction strategy can greatly influence the region over which both risk models are binding.
2. Rebalancing a portfolio is a complex activity that goes beyond solving an optimization problem. It entails a systematic attempt by the PM to understand the impact of constraints on the choice of the optimal portfolio, to evaluate the effect of changing one or more constraint bounds, and to gain insights in the resulting trade-offs. The concept of constraint elasticity gives a natural ordering of constraints to examine

in such an exercise. Heatmaps of portfolio characteristics such as expected return, transfer coefficient, implied beta, etc. derived by perturbing pairs of constraints with comparable elasticities give insight into the terrain of optimal portfolio, and can help discover opportunity pockets that are otherwise difficult to find. Furthermore, a PM can discover portfolios that have multiple desirable characteristics by overlaying the level curves of one characteristic over the heatmap of the other. As our case study illustrated, a systematic application of these ideas can provide access to a class of multiple interesting solutions to the portfolio optimization problem.

References

Ceria, S. (2007). *The Quant Scare*. Presentation to the London Quant Group, October. Available at: http://www.axiomainc.com/research_papers.htm

Foley, T. (2008). Implementing 130/30 equity strategies: diversification among quantitative managers. *Institutional Investor*(Summer), 56–59.

3 Optimal solutions for optimization in practice

Daryl Roxburgh, Katja Scherer and Tim Matthews

Executive Summary

Optimization has been one of the principal techniques to allow quantitative investors to build portfolios. It is not the only technique and there are styles of investment that are not automatically suitable for optimization. Historically, optimization was subject to a detailed criticism by disgruntled users due to e.g. the solutions being excessively sensitive to unstable inputs. It should be noted that optimization has experienced a renaissance in the last five or so years as many high-frequency hedge funds have used it to build portfolios on a daily, or even intra-daily basis. Furthermore, techniques such as Robust and the innovative gain-loss optimization (GLO) have been developed that have addressed problems associated with alpha forecasting errors and the non-normality of returns; we shall discuss these in this chapter together with methods to combine mean-variance and gain-loss to benefit from both their strengths. We conclude with a practical application of such techniques and with a particular focus on the area of Charities and Endowments where we consider investments with infinite investment horizons.

3.1 Introduction

Optimization has been one of the principal techniques to allow quantitative investors to build portfolios. It is not the only technique and there are styles of investment that are not automatically suitable for optimization. Historically, optimization was subject to a detailed criticism by disgruntled users due to the solutions being excessively sensitive to errors in forecast of alpha; we will not discuss this in any detail as it is covered elsewhere in the book. It should be noted that optimization has experienced a renaissance in the last 5 or so years as many high-frequency hedge funds have used it to build portfolios on a daily, or even intradaily basis. Furthermore, techniques have been developed that have addressed the problems associated with alpha measurement; we shall discuss these later in the chapter.

Doi:10.1016/B978-0-12-374952-9.00003-8.

Our experience with optimizers stems from our association with BITA Risk. BITA Risk provides leading private banks and wealth managers with tools to support client acquisition and retention, through risk profiling, suitability assessment, investment proposal generation, and portfolio monitoring and reporting. In addition to this, institutional investment houses across the world use our applications and risk models for portfolio optimization, quant strategy research, risk analysis, and reporting. Our services include the following services discussed below but BITA Risk is also actively engaged in consultancy projects in areas such as risk and quantitative analytics across the globe.

3.1.1 BITA Star(TM)

BITA Star helps wealth managers understand their clients' true attitude to risk, and provides tools to support asset/product suitability screening, portfolio construction, investment proposal generation, and the ongoing monitoring and reporting of portfolio risk.

3.1.2 BITA Monitor(TM)

A sophisticated risk-monitoring tool that supports compliance with the FSA's Treating Customers Fairly Management Information (TCF MI) guidelines—a bespoke yet rapidly deployed solution, used stand-alone or as part of BITA Star.

3.1.3 BITA Curve(TM)

BITA Curve provides a "quant" workbench including optimization, analysis and risk reporting, and data management. This enables users to create and test new strategies, analyze, manage, and report on a portfolio using their own universe of data. BITA Curve gives fund managers unparalleled flexibility, access to data, integration with analytics libraries, and control.

3.1.4 BITA Optimizer(TM)

Developed over the last 14 years, with innovative derivative versions—Robust Charity and Gain Loss (GLO(TM))—BITA Quadratic Optimizer is known for its flexibility, speed of processing, and capacity to handle large complex problems.

This chapter is prefaced with an introduction to optimization in the context of portfolio construction illustrated with a number of applications. Section 3.3 describes mean–variance optimization, which is the most widely implemented optimization methodology. In Sections 3.4–3.6, issues with this now standard technique are identified and alternative methodologies described together with their associated benefits. This exploration completes with implementations of the innovative (GLO) Gain/Loss Optimizer, also implemented by BITA Risk, to address these inherent challenges.

Section 3.7 discusses the application of optimization to charity and endowment investments, which invest in perpetuity in comparison to the shorter-term objectives of traditional investment funds. In Section 3.8, we illustrate a number of bespoke optimization procedures to deal with the practical requests of clients. The chapter concludes in Section 3.9 with a summary of current and future optimization techniques.

3.2 Portfolio optimization

3.2.1 The need for optimization

Consistently adding value through selecting stocks, factors, and "trends" is an obvious challenge for even accomplished asset managers. Many fail, not least when trying to deliver performance net of management and trading costs; this has led to the search for an alternative to "alpha" as a sales story.

Meanwhile, the "cost savings" story has been gaining momentum as "productized" passive funds have gained significant share of the marketplace. This generic "asset product" worldview is evidenced in the active world as well, e.g., active funds' mandates often being defined in terms of the generic asset product they are trying to beat. Most active funds mitigate the risk of excess underperformance by implementing strict constraints on a generic tracking error. This has led to a proliferation of the so-called active tilt funds.

Historically, active tilt funds are run using an optimizer to maximize alpha while constraining tracking error, enforcing the "tilt" either by alpha determination or by systematically over-/underweighting some industry or factor, whilst simultaneously constraining turnover or explicitly subtracting a cost term from the objective function to control the cost of obtaining the strategic portfolio.

3.2.2 Applications of portfolio optimization

An optimizer is designed to create portfolios and trades that are optimal in terms of risk, return, and transaction costs, subject to a wide range of real-world trading and investment constraints. The most important applications of the BITA Optimizer are in portfolio construction and rebalancing. Some examples are shown below.

3.2.3 Program trading

This form of optimization is only concerned with risk and transaction cost. Given an initial position in stocks and index futures as the result of a program trade, the problem is to trade off risk against costs to achieve an overall position (which will include long and short positions in stocks and futures) that has the desired total exposure (measured as total risk or value at risk).

3.2.4 Long–short portfolio construction

Many equity-based portfolio traders in hedge fund and trading firms, as well as institutional fund managers in certain institutions, are allowed to take short positions in futures, indices, currencies, or stocks. For proprietary trading desks and "market neutral" hedge funds, a common constraint is that the long and short sides are equal in value, or that the overall (absolute) risk is constrained. If the objective is to minimize the absolute risk, then the problem is known as optimized hedging.

3.2.5 Active quant management

The active quant is measured against an index or benchmark, but has nonzero alphas for all the stocks of interest, and attempts to maximize the relative return (outperformance) while staying within a given risk budget (constraint on tracking error).

3.2.6 Asset allocation

Asset allocation is characterized by a relatively small number of variables (typically less than 50) and long-only allocations of weights that must sum to unity. Asset allocation is an essential part of the portfolio construction for all pension funds, and for private client investment management.

3.2.7 Index tracking

Index tracking is a form of passive investing in which the benchmark is defined by a particular index (e.g., the S&P 500) and the problem is to minimize the active risk or "tracking error" of a fund. Typical constraints are to be long-only in the stocks of the benchmark, hold little or no cash, and to find the minimum number of names in the tracking fund to stay within a certain tracking error. The alpha (expected return) term does not enter into the optimization.

3.3 Mean–variance optimization

3.3.1 A technical overview

The expected return of a portfolio is a linear combination of the "alphas" of each asset within the portfolio, while the risk (absolute risk/volatility or active risk/tracking error) is most usually expressed as a quadratic expression based on the covariance matrix of asset returns. Transaction costs for a rebalance depend on the size and direction of the trades, and may include both linear and nonlinear terms, to model both commission and market impact of trading.

When creating or rebalancing a portfolio, the optimizer must trade off both risk and cost against expected return—this is achieved by minimizing a "utility function" that includes positive risk and cost terms and negative return terms. The general mathematical form of this utility function, when there is a known benchmark and the risk is defined as active risk, can be expressed as:

$$
\begin{aligned}
U(w, \gamma, \kappa) = & -\frac{\gamma}{1-\gamma} \cdot \alpha^T (w - w_B) \\
& + \frac{\kappa}{1-\kappa} \cdot C(w - w_I) \\
& + \frac{1}{2}(w - w_B)^T \cdot Q \cdot (w - w_B)
\end{aligned}
$$

subject to

$$Aw = d$$

This form includes a general transaction cost function C, where:

α = vector of expected returns, derived from an alpha model;
w = vector of portfolio weights;
w_B = vector of benchmark portfolio weights;
w_1 = vector of initial portfolio weights;
Q = covariance matrix, derived from a risk model;
γ = "risk tolerance" (takes values from 0 to 1);
κ = "cost tolerance" (takes values from 0 to 1);
A = matrix of constraints;
d = vector of limits.

The transpose of any vector is denoted by ^{T}.

As a special case of this, we can consider a piecewise linear form of the transaction cost function, which is defined by a set of buy-cost coefficients b_i and a set of sell-cost coefficients s_i:

$$
\begin{aligned}
U(w, \gamma, \kappa) = & -\frac{\gamma}{1-\gamma} \cdot \alpha^T (w - w_B) \\
& + \frac{\kappa}{1-\kappa} \sum^{\textit{piecewise}} (b_i \max(0, w - w_I) - s_i \min(0, w - w_I) \\
& + \frac{1}{2}(w - w_B)^T \cdot Q \cdot (w - w_B)
\end{aligned}
$$

This form is the most commonly used to rebalance a portfolio when the trading costs include both commission-based and market impact terms.

3.3.2 The BITA optimizer—functional summary

The mean–variance optimizer from BITA Risk delivers a solution with substantial scope and functionality; a summary is given below.

- High-performance quadratic optimization, using the *Active Set* method
- Optimization from cash or portfolio rebalancing
- Linear and piecewise linear cost penalties
- General linear constraints, e.g., asset weight, sector weight, factor weight, asset revision weight, portfolio return, portfolio beta, all of which can be set as hard or soft constraints

$$L_i \le \sum_j A_{ij} w_j \le U_i$$

where L and U are the lower and upper bounds and A is the matrix of constraints.

- Nonlinear constraints including portfolio risk expressed as absolute risk (relative to cash) or tracking error (relative to a benchmark) and turnover
- Threshold constraints, e.g., minimum holding, minimum trade
- Basket (integer number) constraints, e.g., number of stocks, number of trades
- Absolute value constraints, e.g., absolute factor exposure, portfolio gross value

$$L_i \leq \sum_j A_{ij} |w_j| \leq U_i$$

- Lot size (Normal Market Size) trading constraints
- Long/short ratio constraints

$$L \leq S/L \leq U$$

where L and U are the lower and upper bounds and the ratio S/L is always defined.

- Use of composite assets (e.g., index futures, exchange-traded funds)
- Calculation of portfolio utility, and utility per stock
- Calculation of portfolio ex ante risk, and marginal contributions from assets and factors
- Calculation of portfolio transaction cost, and cost per stock
- Stand-alone .dll that is platform-independent, enabling easy integration through its API interfacing through MatLab, .Net C, Java, VB, Perl, Python, and with all major data integrators.

3.4 Robust optimization

3.4.1 Background

Robust optimization can help eliminate some of the instability/cost associated with the traditional approach by reducing the optimizer's need to trade, as opposed to directly controlling the amount of trading it does. This reduction is achieved by reducing the influence of alphas on stocks that have historically been wrong, volatile, or both.

3.4.2 Introduction

BITA Risk developed a new optimizer (BITA Robust) that focuses on managing uncertainty in the expectations that input into portfolio construction. That is, it copes with ranges of inputs for key factors such as expected returns. Traditional mean–variance portfolio optimization and quadratic optimization solutions are notoriously sensitive with respect to inputs. It is also well known that errors in forecasts lead to unrealistic portfolios. BITA Robust accomplishes a more stable solution, which requires less trading to maintain the portfolio's expected risk/return trade-off, thus reducing turnover and costs.

As mentioned above, it is generally accepted that the most unstable input to an optimization is the vector of expected returns. BITA Robust addresses instability in alpha using its enhanced quadratic constraint functionality to set nonlinear boundaries on forecast error (FE) mean and variance. In essence, the process discounts the impact of past FE in alphas, thereby producing a more stable output-reducing turnover.

As mentioned, BITA Robust employs quadratic constraints (not to be confused with the basics of quadratic optimization that uses linear constraints), which allow a number of key investment issues to be addressed:

- Uncertainty in forecast:
  - Expected alpha entered as a range with a degree of confidence.
  - Ranked expected returns.
  - Minimizing expected FE based upon historic FE.
- Risk budgeting
  - Allowing factor, or asset, or sector risk contributions to be set as constraints, enabling detailed risk budgeting within the portfolio construction.

This is considered, and has indeed been proved to offer a better approach than other variations on standard mean–variance optimization such as resampling and bootstrapping.

For mathematical details on BITA Robust, please see Appendix A.

3.4.3 Reformulation of mean–variance optimization

Traditional mean–variance optimization can be couched in terms of maximizing the following expression:

$$U(w, \lambda) = \alpha^T(w - w_B) - \frac{\lambda}{2}(w - w_B)^T Q(w - w_B) \tag{3.1}$$

subject to

$$Aw = d$$

where

U = utility;
w = vector of asset weights in the portfolio;
λ = degree of risk aversion.
α = vector of expected returns;
w_B = vector of asset weights in the benchmark;
C = covariance matrix of asset returns;
A = matrix of linear constraints;
d = vector of limits.

The mean–variance optimization problem is to find the values of w that maximize U given a degree of risk aversion λ, return expectations α, covariance matrix C, and benchmark b, subject to the set of linear constraints.

BITA Robust applies an interior point method to solve a "second-order cone problem". This has enabled the new functionality of applying quadratic constraints. Documentation on the method is in Appendix A. Using BITA Robust Optimizer to apply quadratic inequality constraints, the above can be reformulated as:

$$U(w,\lambda) = \alpha^T(w - w_B) \tag{3.2}$$

subject to

$$Aw = d$$

and the risk constraint (s):

$$\sum\nolimits_{j,k=1}^{n}(w_j - w_{Bj})(w_k - w_{Bk})Q_{ij} < (TE)^2 \tag{3.3}$$

or

$$\sigma^2 = \sum_{j=1}^{m}\sum_{k=1}^{m}\psi_j\psi_k\sigma_{jk}^2 + \sum_{i=1}^{n}\varepsilon_i^2(w_i - w_{Bi})^2 < (TE)^2 \tag{3.4}$$

in the structured factor model case, where:

σ^2 = variance of portfolio p;
ψ_j = sensitivity of portfolio p to factor $j = \sum_{i=1}^{n}\beta_{ij}(w_i - w_{Bi})$;
σ_{jk}^2 = covariance of factor j and factor k;
w_i = weight of security i in the portfolio;
w_{Bi} = weight of security i in the benchmark;
i = error term (residual risk);
TE = benchmark relative portfolio tracking error.

Equations (3.3) and (3.4) are identical, but Equation (3.3) expresses TE in terms of the full covariance matrix, while Equation (3.4) expresses it in terms of factor and residual risk. This eliminates the need for the problematic scaling coefficient λ. Being in the units of "marginal expected return with respect to variance," λ is difficult to evaluate with confidence. This ambiguity in λ is exacerbated by instability in alpha magnitudes when the binding mandate constraint is TE. In practice, users of optimization methods that require an objective function of the form in Equation (3.1) tend to make an initial guess at a value for λ, and subsequently tweak the value to achieve the tracking error required. Many products, including BITA Optimizer, mechanize this by putting an optimization script in place that automates this process.

Using a risk constraint rather than a risk term in the objective function enables the user to put a straightforward constraint on tracking error and the user will know, prior to execution, whether or not the problem is feasible.

3.4.4 BITA Robust applications to controlling FE

Regardless of whether risk is a term in the objective function, or applied as a constraint, both Equations (3.1) and (3.2) suffer from robustness issues. Both posit that the precise weights, w, can be extracted as outputs from this process given the uncertain vector of alphas, α. As α is not known with certainty and is subject to frequent and dramatic change, the output portfolio composition, w, is disproportionately unstable. In practical situations, a portfolio manager would still need to implement a turnover constraint to both methods to attenuate the effect of changing inputs on the output portfolio.

3.4.5 FE constraints

As one potential solution to this issue, we propose to introduce the concept of constraining FE. FE is the difference between the ex ante expectation of return and the realized return in the subsequent corresponding period. The FE on a security is easily measured and is the most straightforward statistic for how well an alpha predicts actual return of a security. It is not dependent on alpha building methodology—whether an alpha is built via multifactor model or expert opinion, a FE can be calculated. Even in cases where the alpha is a ranking or a sign, an analogous term can be constructed.

Observing FE for a historic period of at least 30 observations is recommended. This can be days, weeks, months, or any other period, but should correspond to the investment horizon of the strategy.

In parametric statistics, mean and standard deviation go a long way toward describing the behavior of a distribution. In a "perfect world" where alphas always correspond to subsequent returns, the mean and standard deviation of the FE sample would both be zero.

In the "real world," Robust Optimization is about constraining these two parameters in the realized portfolio to be below the respective constants, which we denote by M and S:

$$\sum\nolimits_{k=1}^{n}\left((w_{Ik} - w_{Ok})\mu[FE_k]\right)^2 < M^2 \tag{3.5}$$

and

$$\sum\nolimits_{j,k=1}^{n}(w_{Ij} - w_{Oj})(w_{Ik} - w_{Ok})\sigma[FE_j, FE_k] < S^2 \tag{3.6}$$

where

w_I = initial weight;
w_O = optimal weight;
$\mu[FE_k]$ = mean FE;
$\sigma[FE_j, FE_k]$ = covariance between FE j and k.

There are two things to note: (1) both are expressed in terms relative to the initial position and (2) both of these constraints are quadratic. By couching the constraints in terms of initial positions, we draw an explicit link between the concepts of robustness and turnover. An alternate way of thinking about Equations (3.5) and (3.6) is as "smart turnover" constraints—they both limit the optimizer's ability to deviate from the initial position, except for the case where there is a good reason (stable, historically accurate alpha) to do so.

Equation (3.5) is quadratic in order to penalize greater deviation from the initial position for stocks with a large FE—in this way, it largely resembles a quadratic penalty, levied as a constraint, rather than an objective function term. Equation (3.6) is quadratic because that is the nature of a covariance term, analogous to risk.

3.4.6 Preliminary results

Traditional methods used to stabilize portfolios, such as applying turnover constraints, merely serve to limit the damage of incorrect data being input into optimization techniques. These methods do nothing to differentiate between the "informative" and "noisy" data that goes into the process.

Controlling FE exposure and variance improves portfolio stability more than traditional methods. It directly addresses the main cause of instability: the instability in returns expectations, by limiting the effect of the "noisy" and allowing the "informative" to add value to the process.

To test this, we defined the following research project:

- US mid- and large-market capitalization equity universe limited as detailed below;
- Alphas from multifactor cross-sectional model (Value, Momentum);
- Limit universe to securities for which 30 months of data (returns, FE) are available—in most cases, this limited the universe to about 500 stocks due to the limited alpha universe during any given month.
- Backtest run on a monthly basis, starting January 1997–November 2005;
- Portfolio weights were adjusted at the end of each month for changes in price and optimized using one of two strategies using the market cap weighted universe as the benchmark:
  - Robust—constraining both M and S from Equations (3.5) and (3.6) to 1% each and tracking error to 5%;
  - Control risk and turnover—tracking error constrained to 5%, turnover constrained to realized turnover from robust strategy.

The resultant strategy performance is clearly dependent on how much it costs to trade. Two sets of results follow:

1. No cost to trade
2. 30 bps per unit of currency turned over

First, the "no cost to trade" results (Figure 3.1).

Now the 30 bps to trade results (Figures 3.2–3.4).

From the "net of control strategy" chart, it is clear that for the alphas provided in this test, the robust strategy outperformed the control strategy over

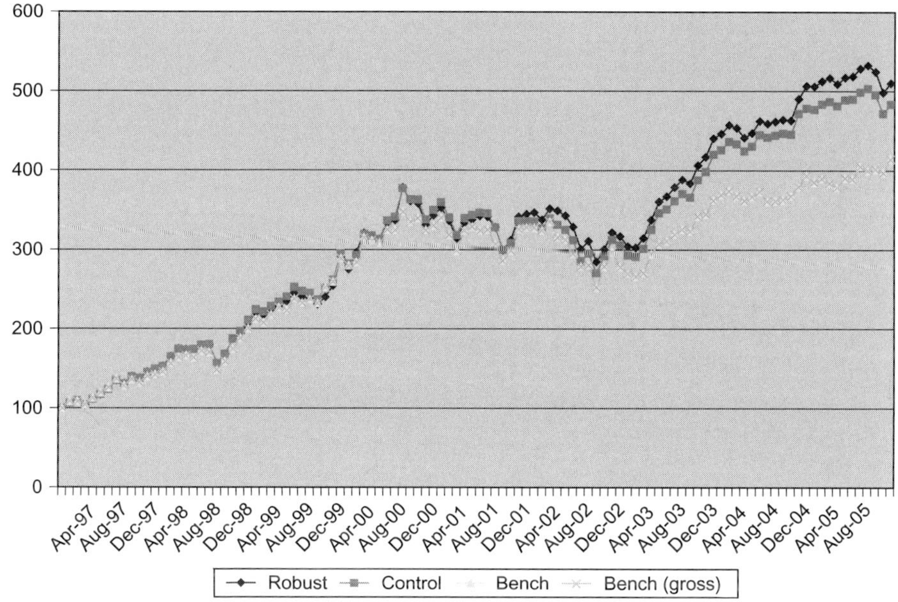

Figure 3.1 Strategy performance (no cost to trade).

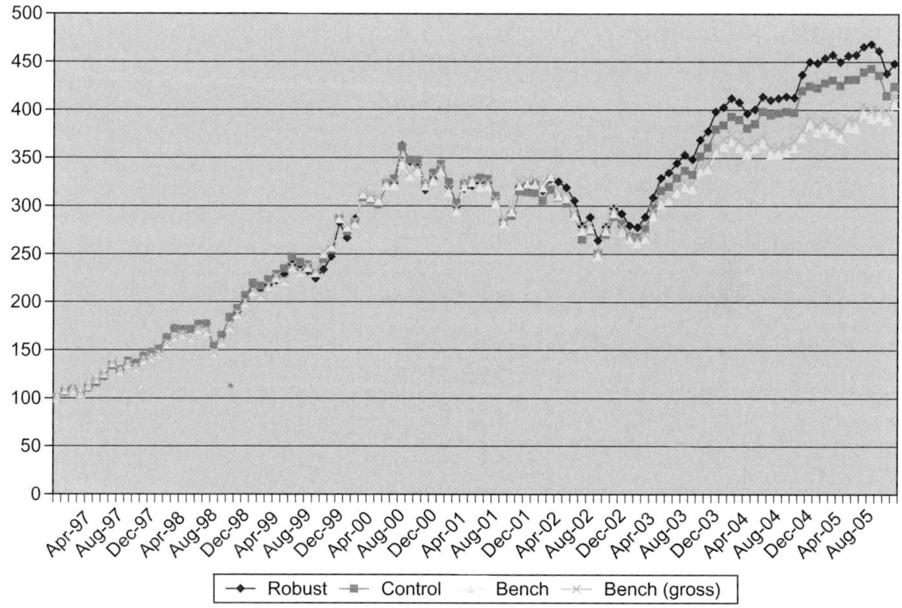

Figure 3.2 Strategy performance (30 bps to trade).

the run of the backtest. Any place where the line is going up, the robust strategy is actively outperforming the control strategy. Curiously, the two strategies tracked each other fairly well during the first third of the period with the robust strategy underperforming mildly just prior to the technology bubble

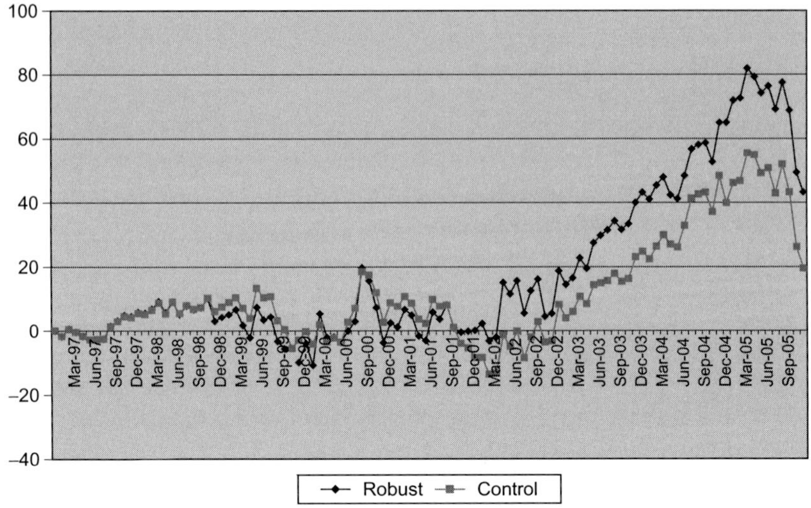

Figure 3.3 Net of benchmark performance (30 bps to trade).

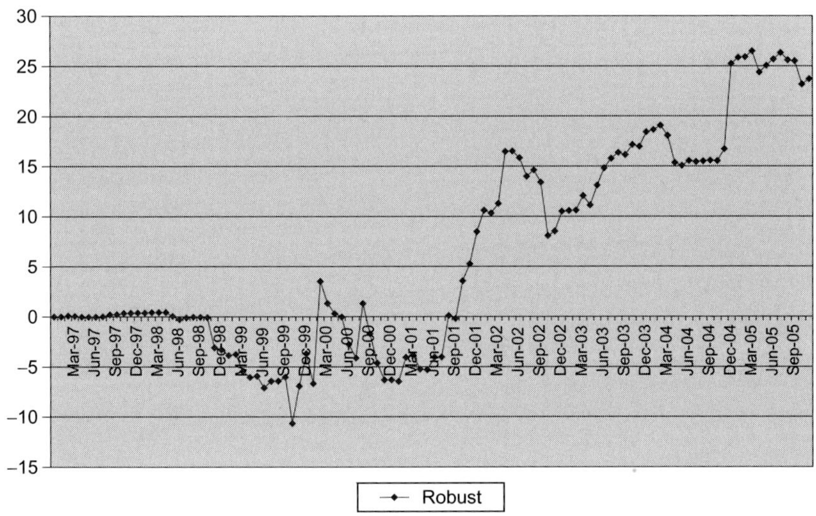

Figure 3.4 Net of control strategy performance (30 bps to trade).

bursting (presumably while the alphas worked). Starting in December 1999, however, just prior to the Dow's peak in January 2000, the robust strategy started to outperform. Furthermore, the outperformance was more or less consistent post bubble, starting in the fall of 2001.

Given that the sample period for the FE means and covariances is 30 months, it is not unreasonable to conclude that once the history contains an extremely volatile market event, the robust strategy starts to differentiate itself from the control strategy.

3.4.7 Mean forecast intervals

Following a suggestion by one of their clients, the BITA team incorporated mean forecast intervals into optimization processes. An example of such a forecast would be, "We assume that the mean return for equities lies between 6% and 8% p.a." The forecast uncertainty is captured by the interval around the expected mean (7% in the example forecast). Consistently, this allows assuming zero FE for the expected mean.

The formulation of this problem is comparable to the application of FE constraints. Equation (3.5), however, becomes redundant with regard to the assumption that uncertainty is completely captured by adding a forecast interval. Further assuming that FE across asset classes are independent, we can reduce Equation (3.6) to:

$$\sum_{k=1}^{n}(w_{Ik} - w_{Ok})^2\sigma^2[FE_k] < S^2 \tag{3.7}$$

where

w_I = initial weight;
w_O = optimal weight;
$\sigma^2[FE_k]$ = variance of FE k.

To compute the FE standard deviation, we map the mean forecast interval to a 95% confidence interval.[1] Assuming that asset returns follow a normal distribution, we can induce the standard deviation from the relationship:

$$d_k = 1.96\sigma[FE_k]$$

where

d_k = error from the estimation (half the interval).

For the example given above, we would have $d_k = 0.01$ and $\sigma[FE_k] = 0.51\%$.

Hence, adding Equation (3.7) as a constraint to the general mean–variance optimization problem in Equation (3.2), where we use the interval mean as the alphas, provides us with an additional application for BITA Robust. The importance of this approach relative to FE constraints is that it needs much lower data requirements. It still, however, allows the additional benefit of robust optimization to add value.

3.4.8 Explicit risk budgeting

BITA Robust enables a generic functionality to apply quadratic constraints to an objective function. The FE framework proposed above is not the only potential

[1] Since the forecast interval can be as wide or narrow as the investor wishes (depending on his or her confidence), we keep the confidence interval fixed at 95%. However, it would, in general, be possible to vary the level to fit the user's degree of confidence.

application of this enhanced functionality to stabilize optimization results. In particular, a further area of research is explicit risk budgeting via constraints.

In the context of a structured factor risk model, this can be implemented as separate constraints on Factor Contributions to Total Factor Risk or to Total Risk. In addition, it would be possible to separately constrain asset-specific risk contribution to Total Risk.

Constraining Contributions to Total Risk enables the user to dictate where exposure to risk is taken, in essence "budgeting" risk exposure. This is useful in hedging risk exposure in areas where alpha expertise is not claimed, e.g., if the manager is taking a systematic bet, the contribution to risk constraint to the relevant risk factor can be loosened. Likewise, if the manager is making asset-specific bets, exposure to asset-specific risk can be expanded. This is different from the traditional factor exposure constraints in that it is constraining the resultant (output) contribution to risk, not the input exposure. In addition, this procedure tells the user a priori whether the goal is feasible, eliminating the need to fiddle with inputs in vain attempts to reach an impossible solution.

Alternatively, if the risk model structure does not represent a relevant categorization upon which to levy risk constraints, it is possible to use any external, arbitrary categorization system and apply risk constraints to each category. For example, if the risk model does not contain sector exposures, and it is important to budget for risk at the sector level, it is possible to levy risk constraints for each of the relevant sectors of the type in Equation (3.4). Effectively, this means splitting the portfolio into several subsections corresponding to sector classifications and constraining the tracking error of each of those. This would be useful in producing a more stable result by ensuring that risk is taken evenly across the investment universe. For example, if the alphas are skewed toward the technology sector, such an approach would prevent the optimizer from loading up on volatile technology assets at the expense of volatile assets in other sectors, in effect forcing the optimizer to diversify within sectors as opposed to across sectors.

3.5 BITA GLO$^{(TM)}$-Gain/Loss optimization

3.5.1 Introduction

Mean–variance investing is arguably the most popular framework to construct optimal portfolios. However, its consistency with expected utility maximization requires that returns are normally distributed and/or the investor's preferences are quadratic.

First, the assumption of normality in return distributions is breached for many asset classes. Second, an investor's utility might rather be driven by the impact of expected gains and losses than by expected return and variance. This is strongly supported by experimental evidence in Kahneman and Tversky (1979). Considering expected gains and losses (i.e., expected returns relative to a target) is part of the prospect theory framework. The theory also includes that an investor is actually concerned about changes in wealth (and not about the absolute level of wealth) and that he or she experiences a higher sensitivity to losses than to gains as described in Cuthbertson and Nitzsche (2004). The latter is expressed by a factor lambda (>1) that addresses the loss aversion within a prospect utility function. For such an investor, who faces asymmetric distributions and defines his or her utility through a functional combination of expected gains and losses, we have developed a suitable (Gain/Loss) optimizer.

3.5.2 Omega and GLO

Omega (Ω), defined as expected gains divided by expected losses, seems to be a natural ingredient of a prospect utility function and became very popular since its introduction in Cascon, Keating, and Shadwick (2002) and Keating and Shadwick (2002a,b). It also "takes the whole information set including higher moments into account" as in Passow (2005) and, hence, does not base on any assumptions in this regard (as normality). When it comes to optimization, however, it appears more sensible to formulate a general utility function as $E[gains] - (1 + l)E[losses]$, where l = loss aversion and "gains" are defined as $(\omega - R)|\omega > R$ and "losses" as $(R - \omega)|\omega \leq R$, where ω and R denote (final) wealth and target rate, respectively. The derivation of this formulation compares to the Sharpe ratio (excess return divided by standard deviation) and its derived utility function $\mu - \frac{\lambda}{2}\sigma^2$.

It is noted that the formulation as "expected gains *minus* loss aversion times expected losses" contrasts with a ratio such as Ω. In fact, it is currently fashionable to treat Ω as a risk metric, which can be computed against the returns of the chosen portfolio. However, there is much that is wrong about this procedure. To start with, Ω does not seem to be a globally concave function of portfolio weights. This means that we will constantly rely on constraints to find the optimum; indeed generically there should be many optima. Furthermore, it seems that Ω is a return metric rather than a risk metric. That is to say, if we could compute our expected utility as a function of Ω, we would expect the partial derivative of expected utility with respect to Ω to be positive.

A further reference capturing the advantages of expected shortfall (ES) or CVaR optimization (which is mathematically equivalent to GLO) is Ho, Cadle, and Theobald (2008) whose conclusions in comparing mean–variance, mean–VaR, and mean–ES portfolio selections are cited below.

> *"We find that the mean–ES framework gives a reasonable and intuitive solution to the asset allocation problem for the investor. By focusing upon asset allocations during the recent "credit crunch" period, we are able to develop insights into portfolio selection and downside risk reductions during high volatility periods.*

> *The expected shortfall values of the optimal portfolio generated by the mean–ES framework are reduced relative to those of the traditional mean–variance and mean–VaR approaches. Furthermore, when portfolio returns are assumed to be normally distributed, the optimal portfolios selected within the mean–VaR and mean–ES frameworks are identical to those generated by the traditional mean–variance approach. Our results are consistent with a growing risk management literature which suggests that, when dealing with fat-tailed distributions, there is a need to focus on the tail region, and further, that expected shortfall is a better framework to base decisions upon than Value-at-Risk."*

Of course, the computation of Ω corresponding to our maximized expected utility remains possible.

3.5.3 Choice of inputs

In order to present the behavior of this GLO, it seems appropriate to compare its outcomes with mean–variance optimization (expected return minus lambda times variance) for various parameterizations. The asset classes chosen are subject to the constraint of long-only investments. We fix the target rate or threshold R that decides if a return is treated as a gain or a loss to 4.8%, which is comparable to the current annual interest rate.[2]

In order to nominate a value for l for the GLO, we confront a virtual investor with a gamble comparable to the Basic Reference Lottery.[3] This earns him or her either a utility rate U (probability = ½) or he or she loses U (also probability = ½), so that his or her total utility equals:

$$\frac{1}{2}U - \frac{1}{2}(1 + l)U$$

The insurance D he or she would pay to avoid the gamble (i.e., the difference between the gamble and a certainty equivalent) reveals his or her loss aversion:

$$\begin{aligned} -(1 + l)D &= \frac{1}{2}U - \frac{1}{2}(1 + l)U \\ <=> l &= \frac{2D}{U - 2D} \end{aligned}$$

Giving up 3% of one's wealth to avoid the uncertainty of a 10% gain or loss with equal probability would, hence, imply $l = 1.5$. Evaluating this as a fairly high degree of loss aversion, we choose a more conservative value of $l = 1.0$. Another procedure to capture loss aversion, which might be more suitable for an institutional investor, is described in section 6.

[2] At the time of the analysis (early 2008).

[3] Gains/Losses utility is linear in wealth and, hence, independent of an investor's wealth level. This allows calculations in terms of rates.

Table 3.1 Summary of optimization results: maximum holding = 100%

	Mean%	StdDev%	Skewness	Kurtosis	Gain%	Loss%	Omega	Probability
(a) G/L $l = 1$	12.63	5.73	−0.79	3.62	12.68	4.85	2.614	0.72
(b) M-V $\lambda = 2$	24.40	20.65	−0.92	4.41	38.65	19.05	2.029	0.66
(c) M-V $\lambda = 10$	15.14	8.55	−0.90	3.72	18.08	7.74	2.336	0.70

Table 3.2 Summary of optimization results: maximum holding = 25%

	Mean%	StdDev%	Skewness	Kurtosis	Gain%	Loss%	Omega	Probability
(a) G/L $l = 1$	13.66	7.73	−0.94	3.86	15.67	6.82	2.298	0.70
(b) M/V $\lambda = 2$	17.51	11.85	−0.96	3.71	23.99	11.27	2.129	0.70
(c) M/V $\lambda = 10$	15.37	9.37	−0.89	3.57	19.38	8.81	2.200	0.67

In order to determine values for risk aversion, we consider an investor who holds a portfolio of cash and FTA only. On the basis of the constant absolute risk aversion MV framework, his or her expected utility is expressed by $E[U(w)] = w\mu + (1 - w)r_f - \frac{\lambda}{2}w^2\sigma^2$, where $w,\mu,\lambda,\sigma^2,r_f$ are the optimal proportion in equity, the expected rate of return, the coefficient of absolute risk aversion, the variance of the FTA, and the riskless rate return, respectively. Maximizing this utility yields $w = \frac{\mu - r_f}{\lambda\sigma^2}$ or $\lambda = \frac{\mu - r_f}{w\sigma^2}$. Considering current values ($\mu = 9\%$, $\sigma = 12\%$, $r_f = 4.6\%$) reduces this to the trade-off $\lambda = \frac{3}{w}$. The proportion of equity investment for institutional investors is indicated by peer-group benchmarks as WM to be 70%. This corresponds to $4 \leq \lambda \leq 5$ as an adequate range for the average institutional investor but probably inappropriate for high net investors. On the basis of these calculations within our mean–variance analysis, we obtain $\lambda = 2$ and $\lambda = 10$ to represent high and low risk aversion. These values would fit portfolios of cash and FTA only with equity positions of 150% and 30%, respectively.

3.5.4 Analysis and comparison

We divide our analysis into four parts, each representing a change in constraints, parameters, or the specification of the utility function; the main characteristics are summarized.

3.5.5 Maximum holding = 100%

a. Gain/Loss optimization: $E[U_{G/L} = E[gains] - (1 + l)E[losses], l = 1$

b. Mean–variance optimization: $E[U_{M/V, \lambda=2}] = \mu - \dfrac{\lambda}{2}\sigma^2, \lambda = 2$

c. Mean–variance optimization: $E[U_{M/V, \lambda=10}] = \mu - \dfrac{\lambda}{2}\sigma^2, \lambda = 10$

All three optimizations include emerging markets in their optimal portfolio. However, GLO obviously considers that asset class' negative skewness and high kurtosis and chooses a much lower weight (13% versus 100% and 25% for low and high risk aversion mean–variance optimization, respectively). GLO and high risk aversion choose the same three asset classes, while the weights differ depending on skewness and kurtosis, which are considered by GLO. Hence, the GLO portfolio has least negative skewness and least kurtosis, suggesting that higher moments are dealt with. Naturally, it also shows the largest Ω.

Broadly speaking, portfolios that control expected losses unsurprisingly have lower expected gains, lower expected losses, and a higher probability of achieving a target. Contrary to popular suspicion, they are not less diversified, they have lower negative skewness, lower kurtosis, lower standard deviation, and lower expected return compared to either high or low risk mean–variance portfolios.

3.5.6 Adding 25% investment constraint

For investors who seek for more diversification, we repeat the analysis with the additional constraint that only 25% may be invested in each asset class.

Again, the GLO seems to be more diversified than either mean–variance solution. The extent to which the constraints bind can be assessed by the number of holdings of 25%. For GLO, we have 3 with three other assets. For high risk tolerant mean–variance, we have 4 and for low risk tolerant mean–variance we have 2 with three other assets.

Again, the same asset classes are considered, as high risk aversion mean–variance adds one class to the 4 chosen by low risk aversion mean–variance and GLO adds an additional one to those 5. That class only picked by GLO, symptomatically shows positive skewness and kurtosis <3.

Now, the improvements in terms of moment behavior are less, skewness is ranked second for GLO, and kurtosis is highest. However, the standard deviations are lowest, probability of achieving the target is highest, and the expected loss is lowest.

3.5.7 Down-trimming of emerging market returns

It might be that the emerging market returns are epoch-dependent and we will get a clearer picture if we trim them down. We do this by subtracting 1% per month off each historic return and repeat our analysis for our three optimizations, both 100%-constrained and 25%-constrained (Tables 3.3 and 3.4).

The GLO (100%-constrained) excepts one asset class that is included by both mean–variance optimizations owing to its negative skewness and higher kurtosis. Otherwise, the results are broadly similar to those in Tables 3.1 and 3.2; emerging markets are only considered by 3.4.

3.5.8 Squared losses

An investor whose disutility of a double unit loss is more than twice a single unit loss might find a utility function that includes the square of expected losses suitable. We also constrain property (IPD) to 10% in all cases (Table 3.5).

a. GLO: $E[U_{G/L}] = E[gains] - (1 + l)E[losses],\ l = 1$

b. GLO: $E[U_{G/L}] = E[gains] - \frac{l}{2}E[(losses)^2],\ l = 20$

c. GLO: $E[U_{G/L}] = E[gains] - \frac{l}{2}E[(losses)^2],\ l = 4$

Table 3.3 Summary of optimization results: maximum holding = 100%, trimmed EM

	Mean%	StdDev%	Skewness	Kurtosis	Gain%	Loss%	Omega	Probability
(a) G/L $l = 1$	10.89	4.07	−0.21	3.01	9.27	3.17	2.924	0.75
(b) M/V $\lambda = 2$	18.17	14.39	−0.53	2.85	28.24	14.87	1.900	0.67
(c) M/V $\lambda = 10$	13.50	7.22	−0.66	3.17	15.46	6.76	2.317	0.69

Table 3.4 Summary of optimization results: maximum holding = 25%, trimmed EM

	Mean%	StdDev%	Skewness	Kurtosis	Gain%	Loss%	Omega	Probability
(a) G/L $l = 1$	11.27	5.57	−0.68	3.37	11.54	5.08	2.271	0.72
(b) M/V $\lambda = 2$	14.48	11.70	−0.96	3.70	21.69	12.01	1.805	0.70
(c) M/V $\lambda = 10$	12.99	7.29	−0.68	3.30	15.11	6.92	2.183	0.69

Table 3.5 Summary of optimization results: IPD constrained to 10%

	Mean%	StdDev%	Skewness	Kurtosis	Gain%	Loss%	Omega	Probability
(a) G/L $l = 1$	7.74	4.80	−0.20	3.61	8.00	5.06	1.581	0.60
(b) G/L^2 $l = 20$	7.42	4.59	−0.32	3.69	7.50	4.88	1.537	0.59
(c) G/L^2 $l = 4$	7.60	4.70	−0.31	3.70	7.75	4.96	1.563	0.61
(d) M/V $\lambda = 2$	10.21	12.30	−0.21	4.55	18.59	13.17	1.412	0.56
(e) M/V $\lambda = 10$	8.12	5.30	0.00	3.40	9.07	5.75	1.577	0.58

d. Mean–variance optimization: $E[U_{M/V,\lambda=2}] = \mu - \frac{\lambda}{2}\sigma^2, \lambda = 2$

e. Mean–variance optimization: $E[U_{M/V,\lambda=10}] = \mu - \frac{\lambda}{2}\sigma^2, \lambda = 10$

It is interesting that the optimal GLO portfolio with $l = 4$ is almost identical to the mean–variance portfolio with $\lambda = 10$. They differ in weights by 1–4 basis points only. Increasing 1 to 20 changes the GLO portfolio substantially. Another 38% is shifted to the asset class with the least negative skewness and least kurtosis (resulting in a total weight of 90%). Also, a fourth asset class is added, with even lower negative skewness and lower kurtosis.

3.5.9 Conclusions

We investigate the extent to which our GLO adds value to mean–variance optimization. It became obvious that both optimization methods can lead to very similar results. Differences are mainly caused by higher loss aversion and asymmetric return distributions. In contrast to mean–variance optimization, the GLO accounts for higher moments and, hence, captures asymmetries in return distributions. In summary, GLO seems to deliver results similar to mean–variance optimization when it is applied to relatively normally distributed returns. Facing asymmetries instead, the GLO is more preferable as it can distinguish by considering higher moments where mean–variance optimization has to assume normal distributions.

To decide which optimization method is most adequate for an investor, he or she has to consider his or her utility function and the distribution of returns of his or her choice of asset classes. Variance is a measure of risk treating both sides of the mean equally. As prospect theory and empirical evidence hypothesize, there are investors who experience downside risk differently to upside potential. Those would define their utility function rather through expected gains and losses. If an investor cannot assume normal return distributions (and chooses another than the well-known quadratic utility function) he or she has to take care of higher moments. Such investors will find our GLO beneficial.

3.6 Combined optimizations

3.6.1 Introduction

Mean–variance (MV) analysis has delivered less interesting portfolios for some time. Mainly the sensitivity of optimal portfolio weights to estimated inputs inspired investors to look for "more robust" portfolios. Closest to MV, obviously, as one specific MV-efficient portfolio is the minimum variance portfolio. Being the only MV portfolio that does not incorporate return expectations, it delivers the desired robustness. However, it still suffers from the second drawback of MV optimization, namely overly concentrated, i.e., not at all diversified, optimal portfolios. Benartzi and Thaler (2001) and Windcliff and Boyle (2004) show that a heuristic approach that seems to naturally solve this diversification weakness is the application of equal weights to the investment universe, i.e., $w = 1/n$ where n is the total number of assets. This also provides robustness as it does not build on estimated inputs and has been shown by DeMiguel, Garlappi, and Uppal (2007) to be efficient under certain circumstances. While achieving diversification with regard to weights, this approach could come with concentration of risks. Hence, portfolios of equal risk contributions of underlying assets were a natural consequence as shown in Maillard, Roncalli, and Teiletche (2009). This, obviously, is also at best a compromise and not based on optimization processes. A second reason for the consideration of alternatives to MV is the recognition that MV utility is inadequate in many situations, while alternatives, such as GLO, seem to come as a remedy. We detail the argument in the discussion below.

These considerations lead us to propose a new utility function that is an "amalgam" (linear combination of MV and GLO). As argued by Brandt (2009), choosing the appropriate objective function is arguably the most important part of the entire portfolio optimization problem. Although many different objective functions have been suggested by practitioners and academics alike, there is little consensus on the appropriate functional form. Consequently, we can choose a finite number of "potential" objective functions and assign larger weights to the objective functions that we believe are most representative of the latent investor preferences. Indeed, we could even go further and nest this idea within a Bayesian framework whereby we formally model these weights as prior probabilities using probability distributions defined upon the unit simplex. Interpreted from this perspective, our utility function parallels estimation under model uncertainty, i.e. the portfolio builder is not clear which is the true objective function but has some intuition as to which objective functions are more reasonable than others.

Alternatively, it may also be possible to derive such a utility function from aggregating the utility functions of individual investors. These "aggregate" utility functions were first explored by Harsanyi (1955) who argued that if both social welfare and individual utility functions take the von Neumann–Morgenstern form, then any social welfare function that satisfies the Pareto principle must be a weighted sum of individual utility functions. What is more the notion of an 'individual' can be made quite general depending on the context; for instance, we can endow a pension fund with one utility function reflecting short run investment and another reflecting long run investment.

We discuss the literature issues in the next section, followed by a discussion of the model, and concluded by a discussion of the incorporation of alpha and risk information.

3.6.2 Discussion

Von Neumann and Morgenstern (1944, hereafter vNM) define utility as a function $U(W)$ over an investor's wealth W. To make use of this decision theoretic foundation in portfolio construction, Markowitz approximated the vNM expected utility theory by his well-known MV utility function from Levy and Markowitz (1979). This is justified for normal return distributions or quadratic utility functions. However, as demonstrated by Mandelbrot (1963), the assumption of normal return distributions does not hold for many assets. Also, many investors would not describe their perception of risk through variance. They relate risk rather to "bad outcomes" than to a symmetrical dispersion around a mean. Sharpe (1964) shows that even Markowitz suggested a preference of semivariance over variance due to computational constraints at the time.

One attempt, Jorion (2006), to describe risk more suitably was the use of VaR. While focusing on the downside of a return distribution and considering nonnormal distributions, it comes with other shortcomings. VaR pays no attention to the loss magnitude beyond its value, implying indifference of an investor with regard to losing the VaR or a possibly considerably bigger loss, and it is complicated to optimize VaR. Using VaR for nonnormal distributions can cause the combination of two assets to have a higher VaR value than the sum of both assets' VaR, i.e., VaR lacks subadditivity. Subadditivity is one of four properties for measures of risk that Artzner, Delbaen, Eber, and Heath (1999) classify as desirable. Risk measures satisfying all four are then called coherent.

Contrary to VaR, lower partial moments are coherent. Fishburn (1977) introduced these and used them as the risk component in his utility function. Here, an investor would formulate his or her utility relative to target wealth, calling for sign dependence. Final wealth above the target wealth has a linearly increasing impact on utility, while outcomes below the target wealth decrease utility exponentially.

Closely related is the expected utility function Kahneman and Tversky (1979) proposed under another descriptive model of choice under uncertainty, as an alternative to vNM. They provide strong experimental evidence for the phenomenon that an investor's utility is rather driven by the impact of expected gains and losses than by expected return and variance. Considering expected gains and losses (i.e., expected returns relative to a target) is part of their prospect theory framework. The theory also includes that an investor is actually concerned about changes in wealth (and not about the absolute level of wealth) and that he or she experiences a higher sensitivity to losses than to gains. The latter is expressed by a factor (>1), which addresses the loss aversion within a prospect utility function.

Hence, it appears that GLO is more amenable to modeling individual utility and is valid for arbitrary distributions; however, it is not set up for using the great deal of valuable information that active managers have accumulated over years of running successful funds. This information is usually in the form of a history of stock or asset alphas and a history of risk model information. Whilst it is possible to use this to improve GLO, it is obvious that gain and loss is a different reward and risk paradigm from the more traditional mean and variance.

3.6.3 The model

We thus propose a utility function V, where:

$$V = \theta(\mu_p - \lambda\sigma_p^2) + (1 - \theta)(g_p - (1 + \varphi)l_p) \tag{3.8}$$

where

μ_p = mean of portfolio p;
σ_p^2 = variance of portfolio p;
g_p = gain of portfolio p relative to some target t;
l_p = loss of portfolio p relative to some target t;
λ = risk aversion coefficient in the MV framework;[4]
ϕ = loss aversion coefficient in the GL framework.

The parameter θ is a weight parameter that takes values between 0 and 1. If $\theta = 1$, we have MV and if $\theta = 0$, we have GLO. This can also be set to resolve scale issues that arise from using alphas in the mean–variance that are not exactly expected annualized rates of return. To determine ϕ, a number of possible approaches can be considered. One we investigate in this chapter is based on the idea that the maximum drawdown on the resulting optimal strategy should be less than 2.5% per month. This approach is described in the final part of this section.

We demonstrate that the above model is equivalent to subtracting variance from a GLO problem. Using the fact that $\mu_p - t = g_p - l_p$ where t is the target rate, it follows that Equation (3.8) becomes:

$$V = g_p + \theta t - (1 + (1 - \theta)\varphi)l_p - \theta\lambda\sigma_p^2 \tag{3.9}$$

[4] We understand that the loss aversion coefficient and the risk aversion coefficient will probably change if they are combined in a mutual utility function. However, this matter will be excluded from this chapter.

It follows that, mathematically, our problem is the same if we simply subtract a portfolio variance term from a GLO. There are a number of considerations that make Equation (3.8) more appealing than Equation (3.9) notwithstanding their mathematical equivalence. These are the incorporation of risk model and alpha information into the problem. We shall deal with each of these issues next.

3.6.4 Incorporation of alpha and risk model information

Incorporating risk model information

In the combined model given by Equation (3.8), risk model information can be directly incorporated into backtesting in the usual way. In the special but important case when we only have gain–loss, the method for incorporating risk model information becomes much more complicated and less obvious. One possible solution is to use the risk model's linear factor structure combined with the stock alphas plus Monte Carlo simulations to create artificial histories of data, which can be used as inputs in the optimization.

Incorporating alpha information

One major difficulty that is virtually universal is trying to identify how long in the future alphas will be effective. This can be partially addressed by information coefficient (IC) analysis, which computes the correlation coefficient either cross-sectionally or as a time series between alphas and future actual returns. Alternatively, one can regress current returns on lagged alphas, this having the advantage that the regression effectively rescales the alphas, saving one the need to do it separately.

One could then include this forecasted return along with K historical returns in the GLO. The difficulty with this is that the alpha information will not enter into the overall data very much. A more ad hoc approach is to use a weighted combination of forecasted past returns with actual past returns. Whilst this may seem ad hoc, it should be able to be justified by some variant of a Bayesian argument much as the one that underpins Black–Litterman, as discussed by Scowcroft and Satchell (2000) and Meucci (2008).

Determination of the loss aversion coefficient

As mentioned above we make use of a drawdown constraint to determine the loss aversion coefficient ϕ. In particular, we consider a maximum drawdown of 2.5% per month. To find a suitable ϕ, we try values between 0.0 and 5.5 in our backtest and look at the resulting drawdowns. Depending on the investor's requirements, ϕ could be chosen as the value that strictly leads to drawdowns smaller than the defined maximum drawdown. Or, for a less constrained investor, the distribution of drawdowns over the backtest period per trial value of ϕ could be examined. One could apply a confidence level and choose ϕ as the value that leads to drawdowns smaller than the defined maximum drawdown given a particular confidence level. This would allow smaller values for ϕ than the first approach.

We consider a subuniverse of our global stocks universe of 250 randomly drawn stocks (with complete history) over the last 4 years, measured weekly. The only constraint we apply here is:

$$-0.02 \leq w_i \leq 0.02$$

The measurement of drawdowns allows certain flexibility with regard to the horizon over which it is calculated. Our analysis starts with the following definitions:

$$D(\tau) = \text{Max}[0, -(R(\tau) / \text{Max}_{t \in (0,\tau)} R(t) - 1)] \tag{3.10}$$

where $\tau \in (0,T)$.

The maximum drawdown (MDD) as the worst of all drawdowns then is:

$$\text{MDD}(T) = \text{Max}_{\tau \in (0,T)}[-(R(\tau) / \text{Max}_{t \in (0,\tau)} R(t) - 1)] \tag{3.11}$$

or—using the results of calculations based on the first definition

$$\text{MDD}(T) = \text{Max}_{\tau \in (0,T)} D(\tau) \tag{3.12}$$

While these definitions allow drawdowns over horizons up to T, we limited this horizon subsequently to any period within a calendar month and then any period within 4 weeks, so:

$$D(m) = \text{Max}[0, -(\text{Min}_{t \in (m_s, m_e)} R(t) / R(m_s) - 1)] \tag{3.13}$$

where m_s and m_e are the beginning and end of a month m (1 ... 35), respectively, and

$$D_1(4w) = \text{Max}[0, -(\text{Min}_{t \in (w, w+4)} R(t) / R(w) - 1)] \tag{3.14}$$

where w (1 ... 152) is a week of the cumulative return series.

In both cases, the maximum drawdown would be calculated similar to Equation (3.12).

Finally, the "drawdown" could be calculated over a fixed period of 4 weeks, which is effectively the 4 weeks return, i.e.,

$$D_2(4w) = R(w + 4) / R(w) - 1 \tag{3.15}$$

Please note that in this case, the "drawdown" is any 4 weeks return, i.e., in contrast to the other drawdown definitions, definition (3.15) results in *negative* drawdown figures for negative returns.

In our attempt to find a suitable loss aversion coefficient that would allow us to meet the drawdown target, we focused on the last definition (3.15). However, analysis for other drawdown definitions is available from the authors as well. We used loss aversion coefficients ranging from 0.0 to 5.5 in 0.5 steps (i.e., 12 backtests). The resulting returns for a loss aversion coefficient of 0 and 3.5 based on definition (3.15) are shown in the appendix. On the basis of these results, we propose a loss aversion coefficient ϕ of 3.5. This allows a few cases of exceeding the drawdown limit of 2.5%; however, the vast majority is just below this limit.

3.7 Practical applications: charities and endowments

3.7.1 Introduction

Endowments present interesting challenges, both to the determination of an appropriate investment policy and its implementation. These challenges differ from those generally encountered by private investors and pension funds because of the unusual characteristic of the majority of endowments, which is their perpetual constitution or effectively infinite investment horizon. The practical expression of this characteristic is the balancing of intergenerational interest in the endowment, or, put another way, sustaining both annual income and long-term real residual capital value.

In this section, we will explore these issues using the concept of the Utility of the Endowment. This concept will be defined and developed later. We will begin with an overview of UK endowments and those aspects of their operation that differentiate them from other investment funds. Although we will not discuss the principles of total return investment in this section, we will assume that income may be a combination of dividends and interest (conventional income) and realized capital.

3.7.2 Why endowments matter

By endowment we mean the investment funds and property bequeathed to charitable organizations that we shall call Foundations, which function in large part or entirely on the returns generated from those assets. Such organizations vary greatly in character and include religious orders, schools, universities and colleges, research bodies, hospital trusts, and grant-making charities that in turn support the work of other charities and good causes.

In the UK, charities exercise a significant influence across a wide range of social, scientific, educational, religious, sporting, cultural, and environmental enterprises. The UK Voluntary Almanac 2009[5] identifies an annual income of £33.2 billion, total assets of £91.3 billion, and a paid workforce of at least 634,000 under the control of General Charities.[6] Significantly, NCVO estimates in its 2008 Almanac that two-thirds of the income is generated by just 2% (3,200) of the total number of organizations in the universe, demonstrating the disproportionate influence of the largest. The income attributable to General Charities is derived from a combination of investment income, donations, grants from government or other charities, and from fees for services provided.

[5] Published by the National Council for Voluntary Organisations (NCVO).

[6] General Charities as defined by NCVO do not include housing associations, independent schools, and government-controlled charities such as NHS charities or religious organizations.

The true scale of Foundations is impossible to quantify but the Charities Aid Foundation identifies the Top 500 UK Trusts in 2006[7] with aggregate assets of £33.3 billion, generating an income of £4.5 billion. The Top 500 Trusts includes a spectrum of research and grant-making Foundations but does not include community foundations, schools, universities, or religious organizations. The scale and variety of Foundations and their endowments is illustrated in Table 3.6, which has been drawn from the Top 500 UK Trusts in 2006.

3.7.3 Managing endowments

The benefactors of Foundations often stipulate criteria within which trustees must operate and, not infrequently, this will compel the trustees to manage the bequest in perpetuity. Even where the obligation to maintain the endowment in perpetuity is not explicitly specified, many institutions depend on their endowment to sustain their operation through time. In all such instances, the requirement to balance the desire for maximum current income and future sustainability is foremost in the operation of an appropriate investment policy. The balancing of intergenerational interest has been highlighted by Tobin (1974), writing on the nature of permanent endowment income:

> *The trustees of an endowed institution are the guardians of the future against the claims of the present. Their task is to preserve equity amongst generations. The trustees of an endowed university like my own [Yale University] assume the institution to be immortal.*

Table 3.6 UK foundations

Organization	Founded	Objective	Assets (£ million)
The Wellcome Trust	1936	Biomedical research	11,674
Garfield Weston Foundation	1958	Grant-making to a range of projects spanning arts, community, education, welfare, medical, social, religion, youth, and environment	3,196
The Leverhulme Trust	1925	Financial support for research & education	1,014
The Henry Smith Charity	1628	Releasing people from need and suffering	622
The Royal Society	1660	The pursuit of science	130
The Clothworkers' Foundation	1977	Capital grants to charities engaged with young people, the elderly, social inclusion, disability, visual impairment, and textiles	66

[7] CAF Charity Trends 2006. Published by Charities Aid Foundation and Caritas Data Ltd.

Tobin eloquently encapsulates the constraint, both practical and fiduciary, that governs organizations that expect or intend to perpetuate through the ages. Such Foundations we shall define as "infinitely lived." Implicit in the decision to limit present consumption for future consumption is the assumption that the future good derived from the endowment is roughly equal on some measure of benefit to the good derived in the present. What will exercise trustees is the comparative cost of delivering the same level of benefit over time (assuming an inflationary impact on goods and wages), hence the preoccupation with the real returns on residual capital after current consumption. Tobin, again, expresses this in the following terms:

> *They [the trustees] want to know, therefore, the rate of consumption from endowment which can be sustained indefinitely. Sustainable consumption is their conception of permanent endowment income.*

Trustees in other circumstances may take a very different approach to distributing an endowment. Assume it is possible to eradicate a disease by pumping money into a research program. The trustees will have a strong subjective bias to present over future benefit and could justify disbursing the entire endowment over a finite time to achieve their objective. As such, three types of endowment can be distinguished:

- Infinitely lived—where rates of present consumption must not compromise future sustainability.
- Finite certain—where the trustees have a fixed investment horizon and wish to maximize returns through that time.
- Finite uncertain—where through unforeseen circumstances or the need to redirect finances to an alternative objective, the maintenance of the residual real value of capital can no longer be a priority.

3.7.4 The specification

Part 1—Find the optimum drawdown by maximizing the expected utility function

$$U(C_0, C_1, \cdots, W_T) = \sum_{t=0}^{T-1} U(C_t, t) + B(W_T, T)$$

where $U(C_t,t)$ is utility at time t depending on consumption at time t and $B(W_T, T)$ is the bequest function (realized at the end of the charity's existence), which depends on the remaining wealth W_T left as legacy.

If we consider infinitely lived charities, the utility function is:

$$U(C_0, C_1, \cdots, W_T) = \sum_{t=0}^{\infty} U(C_t, t)$$

The single period utility is defined as:

$$U(C, t) = \delta^t U(C)$$

where δ is the subjective discount rate.

The optimal drawdown depends on the entity's subjective discount rate and level of risk aversion. To link drawdown to asset allocation, we use the following power utility function:

$$U(C,t) = \frac{1}{1-\alpha} C(t)^{1-\alpha}$$

where α = risk aversion.

The optimal drawdown is:

$$m = 1 - (\delta\varphi)^{\frac{1}{\alpha}}$$

where $\varphi = E(\tilde{Z}_i^{1-\alpha})$.

To estimate ϕ, we bootstrap the monthly portfolio returns using 120,000 random draws, and sum to get 10,000 annual real (gross) returns.

We then use these to compute the sample mean:

$$\tilde{\varphi} = \frac{1}{10{,}000} \sum_{i=1}^{10{,}000} \tilde{Z}_i^{1-\alpha}$$

Part 2—Given the optimum drawdown as derived from the power utility function, we distinguish between:

1. Neutrality: future equal to past
2. Optimism: future better than past
3. Pessimism: future worse than past

Starting from historic data, we calculate the portfolio return:

$$R_{P,t} = \sum_{i=1}^{n} w_{it} r_{it} \quad t = 1, \cdots, T$$

and split the results into good (x%), bad (y%), and neutral ($100 - x - y$%).

For example, to build an optimistic scenario, we oversample from the good and undersample from the bad.

We consider more transformations of distributions that can be described for FSD. For an arbitrary positive continuous density $pdf(x)$; we consider two points x_l and x_u and the probabilities:

$$P_l = \int_0^{x_l} pdf(x)dx, \quad P_u = \int_{x_u}^{\infty} pdf(x)dx \quad \text{and} \quad P_{md} = \int_{x_l}^{x_u} pdf(x)dx$$

Clearly,

$$P_u + P_l + P_{md} = 1$$

We can construct a new density by the following shift:

$$P_l' = P_l - \Delta,\ P_u' = P_u + \Delta \text{ and } P_{md}' = P_{md}$$

where $0 < \Delta < \min(P_u, P_l)$ and

$$\begin{aligned} pdf'(x) &= \frac{P_u' pdf(x)}{P_u} x_u < x < \infty \\ &= pdf(x)\ x_l \leq x \leq x_u \\ &= \frac{P_l' pdf(x)}{P_l}\ 0 \leq x \leq x_l \end{aligned}$$

It is easy to check that $pdf'(x)$ is still a well-defined density although no longer continuous at $x = x_l$ or $x = x_u$. Furthermore, the above transformation can be called optimistic in that it transfers probability form the lower tail to the upper tail of the density. A pessimistic transformation can be defined similarly. Note that since we assumed a continuous density with zero probability mass at any point, the discontinuities induced by our transformation will not affect the existence of the integrals.

3.7.5 Trustees' attitude to risk

Investors' attitude to investment risk is generally assessed by their reaction to such qualities as volatility, peak-to-trough drawdown, recovery time, and absolute loss. Trustees of an infinitely lived endowment should in theory be largely indifferent to all temporary variations to asset values (save for irrecoverable losses), except that market volatility will influence how short-term income is achieved.

An efficiently diversified endowment ought to be delivering optimally smoothed sources of return, which allow annual withdrawal without the necessity of realizing assets that are in a temporary downturn. Of much greater concern to the trustees of any long-lived endowment is the level of future returns that might reasonably be expected. Even if volatility is disregarded and diversification has indeed delivered optimally smoothed sources of return, if the current spending policy is based on expectations of future returns that are not achieved, the longer-term sustainable returns from the endowment will clearly be in question. In Swensen (2000), David Swensen, Yale University Chief Investment Officer, establishes the connection between the rate of annual spend and endowment returns as follows:

> *Spending at levels inconsistent with investment returns either diminishes or enhances future endowment levels. Too much current spending causes future*

endowment levels to fall; too little current spending benefits future students at the expense of today's scholars. Selecting a distribution rate appropriate to the endowment portfolio increases the likelihood of achieving a successful balance between demands of today and responsibilities of tomorrow. (pp. 33 and 34)

It is our contention that for infinitely lived endowments, future returns are more significant than volatility, yet this is often not the focus of investment professionals advising trustees on the strategy to pursue. In developing our concept of theta as a measure of investor attitude to risk, we will seek to encapsulate the unusual characteristics of infinitely lived endowments.

By contrast, investors with a finite investment horizon will be greatly concerned with volatility, as the requirement to meet liabilities at an inopportune time in market cycles can have devastating consequences.

3.7.6 Decision making under uncertainty

We proceed to quantify the above discussions and suggest a model for determining the optimal sustainable Endowment Consumption Rate (ECR) of Foundations, subject to their needs and constraints (we defer discussion of ECR optimality until later, but for clarity at this point, it is measured as a percentage of the real value of the endowment). In this context, the expendable endowment is defined as the real value of the endowment multiplied by the ECR; a monetary value that will vary with the real value of the endowment.

Central to our approach has been the idea of Utility, which may be thought of as a formal way of representing the "well-being" of an endowment through time, as a function of its expendable endowment. For example, if a charitable organization is capable of donating £2 million to worthy causes this year, as opposed to £1 million, its Utility will be greater. Within this context, the optimal ECR maximizes the *expected* Utility throughout the *entire* lifetime of the endowment. It should be noted that perpetual endowments gain no Utility from maintaining capital per se, except in so much as it will support future expenditure.

To determine the optimal ECR requires the trustees to have a view on both present and expected future Utility, taking account of the inflationary impact on goods and services and the future investment returns.

We consider the nature of an endowment's Utility to be dependent on the risk averseness of the trustees, an attribute we label theta. We define a value of theta that is equal to zero as representing risk neutrality, with risk averseness increasing proportionally with larger values. We only allow positive values of theta, as trustees are not expected to be indifferent to risk. Generally, the more risk averse we are, the more we will prefer certain consumption now, to uncertain consumption later.

In addition to theta, an endowment's long-term expected Utility will also be dependent on the subjective discount rate (referred to here as delta). When circumstances are such that there is no time preference (when today's Utility is as important to the trustees as tomorrow's Utility), then we define delta to be equal to 1. However, when the Utility of tomorrow becomes of lesser concern than that of today, delta will be less than 1 (and vice versa). For example, a delta greater than 1 would imply that the trustees prefer to delay gratification.

Each trustee is described by his theta and delta. As expected Utility is critically dependent on both theta and delta, the different values represent a different view of optimal Utility and optimal ECR. Thus, the appropriate values for each individual endowment have to be carefully chosen to reflect the judgment of the trustees and their wishes.

Finally, as we are interested in the optimal ECR that maximizes not just today's Utility, nor the Utility that we expect throughout the lifetime of the endowment, but in fact the overall mix of them, we are naturally concerned with the performance of our investments in the future. For example, if we hold the view that the future will yield poor total returns, and hence sustain lower potential consumption, then our expected Utility is designed to reflect this and would adjust the optimal ECR.

Our discussion has not considered donations. However, we can extend it quite naturally to include donations. Conceptually, one may think of this as an additional return to current wealth; this return could be either certain or uncertain.

3.7.7 Practical implications of risk aversion

Using the BITA Endowment model outlined above, we have computed the optimal ECR for an infinitely lived endowment as a function of risk averseness for three views of what future returns will be: pessimistic, neutral, and optimistic. Figure 3.5 illustrates the results of our calculations.

There are two clear regimes present, dependent on whether theta is less than one, or greater than one. These two cases are examined below. In both instances, delta is assumed constant:

Theta less than 1—less risk averse

The optimal ECR is less for an optimistic view of the future than for a pessimistic view, as greater future returns will yield greater potential consumption in the future (when they are realized), and hence greater expected Utility. This is referred to as the Substitution Effect. This is related to the Trustees having relatively high risk tolerance. In this instance, the risk-tolerant investor lowers the ECR when he makes an optimistic forecast, as he will be preserving wealth to invest and consume in the future (the substitution effect).

Theta more than 1—more risk averse

The optimal ECR is less for a pessimistic view of the future than for an optimistic view, as the high risk aversion of the trustees (measured through the magnitude of theta) leads to lower spending when poor returns are forecasted. However, in this instance, the risk-averse investor, susceptible to the uncertainty of future returns, does not wait and increases his or her ECR in the

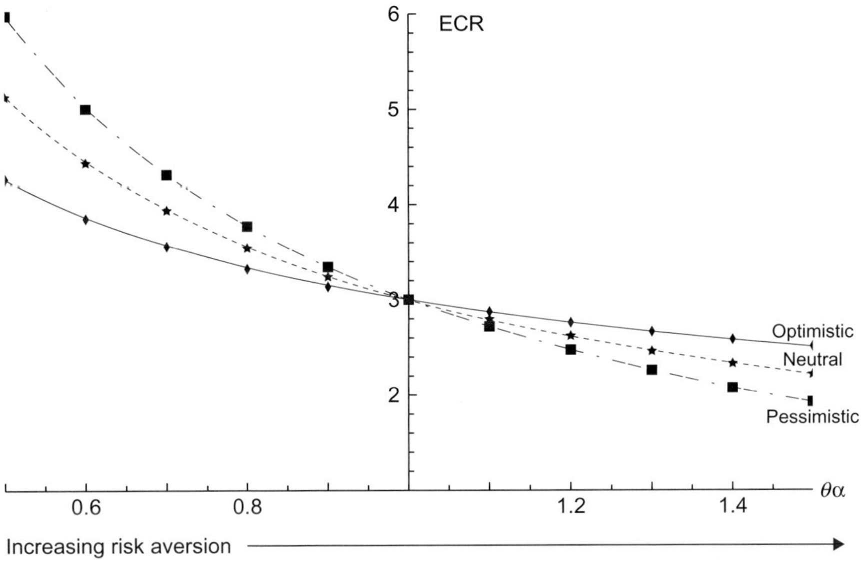

Figure 3.5 The optimal rate of expendable endowment versus the risk-averseness of the endowment's trustees, under three separate future investment scenarios: pessimistic (squares); neutral (stars); and optimistic (diamonds).

face of an optimistic forecast. It is the uncertainty of future consumption that affects this soul, not any self-doubts about his or her future forecast.

When the view of future returns is neutral, then the optimal ECR decreases as theta increases, analogous to the canonical relationship between risk and return.

For the first example, the trustees are advised to consume more when a pessimistic view of future returns is held than when an optimistic view of future returns is held. In the converse situation, this relationship is reversed. The reasons for these choices can be understood in terms of what economists refer to as Income and Substitution effects. We recognize that these ideas are difficult and counterintuitive. Readers who might need further assurance should consult Ingersoll (1987).

This section does not consider empirical analysis. This is partly because historical data on charities and endowments is hard to come by. However, empirical work in financial economics has found values of theta close to 2. Whilst this analysis has been for household investment, it is still probably relevant for the charity and endowments sector. We may consider trustees to be less risk averse than household investors, but it is probably the case that theta should be greater than 1.

Intuition suggests that if we are optimistic about the future, we should increase ECR. However, this is not necessarily true, and if it is, it may not be for the reasons that motivate the intuition.

Sophisticated quantitative analysis has the capacity to assist investors in formulating and implementing appropriate investment strategies. The investment industry has concentrated on providing solutions for private investors with finite investment horizons and liability management strategies for pension funds. In our analysis, we have highlighted characteristics of permanent endowments that require a different approach to reconciling the balance between risk and return. In particular, the requirement for intergenerational equity can be expressed in the technical terms of Utility and we have demonstrated that there can be a counterintuitive relationship between risk aversion and expected future investment return.

Our conclusion is that the trustees of Foundations that intend to endure in perpetuity require a formal expression of Utility that adequately captures both risk aversion (theta) and subjective discounting of consumption over time (delta) to determine a sustainable rate of consumption from the endowment.

3.8 Bespoke optimization—putting theory into practice

At BITA Risk, we see a whole range of tailored requests; the following sections describe two examples. All have been successfully implemented using the range of BITA Risk optimizers.

3.8.1 Request: produce optimal portfolio with exactly 50 long and 50 short holdings

Starting from a stock universe of several hundred stocks, the request was to produce optimal mixes of stocks with exactly 50 long and exactly 50 short positions. Clearly, this requires a heuristic approach since the analytical solution would be infeasible given the computational cost. Our solution was to construct a hybrid methodology of two phases:

(i) On an iterative basis starting with the full candidate stock universe, eliminate a percentage of stocks on the basis of their lowest weight in the preceding optimal portfolio, until the 50 long, 50 short constraint is first violated, i.e., the first step with less than or equal to 50 stocks, both long and short.
(ii) Select and reinsert stocks individually on the basis of the highest marginal utility until the exactly 50 long, exactly 50 short constraint is satisfied.

3.8.2 Request: how to optimize in the absence of forecast returns

A client required a long–short mix of stocks selected on the basis of absolute betas rather than forecast returns, which were not available. In the absence of forecast returns, it is obviously not possible to form a traditional efficient frontier. However, the client could provide measures of beta, measured against the market, for each stock. These absolute betas were used as a constraint in conjunction with more common sector constraints within the optimization. Further, a nonlinear transaction cost function with terms dependent upon both an absolute size factor and the gross value of the portfolio was included. The result is that absolute beta is used to target the desired level of risk.

3.9 Conclusions

We have presented the detailed methodologies associated with a range of optimization techniques including not only standard mean–variance but also Robust to deal with the inherent errors associated with forecast returns, Gain/Loss to deal with nonnormal return distributions and finally a hybrid approach that combines both mean–variance and Gain/Loss methodologies.

We have discussed how optimization can be used in light of the perpetual nature of investments by charities and endowments and offered insights and adjustments to the standard analytical framework to deal with such investments.

Finally, we discussed two of bespoke applications where BITA Risk has successfully implemented some of the more esoteric requirements from clients.

Appendix A: BITA Robust optimization

BITA Robust applies a Second-order Cone Programming (SOCP) problem to the efficient application of quadratic constraints to an objective function.

The standard form in which a second-order cone optimization is expressed is that of minimizing a linear objective subject to a set of linear equality constraints for which the optimization vector variable is restricted to a direct product of quadratic cones K. This is the primal SOCP problem.

$$\begin{aligned} \text{minimize } c^T x \text{ subject to } & Ax = b \\ & x \in K = \prod_{i=1}^{n} \left\{ t : t_{n_i+1}^2 \geq \sum_{j=1}^{n_i} t_j^2, t_{n_i+1} \geq 0 \right\} \end{aligned}$$

where K is the direct product of n quadratic cones of dimension $n_i + 1$, x is represented as a vector of size nn where $nn = \sum_{i=1}^{n} n_i + 1$. A is an nn by m matrix where m is the number of conic constraints, and b is of length m and c nn. We define the dual cone to K as:

$$K_d = \{s : s^t x \geq 0, \forall x \in K\}$$

Associated with the primal problem is the dual SOCP problem:

$$\begin{aligned} \text{maximize } b^T y \text{ subject to } & s = c - A^T y \\ & s \in K_d \end{aligned}$$

where s is a vector with the same size as x and y has the length m. It can be shown that both problems can be solved simultaneously using a primal-dual interior point method and we have implemented the homogenous method described in Andersen, Roos, EA and Terlaky (2003) for the programming, which relies upon the scaling found in Nesterov and Todd (1997). The dual problem may be written in the more convenient form:

$$\begin{aligned} \text{maximize } b^T y \text{ subject to } & \left\| A^T y - c \right\|_1^{n_i} \leq (c - A^T y)_{n_i+1 \forall_i} \forall i \\ & (c - A^T y)_{n_i+1} \geq 0 \forall i \end{aligned}$$

A special case where $A_{j,(n_i+1)}$ $j = 1 \ldots m = 0$ is the problem:

$$\begin{aligned} \text{maximize } b^T y \text{ subject to } & \left\| A^T y - c \right\|_1^{n_i} \leq c_{n_i+1 \forall_i} \forall i \\ & c_{n_i+1} \geq 0 \forall i \end{aligned}$$

which is the maximization of a linear expression subject to upper bounds on the Euclidean norms of general linear terms and theses constraints are equivalent to bounds on positive definite quadratic forms since:

$$(x - b)^T C(x - b) \geq M^2$$
$$\Rightarrow \|Rx - Rb\| \geq M$$

where $C = R^T R$

We can minimize a positive definite quadratic form by noting the following:

$$\lambda a^T x + \frac{1}{2}(x - b)^T C(x - b) = \frac{1}{2}(Rx - Rb + \lambda R^{T-1}a)^2 - \frac{1}{2}\lambda^2(R^{T-1}a)^2 + \lambda a^T b$$

Then, because the last two terms are independent of x,

$$\text{minimize} \;\; \lambda a^T x + \frac{1}{2}(x - b)^T C(x - b)$$

becomes

$$\text{maximize} \;\; -z \;\; \text{subject to} \;\; \left\|Rx - Rb + \lambda R^{T-1}a\right\| \leq z$$

which is an $n + 2$-dimensional cone constraint where z is an extra dummy scalar variable. The two quadratic constraints in our robust optimization become two $n + 1$-dimensional cone constraints and each linear constraint can be written as:

$$\left\|ax - \frac{1}{2}(u + l)\right\| \leq \frac{1}{2}(u + l)$$

which is a two-dimensional cone constraint where a is either a unit vector or a row of A^T, and u and l are upper and lower bounds. We are thus able to program our robust optimization as the dual of the standard SOCP.

Appendix B: BITA GLO

The following relationship holds.

$$\textbf{Expected return} - \textbf{Target} = \textbf{Gain} - \textbf{Loss}$$

Therefore, if we define expected utility by V where $V = \text{Gain} - (1 + \lambda)\text{Loss}$

Then, it follows that:

$$V = \textbf{Expected return} - \textbf{Target} - \lambda\textbf{Loss}$$

It is clear that these two versions are equivalent and that we can optimize either. We choose to maximize:

$$V = \text{Gain} - (1 + \lambda)\text{Loss}$$

It is useful to define an array I_t that contains T integers where T is the number of periods. We assign the value 1 to I_t if the portfolio makes a net gain in period t and 0 to I_t if the portfolio makes a net loss in that period. Note that the values in array I_t depend on the portfolio weights for each asset, the return data, and the target return R

$$I_t = 1 \; if \sum_{i=1}^{n} w_i r_i(t) > R$$

$$I_t = 0 \; if \sum_{i=1}^{n} w_i r_i(t) < R$$

where

w_i = weight of the asset I;
$r_i(t)$ = return of asset i in period t.

The probability that we exceed the target is given by $\left(\sum_{t=1}^{T} I_t\right)/T$. Gain is given by:

$$\sum_{t=1}^{T}\left(\sum_{i=1}^{n} w_i(r_i(t) - R)\right)I_t$$

Loss is given by:

$$\sum_{t=1}^{T}\left(\sum_{i=1}^{n} w_i(R - r_i(t))\right)(I - I_t)$$

Note that both gain and loss are nonnegative numbers. Our utility V appears to be a linear expression in portfolio weights (apart from the dependence of I on w).

With special care, we can maximize V with respect to w and linear constraints on w using an iterative sequential linear programming approach. For the calculations in this section, we just impose nonnegative upper and lower bounds on each weight and constrain their sum to be 1.

References

Andersen, E. D., Roos, C., & Terlaky, T. (2003). On implementing a primal-dual interior point method for conic quadratic optimization. *Mathematical Programming*, 95(2), 249–277.

Artzner, P., Delbaen, F., Eber, J. M., & Heath, D. (1999). Coherent measures of risk. *Mathematical Finance*, 9, 203–228.

Benartzi, S., & Thaler, R. H. (2001). Naïve diversification strategies in defined contribution saving plans. *American Economic Review*, *91*(1), 79–98.

Cascon, A., Keating, C., & Shadwick, W. F. (2002). *The mathematics of the omega measure*, Working Paper, The Finance Development Centre.

Cuthbertson, K., & Nitzsche, D. (2004). Quantitative financial economics. *Stocks, Bonds & Foreign Exchange*, *475*.

DeMiguel, V., Garlappi, L., & Uppal, R. (2007). Optimal versus naïve diversification: How inefficient is the 1/N portfolio strategy? forthcoming. *Review of Financial Studies*.

Fishburn, P. C. (1977). Mean-risk analysis with risk associated with below-target returns. *The American Economic Review*, *67*, 116–126.

Fishburn, P. C. (1982). Foundations of risk measurement. II. Effects of gains on risk. *Journal of Mathematical Psychology*, *25*, 226–242.

Fishburn, P. C. (1984). Foundations of risk measurement. I. Risk as probable loss. *Management Science*, *30*, 396–406.

Ho, L., Cadle, J., & Theobald, M. (2008). Portfolio selection in an expected shortfall framework during the recent "credit crunch" period. *Journal of Asset Management*, *9*, 121–137.

Ingersoll, J. (1987). *Theory of Financial Decision Making*. Savage, MD: Rowman & Littlefield, 245.

Jorion, P. (2006). *Value at risk: The new benchmark for managing financial risk* (3rd ed.). New York: McGraw-Hill.

Kahneman, D., & Tversky, A. (1979). Prospect theory: An analysis of decision under risk. *Econometrica*, *47*(12), 263–291.

Keating, C., & Shadwick, W. F. (2002a). *A universal performance measure*. The Finance Development Centre, London.

Keating, C., & Shadwick, W. F. (2002b). A universal performance measure. *The Journal of Performance Measurement*, 6(3), 59–84.

Levy, H., & Markowitz, H. (1979). Approximating expected utility by a function of mean and variance. *American Economic Review*, *69*(3), 308–317.

Maillard, S., Roncalli, Th., & Teiletche, J. (2009). *On the properties of equally-weighted risk contributions portfolios*, Working Paper.

Mandelbrot, B. (1963). The variation in certain speculative prices. *Journal of Business*, *36*, 394–419.

Meucci, A. (2008). *The Black–Litterman Approach: Original Model and Extensions*. Bloomberg Portfolio Research Paper, August.

Nesterov, Y., & Todd, M. J. (1997). Self-scaled barriers and interior point methods for convex programming. *Mathematical Operator Research*, *22*(1).

Passow, A. (2005). Omega portfolio construction with Johnson distributions. *Risk*, *4*, 85.

Scowcroft, A., & Satchell, S. (2000). A Demystification of the Black–Litterman model: Managing quantitative and traditional construction. *Journal of Asset Management*(September), 138–150.

Sharpe, W. F. (1964). Capital asset prices: A theory of market equilibrium under consideration of risk. *Journal of Finance*, *19*.

Swensen, D. F. (2000). *Pioneering portfolio management—An unconventional approach to institutional investment*. New York: The Free Press.

Tobin, J. (1974). What is permanent endowment income? *The American Economic Review*, Vol. 64, No.2, *Papers and Proceedings of the eighty-sixth Annual meeting of the American Economic Association*, May, 427–432.

Windcliff, H., & Boyle, P. (2004). The 1/n pension plan puzzle. *North American Actuarial Journal*, *9*, 32–45.

4 The Windham Portfolio Advisor

Mark Kritzman

Executive Summary

The Windham Portfolio Advisor (WPA) is a technology platform for efficient portfolio construction and sophisticated risk management. It addresses the overarching investment issues faced by both institutional and individual investors. The WPA is especially distinguished by several unique innovations, including multigoal optimization, which enables investors to optimize portfolios on the basis of both absolute and relative performance simultaneously; within-horizon risk measurement, which assesses a portfolio's exposure to loss throughout the investment horizon rather than just at the conclusion; identification of risk regimes, which partitions historical returns into subsamples representing quiet and turbulent periods; and full-scale optimization, which directly maximizes expected utility, however it is defined, taking into account all features of a return distribution.

4.1 Introduction

The Windham Portfolio Advisor (WPA) is a technology platform for efficient portfolio construction and sophisticated risk management. It is designed to address the overarching investment issues faced by endowment funds, foundations, pension funds, sovereign wealth funds, asset managers, investment consultants, and brokerage firms.[1] It is a highly versatile tool as evidenced by the wide range of portfolio applications described in the Appendix. But it is especially distinguished by several unique innovations, which warrant separate attention, because these features address some of the most vexing problems of investment management. These unique innovations include:

- Multigoal optimization, which enables investors to optimize portfolios on the basis of both absolute and relative performance simultaneously;
- Within-horizon risk measurement, which assesses a portfolio's exposure to loss throughout the investment horizon rather than just at the conclusion;

[1] Windham Capital Management, LLC also offers a companion technology platform called the Windham Financial Planner (WFP). The focus of the WFP is to assist private wealth investors and those professionals who service them.

Doi:10.1016/B978-0-12-374952-9.00004-X.

- Identification of risk regimes, which partitions historical returns into subsamples representing quiet and turbulent periods; and
- Full-scale optimization, which directly maximizes expected utility, however it is defined, from the full sample of returns.

4.2 Multigoal optimization

4.2.1 The problem

Sophisticated investors employ optimization to determine a portfolio's asset allocation. Some investors care about absolute return and risk, while others focus on relative return and risk. Many investors, however, are sensitive to both absolute and relative performance. Typically, they address this dual concern by constraining the asset weights in the optimization process. This approach is inefficient.

4.2.2 The WPA solution

In the years following the introduction of portfolio theory by Harry Markowitz, the investment community, at first reluctantly and then gradually, embraced his prescription for constructing portfolios, but with an important caveat. To the extent that the recommended solution departed from industry norms or prior expectations, investors typically imposed constraints to force a more *palatable* solution. For example, an investor might instruct the optimizer to find the combination of assets with the lowest standard deviation for a particular expected return *subject to the constraint* that no more than 10% of the portfolio is allocated to foreign assets or nontraditional investments and that no less than 40% is allocated to domestic equities. The reason for such constraints is that investors are reticent to depart from the crowd when there is a significant chance they will be wrong, in the sense that they lose money. The matrix shown in Figure 4.1 illustrates this point.

If we accept the notion that investors care not only about how they perform in an absolute sense but also about how their performance stacks up against other investors or a popular benchmark, there are four possible outcomes. An investor achieves favorable absolute returns and at the same time outperforms his or her peers or the benchmark, which would be great as represented by Quadrant I. Alternatively, an investor might beat the competition or benchmark but fall short of an absolute target (Quadrant II). Or an investor might generate a high

Absolute performance	Relative performance	
	Favorable	Unfavorable
Favorable	I. Great	II. Tolerable
Unfavorable	III. Tolerable	IV. Very unpleasant

Figure 4.1 Wrong and alone.

absolute return but underperform the competition or benchmark (Quadrant III). These results would probably be tolerable because the investor produces superior performance along at least one dimension. However, what would likely be very unpleasant is a situation in which an investor generates an unfavorable absolute result and at the same time performs poorly relative to other investors or the benchmark (Quadrant IV). It is the fear of this outcome that induces investors to conform to the norm.

One might argue that investors are driven by a different motivation when they constrain their portfolios to resemble normal industry portfolios; to wit, they lack confidence in their assumptions about expected return, standard deviation, and correlation. It may be the case that, although the optimal result calls for a 50% allocation to foreign equities, the investor lacks confidence in the relevant inputs, and therefore is warranted in overriding the optimal recommendation. This reasoning is a subterfuge for the true motivation, which is fear of being wrong and alone. Consider the following thought experiment. By some form of divine intervention we acquire secret knowledge of the true distributions and correlations of all the assets to be included in our portfolio. With this knowledge, we solve for the optimal portfolio, which again calls for a 50% allocation to foreign equities compared to an industry norm of 10%. Would we be more inclined to accept the recommended solution without constraints knowing that our views about expected return, standard deviation, and correlation are unequivocally true? Probably not, for the following reason. Even if we have perfect foreknowledge of the *expected* returns and the distributions around them, there is about a 50% chance that the actual returns will be either above or below the expected returns during any given period—and there is a substantial likelihood for distributions with high standard deviations that the actual returns will be significantly different from the expected returns. Thus, even though we may be right on average over the long run, we will almost certainly be wrong (due to high standard deviations) and alone (due to an uncommon allocation) over some shorter periods. Most investors, owing either to career or psychological considerations, are unwilling to risk the chance of such an unpleasant outcome.

Investors have sought protection from this unhappy consequence by constraining their portfolios to match typical industry allocations. Although this approach is reasonably effective, it is not the best approach for dealing with the fear of being wrong and alone. Chow (1995) introduced a technique that optimally deals with absolute and relative performance. He simply augmented the quantity to be maximized to equal expected return minus risk aversion times variance minus tracking error aversion times tracking error squared. This measure of investor satisfaction simultaneously addresses concerns about absolute performance and relative performance. Although it includes two terms for risk, it only includes one term for expected return. This is because absolute return and relative return are linearly related to each other. Instead of producing an efficient frontier in two dimensions, this optimization process produces an efficient *surface* in three dimensions: expected return, standard deviation, and tracking error, as shown (Figure 4.2).

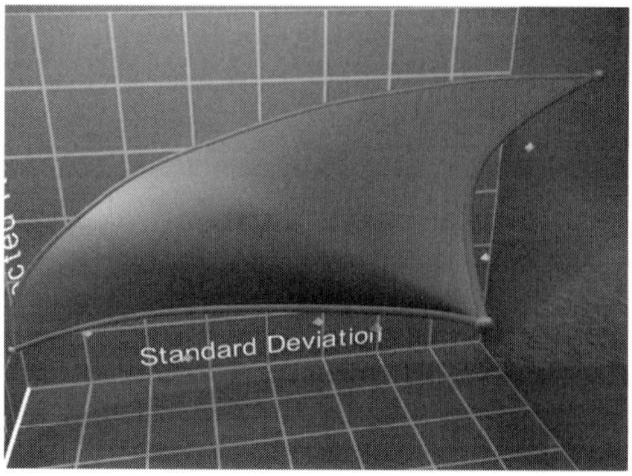

Figure 4.2 The efficient surface.

The efficient surface is bounded on the upper left by the traditional mean–variance efficient frontier, which is composed of efficient portfolios in dimensions of expected return and standard deviation. The leftmost portfolio on the mean–variance efficient frontier is the riskless asset. The right boundary of the efficient surface is the mean–tracking error efficient frontier. It is composed of portfolios that offer the highest expected return for varying levels of tracking error. The leftmost portfolio on the mean–tracking error efficient frontier is the benchmark portfolio because it has no tracking error. The efficient surface is bounded on the bottom by combinations of the riskless asset and the benchmark portfolio. All of the portfolios that lie on this surface are efficient in three dimensions. It does not necessarily follow, however, that a three-dimensional efficient portfolio is always efficient in any two dimensions. Consider, for example, the riskless asset. Although it is on both the mean–variance efficient frontier and the efficient surface, if it were plotted in dimensions of just expected return and tracking error, it would appear very inefficient if the benchmark included high expected return assets such as stocks and long-term bonds. This asset has a low expected return compared to the benchmark and yet a high degree of tracking error.

Multigoal optimization will almost certainly yield an expected result that is superior to constrained mean–variance optimization in the following sense. For a given combination of expected return and standard deviation, it will produce a portfolio with less tracking error. Or for a given combination of expected return and tracking error, it will identify a portfolio with a lower standard deviation. Or finally, for a given combination of standard deviation and tracking error, it will find a portfolio with a higher expected return than a constrained mean–variance optimization. Most of the portfolios identified by constrained mean–variance optimization would lie beneath the efficient surface. In fact, multigoal optimization would fail to improve upon a constrained mean–variance optimization only if the investor knew in advance what constraints were optimal. But, of course, this knowledge could only come from multigoal optimization.

4.2.3 Summary

Investors care about both absolute and relative performance, as revealed by their reluctance to depart significantly from normal industry portfolios. To date, the investment community has addressed aversion to being wrong and alone in an ad hoc fashion by imposing allocation constraints on the mean–variance optimization process. The WPA, by contrast, encompasses both absolute and relative measures of risk in an unconstrained optimization process, which is almost always superior to constrained mean–variance optimization.

4.3 Within-horizon risk measurement

4.3.1 The problem

Investors measure risk incorrectly by focusing exclusively on the distributions of outcomes at the end of their investment horizons. Their investments are exposed to risk during the investment horizon as well. This approach to risk measurement ignores intolerable losses that might occur throughout an investment period, either as an accumulation of many small losses or from a significant loss that later—too late, perhaps—would otherwise recover.

4.3.2 The WPA solution

The WPA provides two related risk measures that enable investors to evaluate a portfolio's exposure to loss all throughout the investment horizon: within-horizon probability of loss and continuous value at risk.

Exhibit 3 illustrates the distinction between risk based on ending outcomes and risk based on outcomes that might occur along the way. Each line represents the path of a hypothetical investment through four periods. The horizontal line represents a loss threshold, which in this example equals 10%. Exhibit 3 reveals that only one of the five paths breaches the loss threshold at the end of the horizon; hence we might conclude that the likelihood of a 10% loss equals 20%. However, four of the five paths at some point during the investment horizon breach the loss threshold, although three of the four paths subsequently recover. If we also care about the investment's performance along the way to the end of the horizon, we would instead conclude that the likelihood of a 10% loss equals 80% (Figure 4.3).

One might argue that calculation of daily value at risk measures a strategy's exposure to loss within an investment horizon, but this is not true. Knowledge of the value at risk on a daily basis does not reveal the extent to which losses may accumulate over time. Moreover, even if daily value at risk is adjusted to account for prior gains and losses, the investor still has no way to know at the inception of the strategy, or at any other point, the cumulative value at risk to any future point throughout the horizon, including interim losses that later recover.

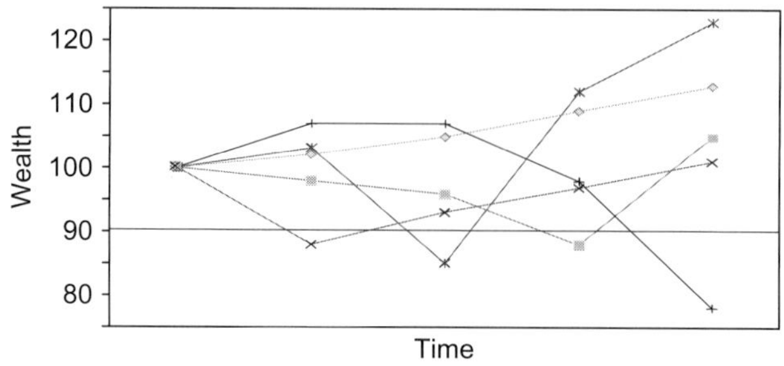

Figure 4.3 Risk of loss: ending versus interim wealth.

End-of-horizon exposure to loss

We estimate probability of loss at the end of the horizon by: (1) calculating the difference between the cumulative percentage loss and the cumulative expected return, (2) dividing this difference by the cumulative standard deviation, and (3) applying the normal distribution function to convert this standardized distance from the mean to a probability estimate, as shown in Equation (4.1).

$$P_E = N[(\ln(1 + L) - \mu T)]/(\sigma\sqrt{T}) \tag{4.1}$$

where

$N[\ ]$ = cumulative normal distribution function;
ln = natural logarithm;
L = cumulative percentage loss in periodic units;
μ = annualized expected return in continuous units;
T = number of years in horizon;
σ = annualized standard deviation of continuous returns.

The process of compounding causes periodic returns to be lognormally distributed. The continuous counterparts of these periodic returns are normally distributed, which is why the inputs to the normal distribution function are in continuous units.

When we estimate value at risk, we turn this calculation around by specifying the probability and solving for the loss amount, as shown:

$$\text{Value at Risk} = -(e^{\mu T - Z\sigma\sqrt{T}} - 1)W \tag{4.2}$$

where

e = base of natural logarithm (2.718282);
Z = normal deviate associated with chosen probability;
W = initial wealth.

Within-horizon exposure to loss

Both of these calculations pertain only to the distribution of values at the end of the horizon and therefore ignore variability in value that occurs throughout the horizon. To capture this variability, we use a statistic called first passage time probability.[2] This statistic measures the probability (P_W) of a first occurrence of an event within a finite horizon. It is equal to:

$$P_W = N[(\ln(1 + L) - \mu T)/(\sigma\sqrt{T})] + N[(\ln(1 + L) + \mu T)/(\sigma\sqrt{T})](1 + L)^{2\mu/\sigma\wedge 2} \tag{4.3}$$

It gives the probability that an investment will depreciate to a particular value over some horizon if it is monitored continuously.[3]

Note that the first part of this equation is identical to the equation for the end of period probability of loss. It is augmented by another probability multiplied by a constant, and there are no circumstances in which this constant equals zero or is negative. Therefore, the probability of loss throughout an investment horizon must always exceed the probability of loss at the end of the horizon. Moreover, within-horizon probability of loss rises as the investment horizon expands in contrast to end-of-horizon probability of loss, which diminishes with time. This effect supports the notion that time does not diversify all measures of risk and that the appropriate equity allocation is not necessarily horizon dependent.

We can use the same equation to estimate continuous value at risk. Whereas value at risk measured conventionally gives the worst outcome at a chosen probability *at the end of an investment horizon*, continuous value at risk gives the worst outcome at a chosen probability *from inception to any time during an investment horizon*. It is not possible to solve for continuous value at risk analytically. We must resort to numerical methods. We set Equation (4.3) equal to the chosen confidence level and solve iteratively for L. Continuous value at risk equals L times initial wealth.

Consider the implications of these risk measures on a hypothetical hedge fund's exposure to loss. Suppose this hedge fund employs an overlay strategy, which has an expected incremental return of 4.00% and an incremental standard deviation of 5.00%. Further, suppose this hedge fund leverages the overlay strategy. Table 4.1 shows the expected return and risk of the hedge fund and its components for varying degrees of leverage.

These figures assume that the underlying asset is a government note with a maturity equal to the specified 3-year investment horizon and that its returns are uncorrelated with the overlay returns. Managers sometimes have a false sense of security, because they view risk as annualized volatility, which diminishes with the duration of the investment horizon. This view, however, fails to consider the likelihood that the fund's assets may depreciate significantly during the investment

[2] The first passage probability is described in Karlin and Taylor (1975).

[3] See Kritzman and Rich (2001) for a discussion of its application to risk measurement.

Table 4.1 Leveraged hedge fund expected return and risk

	Underlying asset	Overlay strategy	Leverage 2	Leverage 4	Leverage 6	Leverage 8	Leverage 10
Expected return	3.50%	4.00%	11.50%	19.50%	27.50%	35.50%	43.50%
Standard deviation	3.00%	5.00%	10.44%	20.22%	30.15%	40.11%	50.09%

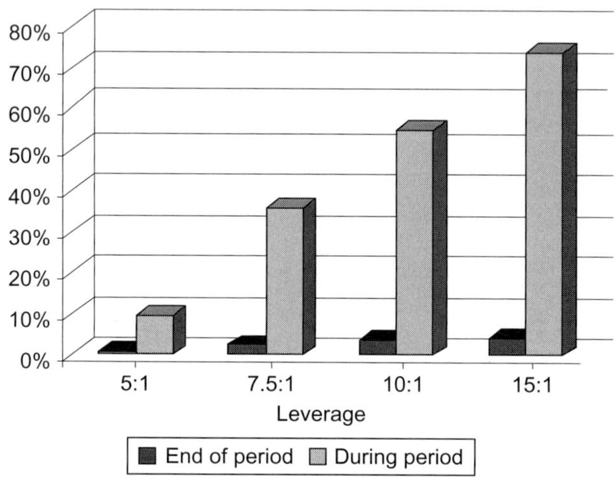

Figure 4.4 Probability of insolvency, 3-year lockup period.

horizon and thereby trigger withdrawals from the fund. Exhibit 6 compares the likelihood of a 10% loss at the end of the 3-year horizon with its likelihood at some point within the 3-year horizon. It reveals that although the chance of a 10% loss at the end of the horizon is low, there is a much higher probability that the fund will experience such a loss at some point along the way, which may trigger withdrawals and threaten the fund's solvency (Figure 4.4).

The same issue applies if exposure to loss is perceived as value at risk. Figure 4.5 shows the hedge fund's value at risk measured conventionally and continuously. Whereas conventional value at risk for leverage ratios less than 10 to 1 is negative (a gain) and still very low for leverage ratios up to 15 to 1, continuous value at risk ranges from more than one-fourth the portfolio's value to nearly half its value.

These examples illustrate rather vividly that investors are exposed to far greater risk throughout their investment horizon than end-of-horizon risk measures indicate. How should investors respond to this reassessment of risk? Investors should not necessarily reduce risk, although such a course of action may sometimes be warranted. Rather, they should at least be aware of the likelihood that they will be unable to sustain their investment strategies. Moreover, investors should be informed of their within-horizon exposure to loss so that

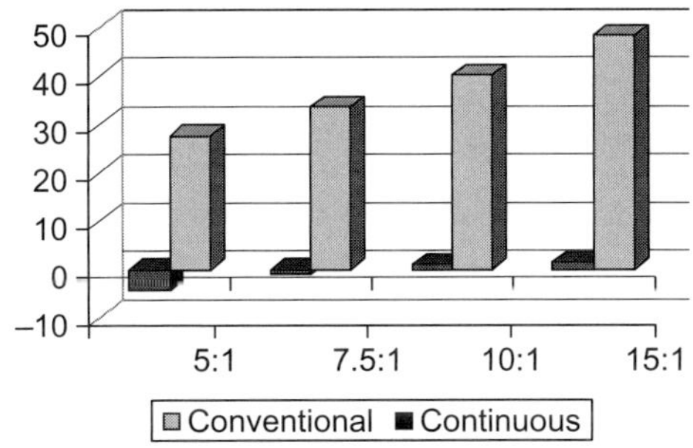

Figure 4.5 5% value at risk, 3-year lockup period.

should an unpleasant loss occur, they will not be unduly surprised and act to reduce risk out of a misguided perception that the nature of their investment strategy has changed.

4.4 Risk regimes

4.4.1 The problem

It is well known that volatilities and correlations are unstable. They may differ significantly depending on the sample used to estimate them. This instability may lead investors to underestimate their exposure to loss and to form portfolios that are not sufficiently resilient to turbulent markets.

4.4.2 The WPA solution

The WPA includes a methodology for partitioning historical returns into distinct samples that are characteristic of quiet markets and turbulent markets.[4] This separation enhances risk management in several ways. It enables investors to stress test portfolios by evaluating their exposure to loss during turbulent conditions. It enables investors to structure portfolios that are more resilient to turbulent markets. And it enables investors to shift their portfolios' risk profiles dynamically to accord with their assessment of the relative likelihood that market conditions will be quiet or turbulent.

The WPA uses a statistical procedure to isolate periods in which returns are unusual either in their magnitude or in their interaction with each other. For example, a period may qualify as unusual because two assets whose returns are positively correlated generate returns that move in the opposite direction. These periods with unusual returns represent statistical outliers, and they are typically associated with turbulent markets. The WPA creates subsamples of these outliers and estimates volatilities and correlations from these turbulent subsamples. It also estimates volatilities and correlations from the remaining returns, which represent quiet markets.

[4] See Chow et al (1999) and Kritzman et al (2001) for a description of this methodology.

How do we determine explicitly whether to classify an observation as "usual" or as an "outlier?" Figure 4.6 shows two independent return series with equal variances presented as a scatter plot.

In order to identify outliers, we first draw a circle around the mean of the data, which is shown as the shaded circle. This shaded circle is the boundary for defining outliers. To determine which observations are outliers, we next calculate the equation of a circle for each observation with its center located at the mean of the data and its perimeter passing through the given observation. If the radius of this calculated circle is greater than the "boundary radius," we define that observation as an outlier. If it is smaller, we define it as an inlier.

This approach is appropriate for a given sample of returns if they are uncorrelated and have the same variance. When the return series have different variances, a circle is no longer appropriate for identifying outliers, as illustrated by Figure 4.7.

Figure 4.7 shows a scatter plot of two positively correlated return series that have unequal variances. Under this condition, an ellipse is the appropriate shape for defining the outlier boundary.[5] As before, we start with the "boundary ellipse," and for each point calculate an ellipse with a parallel perimeter. Then, we compare their boundaries.

These illustrations capture the basic intuition for identifying outliers. However, when the return series are correlated or when the sample is expanded to include more than three return series, we must use matrix algebra for the exact computation of an outlier. This procedure is described below.

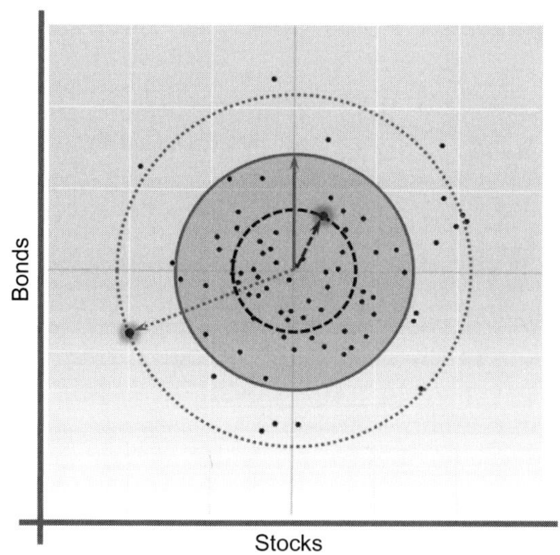

Figure 4.6 Identifying outliers from uncorrelated returns with equal variances.

[5] If we were to consider three return series, the outlier boundary would be an ellipsoid.

The calculation of a multivariate outlier is given by Equation (4.4).

$$d_t = (y_t - \mu)\Sigma^{-1}(y_t - \mu)' \tag{4.4}$$

where

d_t = vector distance from multivariate average;
y_t = return series;
μ = mean vector of return series y_t;
Σ = covariance matrix of return series y_t.

The return series y_t is assumed to be normally distributed with a mean vector μ and a covariance matrix Σ. If we have 12 return series, for example, an individual observation of y_t would be the set of the 12 asset returns for a specific measurement interval. We choose our boundary "distance" and examine the distance, d_t, for each vector in the series. If the observed d_t is greater than our boundary distance, we define that vector as an outlier.

For two uncorrelated return series, Equation (4.4) simplifies to Equation (4.5):

$$d_t = \frac{(y - \mu_y)^2}{\sigma_y^2} + \frac{(x - \mu_x)^2}{\sigma_x^2} \tag{4.5}$$

This is the equation of an ellipse with horizontal and vertical axes. If the variances of the return series are equal, Equation (4.5) further simplifies to a circle.

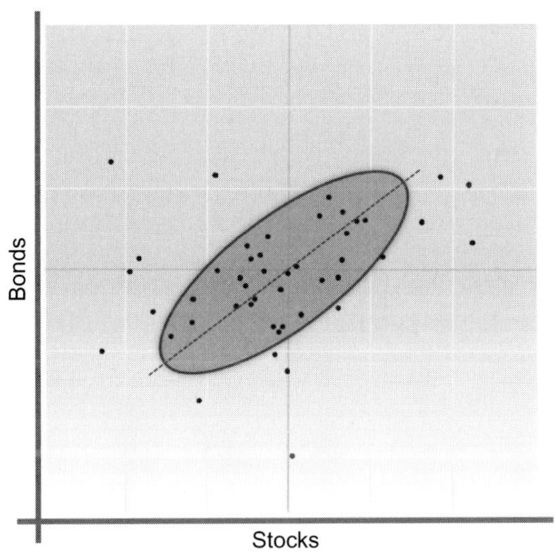

Figure 4.7 Identifying outliers from correlated returns with unequal variances.

For the general n-return normal series case, d_t is distributed as a Chi-Square distribution with n degrees of freedom. Under this assumption, if we define an outlier as falling beyond the outer 25% of the distribution and we have 12 return series, our tolerance boundary is a Chi-Square score of 14.84. Using Equation (4.1), we calculate the Chi-Square score for each vector in our series. If the observed score is greater than 14.84, that vector is an outlier.

4.4.3 Summary

Returns are not typically generated by a single distribution or regime, but instead by distributions associated with quiet and turbulent regimes. The risk parameters that prevail in these separate regimes are often dramatically different than their full sample averages. The WPA includes a procedure for partitioning historical returns into quiet and turbulent subsamples, which can be used to provide better guidance as to a portfolio's exposure to loss during turbulent episodes, and to structure portfolios that are more resilient to turbulence whenever it may occur.

4.5 Full-scale optimization

4.5.1 The problem

Many assets and portfolios have return distributions that display significantly nonnormal skewness and kurtosis. Hedge fund returns, for example, are often negatively skewed and fat-tailed. Because mean–variance optimization ignores skewness and kurtosis, it misallocates assets for investors who are sensitive to these features of return distributions.

4.5.2 The WPA solution

The WPA employs a computationally efficient algorithm called full-scale optimization that enables investors to derive optimal portfolios from the full set of returns given a broad range of realistic investor preference functions. The full-scale approach calculates a portfolio's utility for every period in the return sample considering as many asset mixes as necessary in order to identify the weights that yield the highest expected utility. [6]

Suppose, for example, the aim is to find the optimal blend between two funds whose returns are displayed in Exhibit 10, assuming the investor has log-wealth utility. Log-wealth utility assumes utility is equal to the natural logarithm of wealth, and resides in a broader family of utility functions called power utility. Power utility defines utility as $[(1/\gamma) \times \text{Wealth}\gamma]$. A log-wealth utility function is a special case of power utility. As γ approaches 0, utility

[6] See Cremers et al (2005) for a discussion of this methodology.

approaches the natural logarithm of wealth. A γ equal to 1/2 implies less risk aversion than log-wealth, while a γ equal to -1 implies greater risk aversion.

Utility for each period is calculated as $\ln[(1 + R_A) \times W_A + (1 + R_B) \times W_B]$, where R_A and R_B equal the returns of funds A and B, and W_A and W_B equal their respective weights.

The fund's weights are shifted using a numerical search procedure until the combination that maximizes expected utility is found, which for this example equals a 57.13% allocation to fund A and a 42.87% allocation to fund B. The expected utility of the portfolio reaches a maximum at 9.31%. This approach implicitly takes into account all of the features of the empirical sample, including skewness, kurtosis, and any other peculiarities of the distribution (Table 4.2).

Table 4.2 Full-scale optimization for log-wealth investor

Year	Fund A returns	Fund B returns	Fund A weight	Fund B weight	Portfolio utility
1	10.06%	16.16%	57.13%	42.87%	$\ln[(1 + 0.1006) \times 0.5713 + (1 + 0.1616) \times 0.4287] \times 1/10 = 1.1931\%$
2	1.32%	−7.10%	57.13%	42.87%	$\ln[(1 + 0.0132) \times 0.5713 + (1 - 0.0710) \times 0.4287] \times 1/10 = -0.2317\%$
3	37.53%	29.95%	57.13%	42.87%	$\ln[(1 + 0.3753) \times 0.5713 + (1 + 0.2995) \times 0.4287] \times 1/10 = 2.9477\%$
4	22.93%	0.14%	57.13%	42.87%	$\ln[(1 + 0.2293) \times 0.5713 + (1 + 0.0014) \times 0.4287] \times 1/10 = 1.2367\%$
5	33.34%	14.52%	57.13%	42.87%	$\ln[(1 + 0.3334) \times 0.5713 + (1 + 0.1452) \times 0.4287] \times 1/10 = 2.2533\%$
6	28.60%	11.76%	57.13%	42.87%	$\ln[(1 + 0.2860) \times 0.5713 + (1 + 0.1176) \times 0.4287] \times 1/10 = 1.9375\%$
7	20.89%	−7.64%	57.13%	42.87%	$\ln[(1 + 0.2089) \times 0.5713 + (1 - 0.0764) \times 0.4287] \times 1/10 = 0.8304\%$
8	−9.09%	16.14%	57.13%	42.87%	$\ln[(1 - 0.0909) \times 0.5713 + (1 + 0.1614) \times 0.4287] \times 1/10 = 0.1715\%$
9	−11.94%	7.26%	57.13%	42.87%	$\ln[(1 - 0.1194) \times 0.5713 + (1 + 0.0726) \times 0.4287] \times 1/10 = -0.3777\%$
10	−22.10%	14.83%	57.13%	42.87%	$\ln[(1 - 0.2210) \times 0.5713 + (1 + 0.1483) \times 0.4287] \times 1/10 = -0.6470\%$
Expected utility					9.3138%

Typically, full-scale optimization yields results that are similar to mean–variance optimization for investors with power utility. If investor preferences are instead described by kinked or S-shaped functions, the results of full-scale optimization may differ dramatically from those of mean–variance optimization. We illustrate these points by selecting optimal portfolios for investors with five different preference functions:

1. Log-wealth utility
2. Kinked utility with a -1% threshold
3. Kinked utility with a -5% threshold
4. S-shaped value function with aversion to returns below 0%
5. S-shaped value function with aversion to returns below 0.5%

A kinked utility function changes abruptly at a particular wealth or return level and is relevant for investors who are concerned with breaching a threshold.

$$U(x) = \begin{cases} \ln(1+x), & \text{for } x \geq \theta \\ 10 \times (x-\theta) + \ln(1+\theta), & \text{for } x < \theta \end{cases} \tag{4.6}$$

The kink is located at θ, which in one case is set equal to a monthly return of -1%, corresponding to about an 11% annual loss, and in the second case is set equal to a -5% monthly return, corresponding to a 46% annual loss.

Proponents of behavioral finance have documented a number of contradictions to the neoclassical view of expected utility maximization. In particular, Kahnemann and Tversky (1979) have found that people focus on returns more than wealth levels and that they are risk averse in the domain of gains but risk seeking in the domain of losses. For example, if a typical investor is confronted with a choice between a certain gain and an uncertain outcome with a higher expected value, he or she will choose the certain gain. In contrast, when confronted with a choice between a certain loss and an uncertain outcome with a lower expected value, he or she will choose the uncertain outcome. This behavior is captured by an S-shaped value function, which Kahnemann and Tversky modeled as follows.

$$U(x) = \begin{cases} -A(\theta - x)^{\gamma_1}, & \text{for } x \leq \theta \\ +B(x-\theta)^{\gamma_2}, & \text{for } x > \theta \end{cases} \tag{4.7}$$

subject to:

$$A, B > 0$$
$$0 < \gamma^1, \gamma^2 \leq 1$$

The portfolio's return is represented by x, and A and B are parameters that together control the degree of loss aversion and the curvature of the function for outcomes above and below the loss threshold, θ. In this analysis, the parameters are scaled to match the range of returns one might reasonably expect from investment in hedge funds. In one case, the monthly loss threshold is set at 0%, which implies that investors experience absolute loss aversion. In the second case, the monthly loss threshold is set equal to 0.5%, which corresponds to about a 6% annualized return. This higher threshold implies that investors experience loss aversion relative to a target return such as an actuarial interest rate assumption or a measure of purchasing power.

To illustrate the full-scale approach to optimization, we use monthly hedge fund returns for the 10-year period from January 1994 through December 2003, provided by the Center for International Securities and Derivatives Markets (CISDM). We only use live funds with 10 years of history. These hedge funds deploy four strategies: equity hedge, convertible arbitrage, event driven, and merger arbitrage.

Table 4.3 shows the skewness and kurtosis for each of these funds, and it indicates whether or not the funds passed the Jarque–Bera (JB) test for normality. A normal distribution has skewness equal to 0 and kurtosis equal to 3.

The next step is to identify portfolios of hedge funds in each style category and across the entire sample of funds that maximize utility for each of these utility functions, based on full-scale optimization. This approach reveals the true utility-maximizing portfolios given the precise shape of the empirical return distributions.

Mean–variance optimization is applied to generate the efficient frontier of hedge funds in each category and across the entire sample. Within each category, the mean–variance efficient portfolio is evaluated that has the same expected return as the true utility-maximizing portfolio.

Table 4.4 shows the percentage change in utility gained by shifting from the mean–variance efficient portfolio to the true utility-maximizing portfolio determined by full-scale optimization. For investors with log-wealth utility, mean–variance optimization closely matches the results of full-scale optimization. Mean–variance optimization performs well in these situations because log-wealth utility is relatively insensitive to higher moments.

This result, however, does not prevail for investors who have kinked utility or S-shaped value functions. For investors with these preferences, mean–variance optimization results in significant loss of utility.

Table 4.5 depicts the fraction of the mean–variance efficient portfolio one would need to trade in order to invest the portfolio in accordance with the full-scale optimal weights. Again, mean–variance optimization performs well for log-wealth investors, except in the case in which hedge funds across all four styles are considered. Even in this case, though, the 15% departure from the optimal full-scale weights results in only slight utility loss. In contrast, the mean–variance exposures for investors with kinked utility or S-shaped value functions differ substantially from the true utility-maximizing weights.

4.5.3 Summary

Our analysis reveals that mean–variance optimization performs extremely well for investors with log-wealth utility. This result prevails even though the distributions

Table 4.3 Higher moments of hedge funds

Equity hedge	Skewness	Kurtosis	JB test
1	−0.25	4.98	Failed
2	−0.14	3.61	Passed
3	−0.34	4.69	Failed
4	0.43	4.77	Failed
5	0.07	5.32	Failed
6	0.00	4.75	Failed
7	1.12	8.23	Failed
8	−0.77	4.73	Failed
9	−0.70	5.13	Failed
10	−0.42	3.70	Passed
11	0.38	2.93	Passed
12	0.27	4.23	Failed
13	3.40	26.91	Failed
14	−.087	6.33	Failed
15	−0.15	3.36	Passed
16	−0.17	6.35	Failed
17	−0.02	5.46	Failed
18	0.49	6.35	Failed
19	−0.42	6.46	Failed
20	−0.22	5.72	Failed
21	1.08	7.36	Failed
22	0.56	4.08	Failed
23	0.36	3.46	Passed
24	0.91	8.70	Failed
25	0.10	5.01	Failed
Convertible arbitrage	**Skewness**	**Kurtosis**	
1	−0.45	3.79	Failed
2	−0.39	6.07	Failed
3	−0.84	6.22	Failed
4	−0.11	4.08	Passed
5	−0.35	4.78	Failed
6	0.19	4.44	Failed
7	−0.65	5.04	Failed
8	−0.60	5.25	Failed
9	−1.52	5.83	Failed
10	−1.34	8.72	Failed
Event driven	**Skewness**	**Kurtosis**	**JB test**
1	−1.00	7.60	Failed
2	0.00	4.18	Failed
3	1.43	14.72	Failed
4	0.91	6.87	Failed
5	−1.72	8.89	Failed

(Continued)

Table 4.3 (Continued)

Equity hedge	Skewness	Kurtosis	JB test
6	1.02	8.53	Failed
7	0.16	5.07	Failed
8	−0.50	4.88	Failed
9	0.29	3.92	Passed
10	−0.51	4.57	Failed
11	−0.66	4.28	Failed
12	−0.09	8.26	Failed
13	−0.37	3.75	Passed
14	−3.45	26.50	Failed
15	−0.46	6.26	Failed
16	−1.45	8.12	Failed
17	−0.09	3.98	Passed
18	−0.42	4.53	Failed
19	−0.14	5.30	Failed
Merger arbitrage	**Skewness**	**Kurtosis**	**JB test**
1	−0.30	4.10	Failed
2	0.91	8.40	Failed
3	−0.49	5.25	Failed
4	0.95	6.20	Failed
5	−1.01	6.97	Failed
6	0.17	4.92	Failed
7	0.88	9.75	Failed
Summary	**Average skewness**	**Average kurtosis**	**Percent failing JB**
Equity hedge	0.19	6.10	80%
Convertible arbitrage	−0.61	5.42	90%
Event driven	−0.37	7.38	84%
Merger arbitrage	0.16	6.51	100%
All hedge funds	−0.12	6.44	85%

Table 4.4 Percentage difference in utility

	Log-wealth utility	Kinked utility		S-shaped value functions	
		At −5%	At −1%	At 0%	At +0.5%
Equity hedge	0.01%	20.93%	11.61%	30.76%	61.28%
Convertible arbitrage	0.00%	4.37%	2.90%	9.67%	14.55%
Event driven	0.00%	17.13%	32.62%	12.58%	23.65%
Merger arbitrage	0.00%	7.51%	26.39%	2.41%	6.98%
All hedge funds	0.01%	25.77%	30.41%	13.53%	59.13%

Table 4.5 Turnover required to shift from mean–variance to full-scale efficiency

	Log-wealth utility	**Kinked utility**		**S-shaped value functions**	
		At −5%	**At −1%**	**At 0%**	**At +0.5%**
Equity hedge	4%	21%	31%	40%	53%
Convertible arbitrage	0%	66%	63%	56%	40%
Event driven	0%	84%	30%	31%	42%
Merger arbitrage	0%	36%	42%	12%	14%
All hedge funds	15%	34%	56%	57%	55%

of the hedge fund returns are significantly nonnormal. Moreover, much of this nonnormality survives into the mean–variance efficient portfolios, which implies that log-wealth utility is fairly insensitive to higher moments. However, mean–variance optimization performs poorly for investors with kinked utility or S-shaped value functions.

Appendix—WPA features

Return estimation

The WPA offers several methods for estimating expected returns, including:

- Historical full sample
- Equilibrium based on each asset's beta with respect to a reference portfolio
- Implied returns that assume the current portfolio is optimal
- User-specific views
- Historical blend of historical returns and user views, blended according to user confidence in views
- Equilibrium blend of equilibrium returns and user views about idiosyncratic component of returns, blended according to user confidence in views
- Black–Litterman blend of equilibrium returns and user views about systematic component of returns, blended according to user confidence in views

The user may choose to estimate returns after taxes.

Risk estimation

The WPA offers several options for estimating standard deviations and correlations, including:

- Historical full sample
- Exponential decay factor placing higher weight on recent performance
- Turbulent regime based on periods representing turbulent market conditions
- Quiet regime based on periods representing quiet market conditions
- User-defined weighting of turbulent and quiet periods
- User-specified return or risk thresholds, such as downside risk
- User-defined or imported from other risk model estimation module

The user may choose to estimate risk after taxes.

The WPA checks to ensure that the correlations are positive semidefinite.

Parametric optimization

The WPA enables the user to perform a variety of parametric optimizations, including mean–variance optimization, mean–tracking error optimization, or mean–variance–tracking error optimization.

The user sets the minimum and maximum allowable asset weights.

The user may add other linear or group constraints.

Full-scale optimization

The WPA also allows the user to construct portfolios on the basis of full-scale optimization, which identifies the optimal portfolio by examining a sufficiently large set of asset combinations. Full-scale optimization accommodates a wider range of investor preferences and return distributions than mean–variance optimization. It is especially suitable for assets with nonnormal return distributions, which is common for assets such as hedge funds, private equity, and commodities.

Portfolio attributes

Efficient surface

The WPA shows a three-dimensional plot of the current portfolio, the benchmark portfolio, and 10 optimal portfolios that lie on the efficient surface. The vertical axis is expected return, and the other two dimensions are expected risk (standard deviation) and tracking error.

The user can identify portfolios that have:

1. The same expected return with lower standard deviation
2. The same risk (standard deviation) with higher expected return

Higher moments

The WPA shows a variety of portfolio attributes for the various portfolios, including skewness, kurtosis, and probability of breaching a user-specified threshold. These results are particularly useful for comparing an existing portfolio to a full-scale optimal portfolio.

Probability of loss

The WPA allows the user to evaluate probability of loss in a variety of ways. It shows probability of loss both at the end of the horizon and throughout the horizon. The user has the flexibility to vary the length of the horizon, the percentage loss threshold, and the portfolio value.

The WPA allows the user to estimate probability of loss analytically, by Monte Carlo simulation or by bootstrap simulation.

Value at risk

The WPA shows value at risk measured both at the end of the horizon and throughout the horizon. The user has the flexibility to vary the length of the horizon, the VaR threshold, and the portfolio value.

The WPA allows the user to estimate value at risk analytically, by Monte Carlo simulation or by bootstrap simulation.

Joint probability of loss

Many investors care about both absolute and relative performance. The WPA shows the joint probability of failing to achieve an absolute target and simultaneously underperforming the benchmark.

Risk budgets[7]

The WPA converts optimal portfolio allocations into a risk budget. A risk budget shows the value at risk associated with each portfolio allocation in isolation.

The WPA also shows each allocation's contribution to the total portfolio value at risk.

The WPA also presents the risk budget in terms of continuous value at risk.

Wealth and income analysis

The WPA displays the distribution for wealth and income at the end of the investment horizon, taking into account the expected return and risk of the selected portfolio, the effective tax rate, and cash flows.

The WPA estimates the potential wealth and income of selected portfolios at different dates throughout the investment horizon, for a chosen level of confidence.

All values are shown before and after inflation.

Custom reports

The WPA allows the user to select content for a customized report. The user can select the relevant pages and input the client characteristics. Reports are generated quickly and can be printed or converted to Adobe® PDF format.

References

Chow, G. (1995). Portfolio selection based on return, risk, and relative performance. *Financial Analysts Journal* (March–April).

Chow, G., Jacquier, E., Kritzman, M., & Lowry, K. (1999). Optimal portfolios in good times and bad. *Financial Analysts Journal*(May–June).

Chow, G., & Kritzman, M. (2001). Risk budgets. *The Journal of Portfolio Management* (Winter).

Cremers, J.-H., Kritzman, M., & Page, S. (2005). Optimal hedge fund allocations: Do higher moments matter? *The Journal of Portfolio Management* (Spring).

Karlin, S., & Taylor, H. (1975). *A first course in stochastic processes* (2nd edition). Academic Press.

Kritzman, M., Lowry, K., & Van Royan, A.-S. (2001). Risk, regimes, and overconfidence. *The Journal of Derivatives*(Spring).

Kritzman, M., & Rich, D. (2001). The mismeasurement of risk. *Financial Analysts Journal* (May/June).

[7] See Chow et al (2001) for further discussion of risk budgets.

Section Two

Theory

5 Modeling, estimation, and optimization of equity portfolios with heavy-tailed distributions

Almira Biglova, Sergio Ortobelli, Svetlozar Rachev and Frank J. Fabozzi

Executive Summary

In this chapter, we provide a methodology to solve dynamic portfolio strategies considering realistic assumptions regarding the return distribution. First, we analyze the empirical behavior of some equities, suggesting how to approximate a historical return series with a factor model that accounts for most of the variability and proposing a methodology to generate realistic return scenarios. Then we examine the profitability of some reward–risk strategies based on a forecasted evolution of the returns. Since several studies in behavioral finance have shown that most investors in the market are neither risk averters nor risk lovers, we discuss the use of portfolio strategies based on the maximization of performance measures consistent with these investors' preferences. We first argue the computational complexity of reward–risk portfolio selection problems and then we compare the optimal sample paths of the future wealth obtained by performing reward–risk portfolio optimization on simulated data.

5.1 Introduction

The purpose of this chapter is to model and forecast the behavior of financial asset returns in order to optimize the performance of portfolio choices. In particular, we deal with three fundamental themes in portfolio theory: (1) the reduction of dimensionality of the portfolio problem, (2) the generation of future scenarios, and (3) the maximization of the portfolio performance in a reward–risk plane consistent with investors' behavior. To do so, we suggest a methodology to simulate the joint behavior of future returns to which we apply portfolio selection strategies. For this purpose, we consider the recent historical observations of some equities, paying attention to the modeling of all distributional aspects of the financial series. The empirical analysis performed on these series suggest that they exhibit (1) asymmetry and heavy tailedness

Doi:10.1016/B978-0-12-374952-9.00005-1

and (2) volatility clustering such that calm periods are generally followed by highly volatile periods and vice versa. Moreover, the findings suggest that the dependence model has to be flexible enough to account for the asymmetry of central dependence and, even more importantly, dependence of the tail events ("huge losses go together"). It is no surprise that the Gaussian distributional assumption is rejected for the financial series in our study. In fact, these results are largely confirmed by several empirical studies.[1]

In searching for an acceptable model to describe the dependence structure, we first perform a principal components analysis (PCA) to identify the main portfolio factors whose variance is significantly different from zero. By doing so, we obtain the few components that explain the majority of the return volatility, resulting in a reduction of the dependence structure dimension. In order to simulate realistic future return scenarios, we distinguish between the approximation of PCA-residuals and PCA-factors. The sample residuals obtained from the factor model are well approximated with an ARMA(1,1)-GARCH(1,1) model with stable innovations. As a result, we suggest simulating them independently by the simulated factors. Next, we examine the behavior of each factor with a time-series process belonging to the ARMA-GARCH family with stable Paretian innovations and we suggest modeling dependencies with an asymmetric Student *t*-copula valued on the innovations of the factors.[2] By doing so, we take into account the stylized facts observed in financial markets such as clustering of the volatility effect, heavy tails, and skewness. We then separately model the dependence structure between them.

It is well known that the classic mean–variance framework is not consistent with all investors' preferences. According to several studies, any realistic way of optimizing portfolio performance should maximize upside potential outcomes and minimize the downside outcomes. For this reason, the portfolio literature since about the late 1990s has proposed several alternative approaches to portfolio selection.[3] In particular, in this chapter we analyze portfolio selection models based on different measures of risk and reward. However, the resulting optimization problems consistent with investors' preferences could present more local optima. Thus, we propose solving them using a heuristic for global optimization.[4] Finally, we compare the ex post sample paths of the wealth obtained with the maximization of the Sharpe ratio and of the other performance measures applied to simulated returns.[5]

The chapter is organized as follows. In Section 5.2, we provide a brief empirical analysis of the dataset used in this study. In Section 5.3, we examine a methodology to build scenarios based on a simulated copula. In Section 5.4, we provide

[1] For a summary of studies, see Rachev and Mittnik (2000), Balzer (2001), Rachev *et al.* (2005), and Rachev, Mittnik, Fabozzi, Focardi, and Jasic (2007).

[2] See Sun *et al.* (2008) and Biglova, Kanamura, Rachev, and Stoyanov (2008).

[3] See Balzer (2001), Biglova, Ortobelli, Rachev, and Stoyanov (2004), Rachev *et al.* (2008), Ortobelli, Rachev, Shalit, and Fabozzi (2009).

[4] See Angelelli and Ortobelli (2009).

[5] See Sharpe (1994), Biglova *et al.* (2004), Biglova, Ortobelli, Rachev, and Stoyanov (2009).

a comparison among different strategies. Our conclusions are summarized in Section 5.5.

5.2 Empirical evidence from the Dow Jones Industrial Average components

For purposes of our study, we analyze 30 stocks that were components of the Dow Jones Industrial Average (DJIA) on 10/03/2008.[6] In particular, we investigate the log returns for 1,837 daily observations from 6/14/2001 to 10/03/2008 for each of the 30 stocks. Central theories in finance and important empirical studies assume that asset returns follow a normal distribution. The justification of this assumption is often cast in terms of its asymptotic approximation. However, this can be only a partial justification because the Central Limit Theorem for normalized sums of independent and identically distributed (i.i.d.) random variables determines the domain of attraction of each stable law.[7] Therefore, it is not surprising that when we consider tests for normality such as the Jarque–Bera and Kolmogorov–Smirnov tests (with a 95% confidence level) that the null hypothesis of normality for the daily log returns is rejected for 24 of the 30 stocks. However, if we test the stable Paretian assumption with a 95% confidence level employing the Kolmogorov–Smirnov statistic, we have to reject the null hypothesis only for four of the 30 stocks. Moreover, observing the covariation of the last 2 years of our data (whose time include also the period of the failure of Lehman Brothers), we deduce that the tails of the return distribution should consider (1) asymmetry of dependence and (2) dependence of the tail events. Therefore, the dependence model cannot be approximated with a multivariate normal distribution because it fails to describe both phenomena.[8]

Even from these preliminary tests, it is reasonable to conclude that the assumption of i.i.d. returns and conditional homoskedasticity is not the best model to approximate the return evolution of all equities. Since the prices observed in the market involve information on past market movements, we should consider the return distribution conditioned on information contained in past return data, or a more general information set. The class of autoregressive moving average (ARMA) models is a natural candidate for conditioning on the past of a return series. However, the conditional volatility of ARMA models is

[6] These stocks are: ALCOA INC (AA), AMER EXPRESS INC (AXP), BOEING CO (BA), BK OF AMERICA CP (BAC), CITIGROUP INC (C), CATERPILLAR INC (CAT), CHEVRON CORP (CVX), DU PONT E I DE NEM (DD), WALT DISNEY-DISNEY C (DIS), GEN ELECTRIC CO (GE), GEN MOTORS (GM), HOME DEPOT INC (HD), HEWLETT PACKARD CO (HPQ), IBM, Intel Corporation (INTC), JOHNSON AND JOHNS DC (JNJ), JP MORGAN CHASE CO (JPM), KRAFT FOODS INC (KFT), COCA COLA (KO), MCDONALDS (MCD), 3M COMPANY (MMM), MERCK CO INC (MRK), Microsoft Corporation (MSFT), PFIZER INC (PFE), PROCTER GAMBLE CO(PG), AT&T INC (T), UNITED TECH (UTX), VERIZON COMMUN (VZ), WAL MART STORES (WMT), EXXON MOBIL CP(XOM).

[7] See Zolatorev (1986).

[8] See, for example, Rachev and Mittnik (2000), Rachev *et al.* (2005), and Rachev *et al.* (2007).

independent of past realizations while empirical evidence shows that conditional homoskedasticity is often violated in financial data. In particular, we observe volatility clusters on returns series. Such behavior is captured by autoregressive conditional heteroskedastic models (ARCH)[9] and their generalization (GARCH models),[10] where the innovations are conditionally stable Paretian distributed. Several empirical experiments by Rachev and Mittnik (2000) have reported the typical behavior problem of time-series modeling.

Assume that daily stock returns follow ARMA(p,q)-GARCH(s,u) processes; i.e., assume:

$$\begin{aligned} r_{j,t} &= a_{j,0} + \sum_{i=1}^{p} a_{j,i} r_{j,t-i} + \sum_{i=1}^{q} b_{j,i}\varepsilon_{j,t-i} + \varepsilon_{j,t} \\ \varepsilon_{j,t} &= \sigma_{j,t} z_{j,t} \\ \sigma_{j,t}^2 &= c_{j,0} + \sum_{i=1}^{s} c_{j,i}\sigma_{j,t-i}^2 + \sum_{i=1}^{u} d_{j,i}\varepsilon_{j,t-i}^2 \end{aligned}$$

where $r_{j,t}$ is the daily return of the stock j (j = 1, ..., 30) at day t (t = 1, ..., 1,837). Since several studies have shown that ARMA-GARCH filtered residuals are themselves heavy tailed, then it makes sense to assume that the sequence of innovations $z_{j,t}$ is an infinite-variance process consisting of i.i.d. random variables in the domain of normal attraction of a stable distribution with index of stability α_j belonging to (0,2). That is, there exist normalizing constants $h_j^{(T)} \in R_+$ and $k_j^{(T)} \in R$ such that:

$$\frac{1}{h_j^{(T)}} \sum_{t=1}^{T} z_{j,t} + k_j^{(T)} \xrightarrow{d} S_{\alpha_j}(\sigma_j, \beta_j, \mu_j),$$

where the constants $h_j^{(T)}$ have the form $h_j^{(T)} = L_j(T)\sqrt[\alpha_j]{T}$ and $L_j(T)$ are slowly varying functions as $T \to \infty$. $S_{\alpha j}(\sigma_j, \beta_j, \mu_j)$ is a stable Paretian distribution with index of stability, $\alpha_j \in (0,2]$, skewness parameter, $\beta_j \in [-1,1]$, scale parameter, $\sigma_j \in \Re_+$, and location parameter, $\mu_j \in \Re$.[11] In particular, we can easily test different distributional hypotheses for the innovations of ARMA(1,1)-GARCH(1,1):

$$\begin{aligned} r_{j,t} &= a_{j,0} + a_{j,1} r_{j,t-1} + b_{j,1}\varepsilon_{j,t-1} + \varepsilon_{j,t}; \ \varepsilon_{j,t} = \sigma_{j,t} z_{j,t} \\ \sigma_{j,t}^2 &= c_{j,0} + c_{j,1}\sigma_{j,t-1}^2 + d_{j,1}\varepsilon_{j,t-1}^2 \end{aligned} \tag{5.1}$$

[9] See Engle (1982).

[10] See Bollerslev (1986).

[11] Refer to Samorodnitsky and Taqqu (1994) and Rachev and Mittnik (2000) for a general discussion of the properties and use of stable distributions.

estimated for the equity returns. First, we observe that a simple Ljung–Box Q-statistic for the full model (see Box, Jenkins, and Reindel, 1994) indicates that we cannot reject an ARMA(1,1)-GARCH(1,1) model for all return series. Moreover, once the maximum likelihood estimates of the model are obtained from the empirical innovations $\hat{\varepsilon}_{j,t} = r_{j,t} - \hat{a}_{j,0} + \hat{a}_{j,1} r_{j,t-1} + \hat{b}_{j,1} \hat{\varepsilon}_{j,t-1}$, we can easily get the standardized innovations $\hat{z}_{j,t} = \hat{\varepsilon}_{j,t} / \hat{\sigma}_{j,t}$. We can then test these innovations with respect to the stable non-Gaussian distribution versus the Gaussian one by applying the Kolmogorov–Smirnov (KS) statistic according to:

$$KS = \sup_{x \in R} | F_S(x) - \hat{F}(x)|,$$

where $F_S(x)$ is the empirical sample distribution and $\hat{F}(x)$ is the cumulative distribution function evaluated at x for the Gaussian or stable non-Gaussian fit, respectively. The KS test allows a comparison of the empirical cumulative distribution of innovations with either a simulated Gaussian or a simulated stable distribution. For our sample, KS statistics for the stable non-Gaussian test is almost 10 times smaller (in average) than the KS distance in the Gaussian case.

5.3 Generation of scenarios consistent with empirical evidence

Several problems need to be overcome in order to forecast, control, and model portfolios in volatile markets. First, we have to reduce the dimensionality of the problem, to get robust estimations in a multivariate context. Second, as has been noted in the portfolio selection literature, it is necessary to properly take into consideration the dependence structure of financial returns. Finally, the portfolio selection problem should be based on scenarios that take into account all the characteristics of the stock returns: heavy-tailed distributions, volatility clustering, and non-Gaussian copula dependence.

5.3.1 *The portfolio dimensionality problem*

When we deal with the portfolio selection problem under uncertainty conditions, we always have to consider the robustness of the estimates necessary to forecast the future evolution of the portfolio. Since we want to compute optimal portfolios with respect to some ordering criteria, we should also consider the sensitivity of risk and reward measures with respect to changes in portfolio composition. Thus, the themes related to robust portfolio theory are essentially twofold: (1) the risk (reward) contribution given by individual stock components of the portfolio[12] and (2) the estimation of the inputs (i.e., statistical parameters).[13]

[12] See, for example, Fischer (2003) and Tasche (2000).

[13] See, among others, Chopra and Ziemba (1993), Papp *et al.* (2005), Kondor *et al.* (2007), Rachev *et al.* (2005), Sun *et al.* (2008a,), and Biglova *et al.* (2008).

As discussed by Rachev, Menn, and Fabozzi (2005) and Sun, Rachev, Stoyanov, and Fabozzi (2008), the portfolio dimensional problem is strictly linked to the approximation of statistical parameters describing the dependence structure of the returns. Moreover, Kondor, Pafka, and Nagy (2007) have shown that the sensitivity to estimation error of portfolios optimized under various risk measures can have a strong impact on portfolio optimization, in particular when we consider the probability of rare events. Thus, according to the studies by Papp, Pafka, Nowak, and Kondor (2005) and Kondor *et al.* (2007), robustness of the approximations could be lost if there is not an "adequate" number of observations. In fact, Papp *et al.* (2005) have shown that the ratio v between the estimated optimal portfolio variance and the true one follows the rule:

$$v = \left(1 - \frac{n}{K}\right)^{-1}$$

where K is the number of observations and n is the number of assets. Consequently, in order to get a good approximation of the portfolio variance, we need to have a much larger number of observations relative to the number of assets.

Similar results can be proven for other risk parameter estimates such as conditional value at risk.[14] Because in practice the number of observations is limited, in order to get a good approximation of portfolio input measures, it is necessary to find the right trade-off between the number of historical observations and a statistical approximation of the historical series depending only on a few parameters.

One way to reduce the dimensionality of the problem is to approximate the return series with a regression-type model (such as a k-fund separation model or other model) that depends on an adequate number (not too large) of parameters.[15] For this purpose, we perform a PCA of the returns of the 30 stocks used in this chapter in order to identify few factors (portfolios) with the highest variability. Therefore, we replace the original n ($n = 30$ for our case) correlated time series r_i with n uncorrelated time series P_i assuming that each r_i is a linear combination of the P_i. Then we implement a dimensionality reduction by choosing only those portfolios whose variance is significantly different from zero. In particular, we call *portfolios factors* f_i the p portfolios P_i with a significant variance, while the remaining n–p portfolios with very small variances are summarized by an error ε. Thus, each series r_i is a linear combination of the factors plus a small uncorrelated noise:

$$r_i = \sum_{i=1}^{p} c_i f_i + \sum_{i=p+1}^{n} d_i P_i = \sum_{i=1}^{p} c_i f_i + \varepsilon,$$

[14] See Kondor *et al.* (2007).

[15] See Ross (1978).

Generally, we can apply the PCA either to the variance–covariance matrix or to the correlation matrix. Since returns are heavy-tailed dimensionless quantities, we apply PCA to the correlation matrix obtaining 30 principal components, which are linear combinations of the original series, $r = (r_1,\ldots,r_{30})'$.

Table 5.1 shows the total variance explained by a growing number of components. Thus, the first component explains 41.2% of the total variance and the first 14 components explain 80.65% of the total variance. Because all the other components contribute no more than 1.75% of the global variance, we implement a dimensionality reduction by choosing only the first 14 factors. As a consequence of this principal component analysis, each series r_i $(i = 1,\ldots,30)$ can be represented as a linear combination of 14 factors plus a small uncorrelated noise.

Once we have identified 14 factors that explain more than 80% of the global variance, then we can generate the future returns r_i using the factor model:

$$r_{i,t} = \alpha_i + \sum_{j=1}^{14} \beta_{i,j} f_{j,t} + e_{i,t} \quad t = 1,\ldots,1837;\ i = 1,\ldots,30 \tag{5.2}$$

Table 5.2 reports the coefficients α_i; $\beta_{i,j}$ of factor model (5.2). The generation of future scenarios should take into account (1) all the anomalies observed

Table 5.1 Percentage of the total variance explained by a growing number of components based on the covariance matrix

Principal component	Percentage of variance explained	Percentage of total variance explained	Principal component	Percentage of variance explained	Percentage of total variance explained
1	41.20	41.20	16	1.71	84.11
2	5.33	46.53	17	1.64	85.75
3	4.52	51.05	18	1.53	87.28
4	4.08	55.14	19	1.45	88.73
5	3.82	58.95	20	1.40	90.13
6	3.24	62.19	21	1.37	91.50
7	2.84	65.03	22	1.31	92.81
8	2.72	67.75	23	1.24	94.05
9	2.63	70.38	24	1.15	95.20
10	2.37	72.75	25	1.05	96.26
11	2.14	74.89	26	0.97	97.23
12	2.03	76.92	27	0.88	98.11
13	1.90	78.82	28	0.72	98.83
14	1.83	80.65	29	0.65	99.48
15	1.75	82.40	30	0.52	100.00

Table 5.2 Estimated coefficients

	MMM	AA	AXP	T	BAC	BA	CAT
Alpha	0.001%	−0.017%	−0.004%	−0.010%	0.004%	−0.005%	0.015%
Beta 1	0.528%	0.600%	0.660%	0.501%	0.522%	0.497%	0.552%
Beta 2	−0.061%	0.175%	0.082%	−0.189%	0.116%	0.090%	0.172%
Beta 3	0.079%	0.392%	−0.137%	−0.139%	−0.273%	0.146%	0.212%
Beta 4	0.004%	−0.122%	−0.121%	0.166%	−0.359%	0.031%	−0.078%
Beta 5	0.182%	−0.054%	−0.035%	−0.310%	−0.017%	0.195%	0.080%
Beta 6	−0.014%	0.063%	−0.026%	0.252%	0.009%	0.031%	−0.052%
Beta 7	0.057%	−0.202%	0.019%	−0.221%	0.074%	−0.242%	−0.032%
Beta 8	−0.083%	0.020%	0.009%	−0.068%	0.010%	0.185%	−0.043%
Beta 9	0.118%	−0.058%	−0.028%	0.093%	0.000%	0.114%	0.117%
Beta 10	0.001%	−0.194%	0.070%	−0.020%	0.057%	0.152%	−0.106%
Beta 11	0.118%	0.155%	0.010%	−0.026%	−0.052%	−0.327%	0.148%
Beta 12	0.021%	0.069%	−0.107%	0.026%	−0.002%	0.004%	−0.078%
Beta 13	0.069%	0.058%	−0.076%	0.027%	0.093%	0.015%	−0.020%
Beta 14	0.128%	0.193%	0.010%	0.033%	0.068%	−0.105%	0.097%

	HD	IBM	INTC	JNJ	JPM	KFT	MCD
Alpha	−0.018%	−0.003%	−0.012%	0.006%	0.001%	0.001%	0.016%
Beta 1	0.585%	0.444%	0.670%	0.434%	0.646%	0.287%	0.374%
Beta 2	0.121%	0.067%	0.156%	−0.324%	0.152%	−0.188%	−0.020%
Beta 3	−0.158%	0.017%	0.050%	−0.048%	−0.251%	−0.135%	0.010%
Beta 4	0.085%	0.184%	0.408%	0.032%	−0.239%	−0.036%	0.052%
Beta 5	0.172%	−0.103%	−0.057%	0.060%	−0.129%	0.182%	0.196%
Beta 6	0.001%	−0.043%	−0.043%	−0.142%	0.013%	0.330%	0.047%
Beta 7	0.049%	0.092%	0.267%	0.025%	0.076%	0.050%	−0.223%
Beta 8	−0.209%	0.036%	0.086%	−0.025%	0.047%	0.399%	−0.185%
Beta 9	−0.056%	−0.013%	−0.045%	0.101%	−0.009%	−0.124%	−0.475%
Beta 10	−0.318%	−0.040%	−0.018%	0.045%	0.072%	−0.216%	0.181%
Beta 11	−0.191%	0.087%	0.000%	0.148%	−0.017%	0.060%	0.027%
Beta 12	−0.007%	−0.183%	0.056%	−0.205%	−0.019%	−0.150%	−0.142%
Beta 13	−0.196%	0.066%	0.110%	0.002%	−0.007%	0.033%	0.066%
Beta 14	0.008%	−0.156%	−0.104%	−0.183%	0.100%	0.001%	0.001%

in equity returns; (2) the time evolution of factor $f_{j,t}$ and of errors $e_{i,t}$, and (3) the comovements of the vector of the returns considering the skewness and kurtosis of the joint distribution.

To deal with the third problem, we suggest employing a skewed copula with heavy tails. A *copula function* C associated to random vector $v = (v_1,\ldots,v_n)$ is a probability distribution function on the n-dimensional hypercube, such that:

$$F_v(y_1,\ldots,y_n) = P(v_1 \le y_1,\ldots,v_n \le y_n) = C(P(v_1 \le y_1),\ldots,P(v_n \le y_n))$$
$$= C(F_{v_1}(y_1),\ldots,F_{v_n}(y_n)),$$

alpha and betas of the factor model

CVX	C	KO	DD	XOM	GE	GM	HPQ
0.012%	−0.022%	0.003%	−0.004%	0.013%	−0.019%	−0.045%	0.010%
0.426%	0.643%	0.416%	0.531%	0.531%	0.619%	0.460%	0.427%
−0.165%	0.152%	−0.326%	0.021%	−0.223%	0.029%	0.184%	0.262%
0.411%	−0.248%	−0.053%	0.095%	0.370%	−0.028%	−0.146%	0.071%
−0.289%	−0.280%	0.100%	−0.037%	−0.230%	0.010%	−0.142%	0.331%
−0.180%	−0.104%	0.208%	0.090%	−0.131%	0.028%	0.052%	−0.124%
0.062%	−0.036%	0.135%	−0.012%	0.053%	−0.055%	0.066%	−0.020%
0.148%	0.083%	0.189%	−0.033%	0.156%	−0.013%	−0.085%	0.117%
−0.024%	0.008%	−0.033%	−0.047%	−0.040%	0.019%	−0.018%	0.120%
−0.100%	0.003%	0.027%	0.076%	−0.052%	0.078%	−0.059%	−0.169%
0.016%	0.067%	0.150%	−0.050%	0.007%	0.037%	−0.017%	0.143%
−0.102%	−0.033%	0.072%	0.083%	−0.096%	0.074%	0.169%	0.001%
−0.010%	−0.052%	0.193%	0.088%	−0.004%	−0.082%	0.419%	0.181%
−0.061%	−0.010%	−0.165%	0.087%	−0.040%	0.017%	0.017%	−0.051%
−0.105%	0.072%	0.114%	0.131%	−0.092%	0.025%	−0.438%	0.272%

MRK	MSFT	PFE	PG	UTX	VZ	WMT	DIS
−0.018%	−0.007%	−0.019%	0.019%	0.007%	−0.012%	0.004%	−0.001%
0.309%	0.543%	0.516%	0.411%	0.554%	0.491%	0.469%	0.528%
−0.339%	0.024%	−0.270%	−0.294%	0.138%	−0.165%	−0.014%	0.095%
−0.087%	0.034%	−0.079%	−0.034%	0.140%	−0.109%	−0.093%	0.023%
−0.008%	0.259%	−0.037%	0.050%	0.009%	0.170%	0.105%	0.131%
−0.106%	−0.079%	0.005%	0.184%	0.171%	−0.281%	0.079%	−0.023%
−0.360%	−0.049%	−0.281%	0.063%	−0.026%	0.179%	−0.011%	−0.041%
−0.161%	0.129%	−0.085%	0.094%	−0.132%	−0.199%	0.028%	−0.109%
0.218%	−0.002%	−0.005%	−0.081%	0.101%	−0.092%	−0.167%	0.031%
−0.142%	−0.018%	−0.002%	0.065%	0.160%	0.083%	0.003%	0.057%
−0.108%	−0.017%	−0.051%	0.121%	0.094%	0.021%	−0.205%	0.087%
−0.034%	−0.101%	0.003%	−0.030%	−0.106%	−0.007%	−0.188%	0.170%
0.207%	−0.027%	−0.061%	0.140%	0.018%	0.016%	−0.022%	−0.193%
−0.141%	0.102%	0.197%	0.084%	0.025%	0.019%	0.010%	−0.346%
0.090%	−0.134%	−0.024%	0.052%	−0.016%	0.032%	0.022%	−0.129%

where F_{υ_i} is the marginal distribution of the i-th component (see Sklar (1959)). So once we have generated scenarios with the copula $C(u_1,\ldots,u_n) = F_\upsilon\left(F_{\upsilon_1}^{-1}(u_1),\ldots,F_{\upsilon_n}^{-1}(u_n)\right)$ (where $F_{\upsilon_i}^{-1}$ is the inverse cumulative function of the i-th marginal derived from the multivariate distributional assumption F_υ) that summarizes the dependence structure of returns, then we can easily generate joint observations using the most opportune inverse distribution functions $\bar{F}_{\upsilon_i}^{-1}$ of the single components applied to the points generated by the copula. In particular, we next tackle the general problem of return generation considering a multivariate skewed Student's t-copula for the joint generation of innovations of the 14 factors.

5.3.2 Generation of return scenarios

Let us summarize the algorithm we propose to generate return scenarios according to the empirical evidence. Assume the log returns follow model (5.2). In Step 1 of the algorithm, we approximate each factor $f_{j,t}$ with an ARMA(1,1)-GARCH(1,1) process with stable Paretian innovations. Then, we provide the marginal distributions for standardized innovations of each factor used to simulate the next-period returns. In Step 2 of the algorithm, we estimate the dependence structure of the vector of standardized innovations with a skewed Student *t*. In particular, we first estimate the dependence structure among the innovations with an asymmetric *t*-copula. Then, we combine the marginal distributions and the scenarios for the copula into scenarios for the vector of factors. By doing so, we generate the vector of the standardized innovation assuming that the marginal distributions are α_j-stable distributions and considering an asymmetric *t*-copula to summarize the dependence structure. Then, we can easily generate the vector of factors and in the last step of the algorithm we show how to generate future returns.

The algorithm is as follows.

Step 1. Carry out maximum likelihood parameter estimation of ARMA(1,1)-GARCH(1,1) for each factor $f_{j,t}$ ($j = 1,\ldots,14$).

$$\begin{aligned} f_{j,t} &= a_{j,0} + a_{j,1} f_{j,t-1} + b_{j,1}\varepsilon_{j,t-1} + \varepsilon_{j,t} \\ \varepsilon_{j,t} &= \sigma_{j,t} u_{j,t} \\ \sigma^2_{j,t} &= c_{j,0} + c_{j,1}\sigma^2_{j,t-1} + d_{j,1}\varepsilon^2_{j,t-1}; \\ j &= 1,\ldots,14;\ t = 1,\ldots,T. \end{aligned} \tag{5.3}$$

Since we have 1,837 historical observations, we use a window of $T = 1{,}837$. Table 5.3 reports the maximum likelihood estimates for the ARMA-GARCH parameters for all 14 factors.

Approximate with α_j-stable distribution $S_{\alpha_j}(\sigma_j, \beta_j, \mu_j)$ the empirical standardized innovations $\hat{u}_{j,t} = \hat{\varepsilon}_{j,t} / \sigma_{j,t}$ where the innovations $\hat{\varepsilon}_{j,t} = f_{j,t} - a_{j,0} - a_{j,1} f_{j,t-1} - b_{j,1}\varepsilon_{j,t-1}$ $j = 1,\ldots,14$.[16]

In order to value the marginal distribution of each innovation, we first simulate S stable distributed scenarios for each of the future standardized innovations series. Then, we compute the sample distribution functions of these simulated series:

$$F_{\hat{u}_{j,T+1}}(x) = \frac{1}{S}\sum_{s=1}^{S} I_{\{\hat{u}^{(s)}_{j,T+1} \le x\}},\ x \in \mathbb{R},\ j = 1,\ldots,14 \tag{5.4}$$

[16] For a general discussion on properties and use of stable distributions, see Samorodnitsky and Taqqu (1994) and Rachev and Mittnik (2000).

Table 5.3 Maximum likelihood estimates of ARMA(1,1)-GARCH(1,1) parameters for the 14 factors

Coefficients	Factor 1	Factor 2	Factor 3	Factor 4	Factor 5	Factor 6	Factor 7
$a_{j,0}$	0.02689	0.01643	0.01538	0.00421	0.04306	0.02288	−0.03331
$a_{j,1}$	0.25134	0.12943	−0.20296	0.13580	−0.26676	−0.35932	−0.36273
$b_{j,1}$	−0.32185	−0.03155	0.25612	−0.12642	0.27020	0.40564	0.37829
$c_{j,0}$	0.00656	0.01864	0.01248	0.00470	0.00000	0.17677	0.00294
$c_{j,1}$	0.91919	0.86903	0.90736	0.93934	0.95628	0.71501	0.96956
$d_{j,1}$	0.07633	0.11862	0.07947	0.05868	0.04372	0.10935	0.02775
Coefficients	**Factor 8**	**Factor 9**	**Factor 10**	**Factor 11**	**Factor 12**	**Factor 13**	**Factor 14**
$a_{j,0}$	0.00197	−0.01988	0.00786	0.00034	0.00047	−0.01269	−0.01689
$a_{j,1}$	0.77222	−0.60207	−0.03102	−0.50587	0.94011	−0.60657	0.01424
$b_{j,1}$	−0.79557	0.64574	0.04524	0.46606	−0.93076	0.62172	0.03502
$c_{j,0}$	0.00428	0.00927	0.00353	0.00665	0.01316	0.02258	0.01895
$c_{j,1}$	0.98033	0.95180	0.97129	0.96058	0.94889	0.91950	0.93939
$d_{j,1}$	0.01504	0.03947	0.02580	0.03321	0.03883	0.05879	0.04197

where $\hat{u}_{j,T+1}^{(s)}$ $(1 \leq s \leq S)$ is the s-th value simulated with the fitted α_j-stable distribution for future standardized innovation (valued in $T + 1$) of the jth factor.

Step 2. Fit the 14-dimensional vector of empirical standardized innovations $\hat{u} = [\hat{u}_1,\ldots,\hat{u}_{14}]'$ with an asymmetric t-distribution $\mathbf{V} = [V_1,\ldots,V_{14}]'$ with v degree of freedom; i.e.,

$$\mathbf{V} = \boldsymbol{\mu} + \boldsymbol{\gamma} Y + \sqrt{Y}\mathbf{Z} \tag{5.5}$$

where $\boldsymbol{\mu}$ and $\boldsymbol{\gamma}$ are constant vectors and Y is inverse γ- distributed $IG(v/2;v/2)$[17] independent of the vector $\mathbf{Z}$ that is normally distributed with zero mean and covariance matrix $\Sigma = [\sigma_{ij}]$. We use the maximum likelihood method

[17] See, among others, Rachev and Mittnik (2000).

to estimate the parameters $(v,\hat{\mu}_i,\hat{\sigma}_{ii},\hat{\gamma}_i)$ of each component. Then, an estimator of matrix Σ is given by

$$\hat{\Sigma} = \left(\text{cov}(\mathbf{V}) - \frac{2v^2}{(v-2)^2(v-4)}\hat{\gamma}\hat{\gamma}'\right)\frac{v-2}{2}$$

where $\hat{\gamma} = (\gamma_1,\ldots,\gamma_{14})$ and cov(**V**) is the variance–covariance matrix of **V**. Table 5.4 reports the estimated parameters of the multivariate skewed Student's t-distribution for the 14 factors.

Since we have estimated all the parameters of Y and **Z**, we can generate S scenarios for Y and, independently, S scenarios for **Z**, and using Equation (5.5) we obtain S scenarios for the vector of standardized innovations $\hat{u} = [\hat{u}_1,\ldots,\hat{u}_{14}]'$ that is asymmetric t-distributed. Denote these scenarios by $(V_1^{(s)},\ldots,V_{14}^{(s)})$ for $s = 1,\ldots,S$ and denote the marginal distributions $F_{V_j}(x)$ for $1 \le j \le 14$ of the estimated 14-dimensional asymmetric t-distribution by $F_V(x_1,\ldots,x_{14}) = P(V_1 \le x_1,\ldots,V_{14} \le x_{14})$. Then, considering $U_j^{(s)} = F_{V_j}(V_j^{(s)})$, $1 \le j \le 14$; $1 \le s \le S$, we can generate S scenarios $(U_1^{(s)},\ldots,U_{14}^{(s)})$, $s = 1,\ldots,S$ of the uniform random vector $(U_1,\ldots,U_{14})$ (with support on the 14-dimensional unit cube) and whose distribution is given by the copula:

$$C(t_1,\ldots,t_{14}) = F_V(F_{V_1}^{-1}(t_1),\ldots,F_{V_{14}}^{-1}(t_{14}));0 \le t_i \le 1;1 \le i \le 14.$$

Considering the stable distributed marginal sample distribution function of the j-th standardized innovation $F_{\hat{u}_{j,T+1}}$; $j = 1,\ldots,14$ (see Equation (5.4)) and the scenarios $U_j^{(s)}$ for $1 \le j \le 14$; $1 \le s \le S$, then we can generate S scenarios of the vector of standardized innovations (taking into account the dependence structure of the vector) $u_{T+1}^{(s)} = (u_{T+1}^{(1,s)},\ldots,u_{T+1}^{(14,s)})$, $s = 1,\ldots,S$ valued at time $T+1$ assuming:

$$u_{T+1}^{(j,s)} = (F_{u_{j,T+1}})^{-1}\left(U_j^{(s)}\right);1 \le j \le 14;1 \le s \le S.$$

Once we have described the multivariate behavior of the standardized innovation at time $T+1$ using relation (5.3), we can generate S scenarios of the vector of innovation:

$$\varepsilon_{T+1}^{(s)} = \left(\varepsilon_{T+1}^{(1,s)},\ldots,\varepsilon_{T+1}^{(14,s)}\right) = \left(\sigma_{1,T+1}u_{T+1}^{(1,s)},\ldots,\sigma_{14,T+1}u_{T+1}^{(14,s)}\right), s = 1,\ldots,S,$$

where $\sigma_{j,T+1}$ are still defined by Equation (5.3). Thus, using relation (5.3), we can generate S scenarios of the vector of factors $f_{T+1}^{(s)} = \left[f_{1,T+1}^{(s)},\ldots,f_{14,T+1}^{(s)}\right]$ valued at time $T+1$. Observe that this procedure can always be used

Table 5.4 Maximum likelihood estimate of parameters of the skewed Student's t distribution for the 14 factors

	F1	F2	F3	F4	F5	F6	F7	F8	F9	F10	F11	F12	F13	F14
γ	−7.19%	3.29%	−17.5%	−4.22%	3.95%	9.66%	2.17%	−10.66%	1.72%	0.03%	2.33%	−2.68%	1.06%	−5.87%
μ	−2.11%	−18.3%	39.68%	21.77%	−9.50%	−16.2%	7.76%	7.56%	−6.6%	−5.93%	−11.9%	−9.34%	−0.81%	19.07%
V	5													

Matrix Σ

	F1	F2	F3	F4	F5	F6	F7	F8	F9	F10	F11	F12	F13	F14
1	72.06%	47.54%	−35.8%	−41.1%	−28.0%	8.48%	8.98%	11.54%	−20.9%	6.68%	7.11%	29.45%	9.01%	−14.8%
2	47.54%	70.38%	−33.2%	−34.7%	−11.3%	2.82%	16.89%	3.54%	−10.6%	5.87%	11.99%	23.04%	14.53%	−23.9%
3	−35.8%	−33.2%	71.18%	39.25%	13.87%	10.64%	−12.5%	−3.49%	8.46%	−15.2%	14.68%	−14.2%	11.15%	12.07%
4	−41.1%	−34.7%	39.25%	77.07%	17.14%	0.25%	−2.98%	−4.41%	3.65%	−7.37%	−1.15%	−24.9%	−2.04%	13.57%
5	−28.0%	−11.3%	13.87%	17.14%	77.56%	−1.59%	−18.1%	−21.0%	10.39%	−8.26%	1.86%	−5.26%	−0.69%	−14.5%
6	8.48%	2.82%	10.64%	0.25%	−1.59%	57.56%	−1.83%	−15.8%	5.43%	−1.54%	6.92%	−0.13%	13.64%	−12.7%
7	8.98%	16.89%	−12.5%	−2.98%	−18.1%	−1.83%	70.42%	2.43%	−12.5%	13.65%	−4.07%	−13.4%	7.58%	10.50%
8	11.54%	3.54%	−3.49%	−4.41%	−21.0%	−15.8%	2.48%	62.45%	−8.60%	14.39%	−2.50%	6.44%	−3.28%	10.94%
9	−20.9%	−10.6%	8.46%	3.65%	10.39%	5.43%	−12.5%	−8.60%	61.14%	−6.05%	−5.60%	−14.3%	−0.76%	6.40%
10	6.68%	5.87%	−15.2%	−7.37%	−8.26%	−1.54%	13.65%	14.39%	−6.05%	70.92%	−20.4%	−18.4%	−0.06%	14.32%
11	7.11%	11.99%	14.68%	−1.15%	1.86%	6.92%	−4.07%	−2.50%	−5.60%	−20.4%	66.84%	28.08%	12.74%	−15.2%
12	29.45%	23.04%	−14.1%	−24.9%	−5.26%	−0.13%	−13.4%	6.44%	−14.3%	−18.4%	28.08%	70.98%	8.16%	−45.0%
13	9.01%	14.53%	11.15%	−2.04%	−0.69%	13.64%	7.58%	−3.23%	−0.76%	−0.06%	12.74%	8.16%	63.76%	−15%
14	−14.8%	−23.9%	12.07%	13.57%	−14.5%	−12.7%	10.50%	10.94%	6.40%	14.32%	−15.2%	−45.0%	−15.%	70.70%

to generate a distribution with some given marginals and a given dependence structure.[18]

Step 3. In order to estimate future returns valued at time $T + 1$, we first estimate a model ARMA(1,1)-GARCH(1,1) for the residuals of the factor model (5.2). That is, we consider the empirical residuals:

$$\hat{e}_{i,t} = r_{i,t} - \hat{\alpha}_i - \sum_{j=1}^{14} \hat{\beta}_{i,j} f_{j,t}$$

and then we estimate the parameters $g_{i,0}$, $g_{i,1}$, $h_{i,1}$, $k_{i,0}$, $k_{i,1}$, $p_{i,1}$ for all $I = 1,\ldots,30$ of the ARMA(1,1)-GARCH(1,1):

$$\begin{aligned} \hat{e}_{i,t} &= g_{i,0} + g_{i,1}\hat{e}_{i,t-1} + h_{i,1}q_{i,t-1} + q_{i,t} \\ q_{i,t} &= v_{i,t}z_{i,t} \\ v_{i,t}^2 &= k_{i,0} + k_{i,1}v_{i,t-1}^2 + p_{i,1}q_{i,t-1}^2; \\ \mathrm{i} &= 1,\ldots,30;\ t = 1,\ldots,T. \end{aligned} \tag{5.6}$$

Moreover, as for the factor innovation, we approximate with α_j-stable distribution $S_{\alpha_i}(\sigma_i, \beta_i, \mu_i)$ for any $i = 1,\ldots,30$ the empirical standardized innovations $\hat{z}_{i,t} = \hat{q}_{i,t}/v_{i,t}$, where the innovations $\hat{q}_{i,t} = \hat{e}_{i,t} - g_{i,0} - g_{i,1}\hat{e}_{i,t-1} - h_{i,1}q_{i,t-1}$. Then, we can generate S scenarios α_j-stable distributed for the standardized innovations $z_{i,T+1}^{(s)}$ $s = 1,\ldots,S$ and from Equation (5.6) we get S possible scenarios for the residuals $e_{i,T+1}^{(s)} = v_{i,T+1}z_{i,T+1}^{(s)}$ $s = 1,\ldots,S$. Therefore, combining Step 2 with the estimation of future residuals from factor model (5.2), we get S possible scenarios of returns:

$$r_{i,T+1}^{(s)} = \hat{\alpha}_i + \sum_{j=1}^{14} \hat{\beta}_{i,j} f_{j,t}^{(s)} + e_{i,t}^{(s)} \quad s = 1,\ldots,S. \tag{5.7}$$

The procedure illustrated here permits one to generate S scenarios at time $T + 1$ of the vector of returns.

5.4 The portfolio selection problem

Suppose we have a frictionless market in which no short selling is allowed and all investors act as price takers. The classical portfolio selection problem

[18] See, among others, Rachev *et al.* (2005), Sun *et al.* (2008a), Biglova *et al.* (2008), and Cherubini, Luciano, and Vecchiato (2004) for the definition of some classical copula used in finance literature.

among n assets in the reward–risk plane consists of minimizing a given risk measure ρ provided that the expected reward ν is constrained by some minimal value m; i.e.,

$$\min_x \rho(x'r - r_b)$$
$$s.t.$$
$$\nu(x'r - r_b) \geq m;\ x_i \geq 0, \sum_{i=1}^{n} x_i = 1; \tag{5.8}$$

where r_b denotes the return of a given benchmark, and $x'r = \sum_{i=1}^{n} x_i r_i$ stands for the returns of a portfolio with composition $x = (x_1,\ldots,x_n)'$. The portfolio that provides the maximum expected reward ν per unit of risk ρ is called the *market portfolio* and is obtained from problem (5.8) for one value m among all admissible portfolios. In particular, when the reward and risk are both positive measures, the market portfolio is obtained as the solution of the optimization problem:

$$\max_x \frac{\nu(x'r - r_b)}{\rho(x'r - r_b)}$$
$$s.t.$$
$$x_i \geq 0, \sum_{i=1}^{n} x_i = 1 \tag{5.9}$$

Clearly, there exist many possible performance ratios $G(X) = \nu(X)/\rho(X)$.

A first classification with respect to the different characteristics of reward and risk measures is given in Rachev, Ortobelli, Stoyanov, Fabozzi, and Biglova (2008). The most important characteristic is the isotony (consistency) with an order of preference; i.e., if X is preferable to Y, then $G(X) \geq G(Y)$ $(G(X) \leq G(Y))$ Although the financial literature on investor behavior agrees that investors are nonsatiable, there is not a common vision about the investors' aversion to risk. Thus, investors' choices should be isotonic with nonsatiable investors' preferences (i.e., if $X \geq Y$, then $G(X) \geq G(Y)$).

Several behavioral finance studies suggest that most investors are neither risk averse nor risk loving.[19] Thus, according to Bauerle and Müller (2006), if risk and reward measures are invariant in law (i.e., if X and Y have the same distribution, then $\rho(X) = \rho(Y)$ and $\nu(X) = \nu(Y)$), and the risk measure is positive and convex (concave) and the reward is positive and concave (convex), then the performance ratio is isotone with risk-averse (lover) preferences.

[19] See Friedman and Savage (1948), Markowitz (1952), Tversky and Kahneman (1992), Levy and Levy (2002), and Ortobelli *et al.* (2009).

Rachev *et al.* (2008) and Stoyanov, Rachev, and Fabozzi (2007) have classified the computational complexity of reward–risk portfolio selection problems. In particular, Stoyanov *et al.* (2007) have shown that we can distinguish four cases of reward/risk ratios $G(X)$ that admit unique optimum portfolio strategies. The most general case with unique optimum is when the ratio is a quasi-concave function; i.e., the risk functional $\rho(X)$ is positive convex and the reward functional $v(X)$ is positive concave. As observed above, by maximizing the ratio $G(X)$, we obtain optimal choices for risk-averse investors. In the other cases, when both measures $\rho(X)$ and $v(X)$ are either concave or convex, then the ratio $G(X)$ is isotone with investors' preferences that are neither risk averse nor risk loving. However, in this last case, the performance ratio admits more local optima.

5.4.1 Review of performance ratios

Here, we will review three performance ratios that we will use in the next section when we perform our empirical comparisons: Sharpe ratio, Rachev ratio, and Rachev higher moments ratio.

According to Markowitz' mean–variance analysis, Sharpe (1994) suggested that investors should maximize what is now referred to as Sharpe ratio (SR) given by:

$$SR(x'r) = \frac{E(x'r_{T+1} - r_{T+1,b})}{STD(x'r_{T+1} - r_{T+1,b})}$$

where $STD(x'r_{T+1} - r_{T+1,b})$ is the standard deviation of excess returns. Maximizing the Sharpe ratio, we get a market portfolio that should be optimal for nonsatiable risk-averse investors, and that is not dominated in the sense of second-order stochastic dominance. The maximization of the Sharpe ratio can be solved as a quadratic-type problem and thus it presents a unique optimum. In contrast to the Sharpe ratio, the next two performance ratios (Rachev ratio and Rachev higher moments ratio) are isotonic with the preferences of nonsatiable investors that are neither risk averse nor risk lovers.

The Rachev ratio (RR)[20] is the ratio between the average of earnings and the mean of losses; i.e.,

$$RR(x'r_{T+1}, \alpha, \beta) = \frac{ETL_{\beta}(r_{T+1,b} - x'r_{T+1})}{ETL_{\alpha}(x'r_{T+1} - r_{T+1,b})}$$

[20] See Biglova *et al.* (2004).

where the ETL is the expected tail loss, also known as conditional value at risk (CVaR), is defined as:

$$ETL_\alpha(X) = \frac{1}{\alpha}\int_0^\alpha VaR_q(X)dq,$$

and

$$VaR_q(X) = -F_X^{-1}(q) = -\inf\{x|P(X \leq x) > q\}$$

is the value at risk (VaR) of the random return X. If we assume a continuous distribution for the probability law of X, then $ETL_\alpha(X) = -E(X|X \leq -VaR_\alpha(X))$ and, therefore, ETL can be interpreted as the average loss beyond VaR. Figure 5.1 shows the values of this performance ratio when $\alpha = 0.01 = \beta$ and the components of three assets vary on the simplex:

$$SIMP = \left\{(x_1, x_2, x_3) \in R^3 / \sum_{i=1}^{3} x_i = 1; x_i \geq 0\right\}.$$

As we can see from Figure 5.1, this performance ratio admits more local maxima. In our comparison, we consider the parameters $\alpha = 0.35$ and $\beta = 0.1$.

The Rachev higher moments ratio (RHMR)[21] is given by:

$$RHMR(x'r) = \frac{v_1(x'r_{T+1} - r_{T+1,b})}{\rho_1(x'r_{T+1} - r_{T+1,b})}$$

where

$$\begin{aligned} v_1(x'r - r_b) &= E(x'r - r_b / x'r - r_b > F^{-1}_{x'r-r_b}(p_1)) \\ &\quad + \sum_{i=2}^{4} a_i E\left(\left(\frac{x'r - r_b}{\sigma_{x'r-r_b}}\right)^i \middle| x'r - r_b > F^{-1}_{x'r-r_b}(p_i)\right); \rho_1(x'r - r_b) \\ &= -E(x'r - r_b / x'r - r_b < F^{-1}_{x'r-r_b}(q_1)) \\ &\quad - \sum_{i=2}^{4} b_i E\left(\left(\frac{x'r - r_b}{\sigma_{x'r-r_b}}\right)^i \middle| x'r - r_b < F^{-1}_{x'r-r_b}(q_i)\right), \end{aligned}$$

$\sigma_{x'r-r_b}$ is the standard deviation of $x'r - r_b$, $a_i, \ldots, b_i \in$ R and $p_i, q_i \in (0,1)$.

[21] See Ortobelli *et al.* (2009).

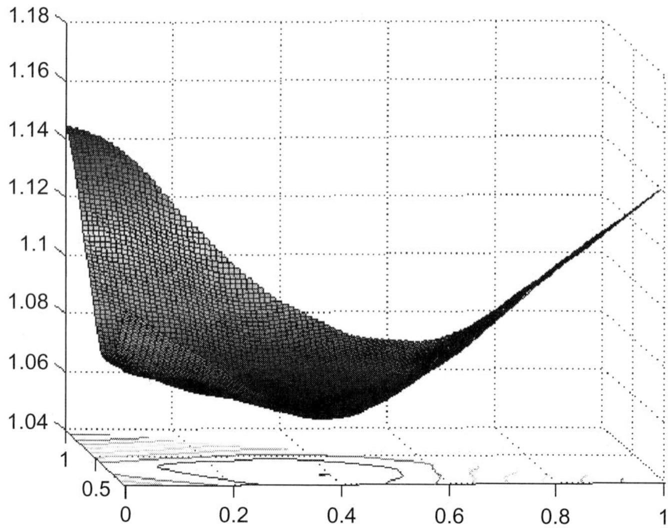

Figure 5.1 Rachev ratio with parameters $\alpha = 0.01 = \beta$ valued varying the composition of three components of DJIA.

This performance ratio was introduced to approximate the nonlinearity attitude to risk of decision makers considering the first four moments of the standardized tails of the return distribution.[22] As we can observe from the definition, the RHMR is very versatile and depends on many parameters. To simplify our analysis in the empirical comparison to follow, we assume $a_1 = b_1 = 1$; $a_2 = b_2 = -1/2$; $a_3 = b_3 = 1/6$; $a_4 = b_4 = -1/24$; $p_1 = 0.9$; $p_2 = 0.89$; $p_3 = 0.88$; $p_4 = 0.87$; and $q_i = 0.35$, $i = 1,2,3,4$. Figure 5.2 shows the values of this performance ratio when the composition of three assets varies on the simplex. As we can see from Figure 5.2, this performance ratio admits more local maxima.

In order to overcome the computational complexity problem for global maximum, we use the heuristic proposed by Angelelli and Ortobelli (2009) that presents significant improvements in terms of objective function and portfolio weights with respect to the classic function "fmincon" provided with the optimization toolbox of MATLAB. Moreover, this heuristic approximates the global optimum with an error that can be controlled in much less computational time than classic algorithms for global maximum such as simulated annealing.

5.4.2 An empirical comparison among portfolio strategies

In order to value the impact of nonlinear reward–risk measures, we provide an empirical comparison among the above strategies based on simulated data. We assume that decision makers invest their wealth in the market portfolio solution

[22] See Rachev *et al.* (2008) and Biglova *et al.* (2009).

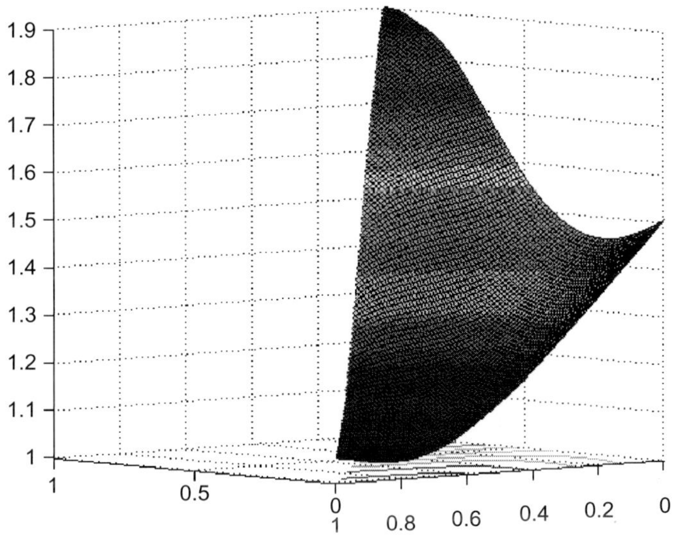

Figure 5.2 Rachev higher moments ratio and Rachev ratio valued varying the composition of three components of DJIA.

given by Equation (5.9) and we consider the sample path of the final wealth and of the cumulative return obtained from the different approaches. We assume that the investor recalibrates the portfolio daily and has an initial wealth W_0 equal to 1 and an initial cumulative return CR_0 equal to 0 (at the date 10/3/2008 when we use $T = 1{,}837$). Since we do not know the future evolution of assets returns from 10/3/2008, we assume that the returns for each future date correspond to those obtained as the mean of the scenarios and the same for the factors and the residuals of the previous factor model, i.e., $r_{i,T+k} = \frac{1}{S}\sum_{s=1}^{S} r_{i,T+k}^{(s)}$; $e_{i,T+k} = \frac{1}{S}\sum_{s=1}^{S} e_{i,T+k}^{(s)}$ for $i = 1,\ldots,30$ and $f_{j,T+k} = \frac{1}{S}\sum_{s=1}^{S} f_{j,T+k}^{(s)}$ for $j = 1,\ldots,14$. Therefore, at the kth recalibration, three main steps are performed to compute the ex post final wealth and cumulative return:

Step 1. Choose a performance ratio. Simulate 3,000 scenarios using the algorithm of the previous section. Determine the market portfolio $x_M^{(k)}$ solution to the optimization problem given by Equation (5.9) that maximizes the performance ratio.

Step 2. The ex-post final wealth is given by:

$$W_{k+1} = W_k\left((x_M^{(k)})'(1+)\right),$$

where r_{T+k} is the vector of returns mean of our scenarios. The ex post cumulative return is given by:

$$CR_{k+1} = CR_k + \left(x_M^{(k)}\right)' r_{T+k}$$

Step 3. The optimal portfolio $x_M^{(k)}$ is the new starting point for the $(k + 1)$th optimization problem given by Equation (5.9).

Steps 1, 2, and 3 were repeated for all the performance ratios 1,000 times so that we forecasted the future behavior of the optimal portfolio strategies in the next 4 years. The output of this analysis is represented in Figures 5.3–5.5. Figure 5.3 compares the sample paths of wealth and of the total return obtained with the application either of the Angelelli–Ortobelli heuristic or of the local maximization function fmincon of Matlab. This comparison shows that if we maximize the Rachev ratio with $\alpha = 0.35$; $\beta = 0.1$ with the function for local maximum of Matlab, we could lose more than 20% of the initial wealth in 4 years. Figure 5.4 compares the sample paths of wealth and of the total return obtained with the Rachev ratio and the Sharpe ratio. In particular, the results suggest that using the Rachev ratio we can increase final wealth by more than 25%. Analogously, Figure 5.5 shows that using the Rachev higher moments ratio we can increase final wealth by more than 15%. Comparing Figures 5.4 and 5.5 we also see the superiority of the Rachev higher moments ratio approach relative to the Rachev ratio during the first 300 days. Then we see a superior performance of the Rachev ratio.

What is clear from all of the comparisons is that the use of an adequate statistical and econometric model, combined with appropriate risk and performers measures, could have a significant impact on the investors' final wealth.

5.5 Concluding remarks

In this chapter, we provide a methodology to compare dynamic portfolio strategies consistent with the behavior of investors based on realistic simulated scenarios after a reduction of dimensionality of the portfolio selection problem.

We first summarize the empirical evidence regarding the behavior of equity returns: heavy-tailed distributions, volatility clustering, and non-Gaussian copula dependence. Then, we discuss how to generate scenarios that take into account the empirical evidence observed for equity return distributions. In particular, we first propose a way to reduce the dimensionality of the problem using PCA. Then, we approximate the returns using a factor model on a restricted number of principal components. The factors (i.e., principal components) and residuals of the factor model are modeled with an ARMA(1,1)-GARCH(1,1) with stable innovations. Moreover, we propose a copula approach for the innovations of the factors. This approach allows us to generate future scenarios. Second, we examine the use of reward/risk criteria to select optimal portfolios, suggesting the use of the Sharpe ratio, the Rachev ratio, and the Rachev higher moments ratio. Finally, we provide an empirical comparison among final wealth and cumulative return processes obtained using the simulated data. The empirical comparison between the Sharpe ratio and the two Rachev ratios shows the greater predictable capacity of the latter.

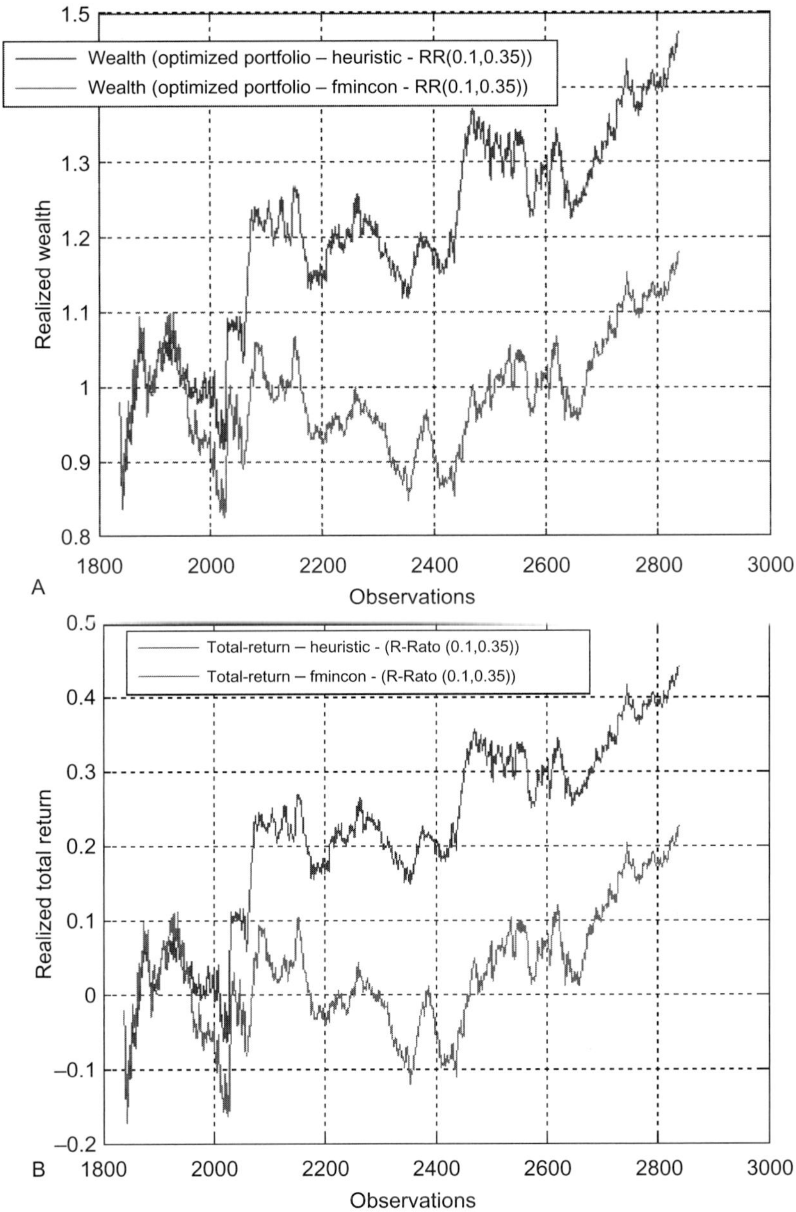

Figure 5.3 Final wealth and total return realized in 1,000 days using the Rachev ratio with parameters $\alpha = 0.35$; $\beta = 0.1$ and maximizing it either with the Angelelli–Ortobelli heuristic or the function fmincon of Matlab.

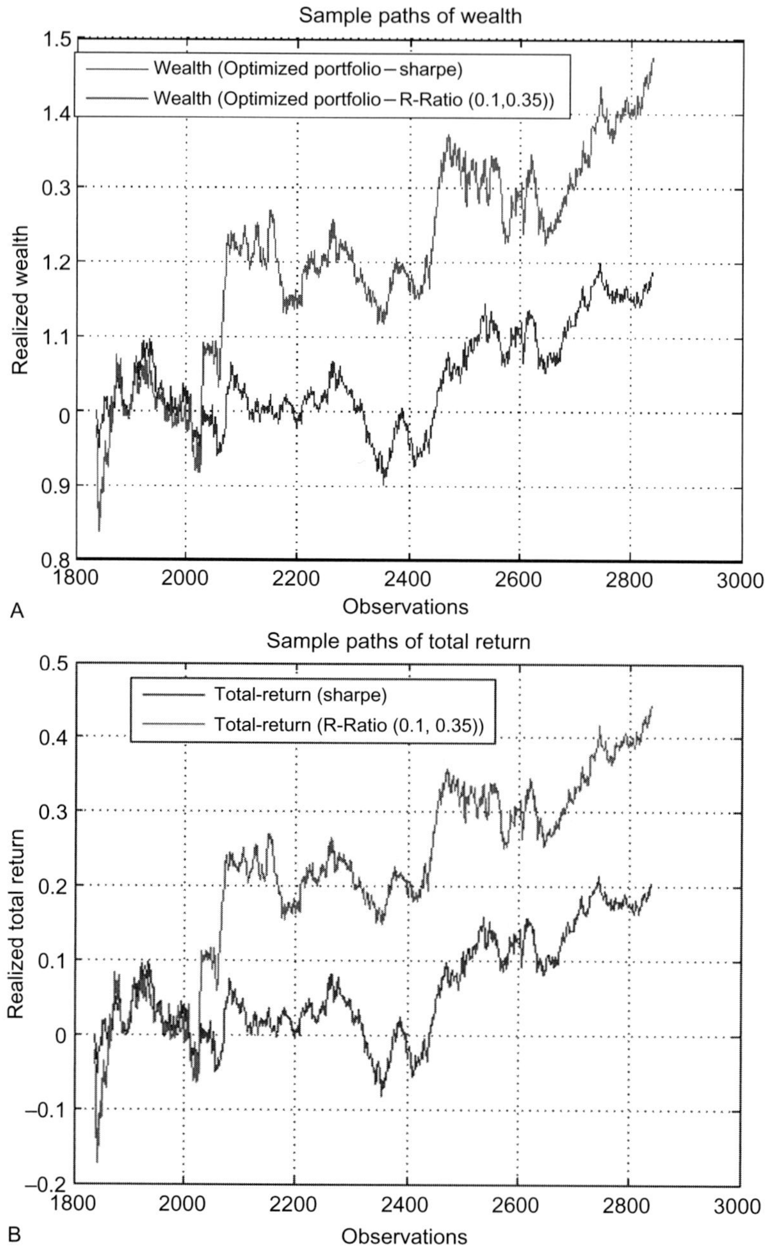

Figure 5.4 Final wealth and total return realized in 1,000 days using either the Rachev ratio with parameters $\alpha = 0.35$; $\beta = 0.1$ or the Sharpe ratio.

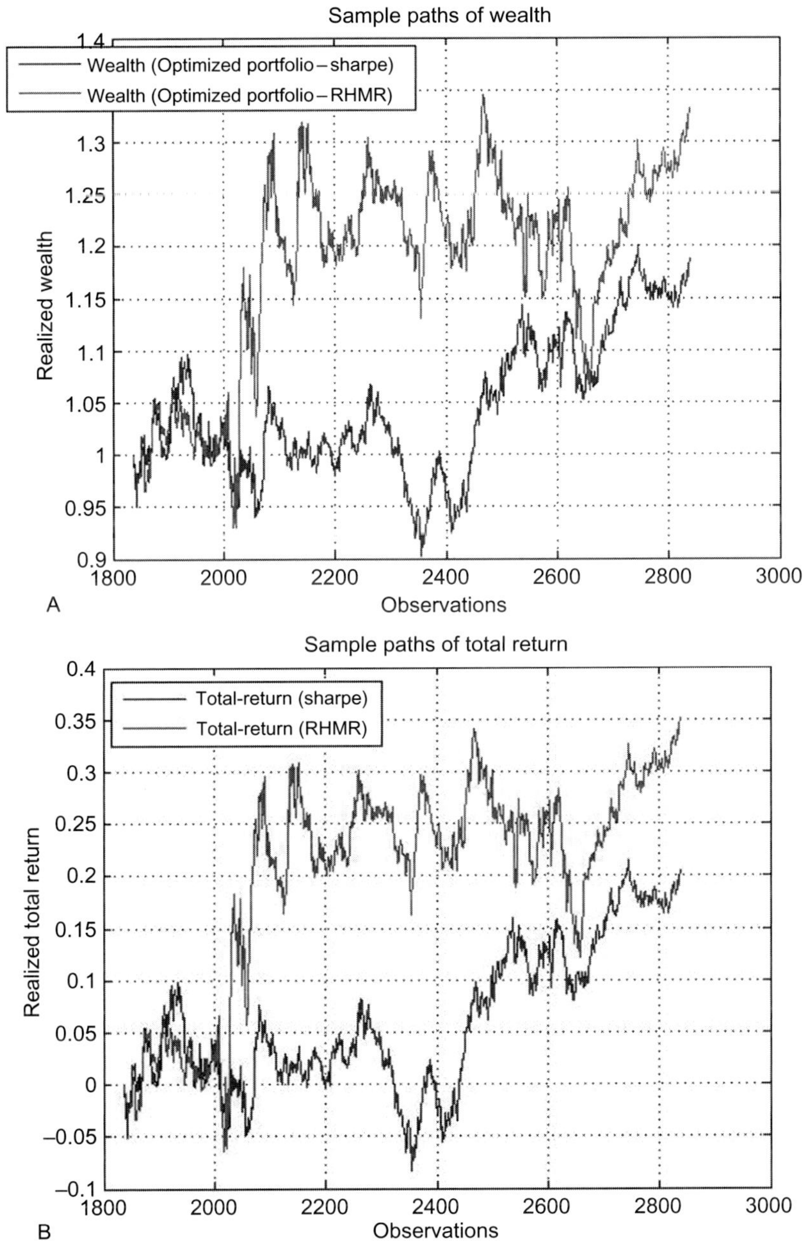

Figure 5.5 Final wealth and total return realized in 1,000 days using either the Rachev high moment ratio or the Sharpe ratio.

References

Angelelli, E., & Ortobelli, S. (2009). American and European portfolio selection strategies: The Markovian approach. To appear as Chapter 5 in *Financial Hedging*. New York: Novascience.

Balzer, L. A. (2001). Investment risk: A unified approach to upside and downside returns. In F. Sortino & Satchell. (Eds.), *Managing downside risk in financial markets: Theory practice and implementation* (pp. 103–156). Oxford: Butterworth-Heinemann.

Biglova, A., Ortobelli, S., Rachev, S., & Stoyanov, S. (2004). Different approaches to risk estimation in portfolio theory. *Journal of Portfolio Management*, *31*, 103–112.

Biglova, A., Kanamura, T., Rachev, S. T., & Stoyanov, S. (2008). Modeling, risk assessment and portfolio optimization of energy futures. *Investment Management and Financial Innovations*, *5*(1), 17–31.

Biglova, A., Ortobelli, S., Rachev, S., & Stoyanov, S. (2009). A note on the impact of nonlinear reward and risk measures. To appear in *Journal of Applied Functional Analysis*.

Bollerslev, T. (1986). Generalized autoregressive conditional heteroskedasticity. *Journal of Econometrics*, *31*, 307–327.

Box, G. E. P., Jenkins, G. M., & Reindel, G. C. (1994). *Time series analysis: Forecasting and control* (3rd ed.). San Francisco: Prentice Hall.

Cherubini, U., Luciano, E., & Vecchiato, W. (2004). *Copula methods in finance*. Hoboken: John Wiley & Sons.

Chopra, C., & Ziemba, W. (1993). The effect of errors in mean variances and covariances on optimal portfolio choice. *Journal of Portfolio Management*, *19*, 6–10.

Engle, R. F. (1982). Autoregressive conditional heteroskedasticity with estimates of the variance of UK inflation. *Econometrica*, *50*, 987–1008.

Fischer, T. (2003). Risk capital allocation by coherent risk measures based on one-sided moments. *Insurance: Mathematics and Economics*, *32*(1), 135–146.

Friedman, M., & Savage, L. J. (1948). The utility analysis of choices involving risk. *Journal of Political Economy*, *56*, 279–304.

Kondor, I., Pafka, S., & Nagy, G. (2007). Noise sensitivity of portfolio selection under various risk measures. *Journal of Banking and Finance*, *31*, 1545–1573.

Levy, M., & Levy, H. (2002). Prospect theory: Much ado about nothing? *Management Science*, *48*, 1334–1349.

Markowitz, H. M. (1952). The utility of wealth. *Journal of Political Economy*, *60*, 151–158.

Ortobelli, S., Rachev, S., Shalit, H., & Fabozzi, F. J. (2009). Orderings and probability functionals consistent with preferences. *Applied Mathematical Finance*, *16*(1), 81–102.

Papp, G., Pafka, S., Nowak, M. A., & Kondor, I. (2005). Random matrix filtering in portfolio optimization. *ACTA Physica Polonica B*, *36*, 2757–2765.

Rachev, S., Ortobelli, S., Stoyanov, S., Fabozzi, F. J., & Biglova, A. (2008). Desirable properties of an ideal risk measure in portfolio theory. *International Journal of Theoretical and Applied Finance*, *11*(1), 19–54.

Rachev, S. T., & Mittnik, S. (2000). *Stable Paretian models in finance*. Chichester: John Wiley & Sons.

Rachev, S. T., Menn, C., & Fabozzi, F. J. (2005). *Fat-tailed and skewed asset return distributions: Implications for risk management, portfolio selection, and option pricing*. New York: John Wiley & Sons.

Rachev, S. T., Mittnik, S., Fabozzi, F. J., Focardi, S. M., & Jasic, T. (2007). *Financial econometrics: From basics to advanced modeling techniques*. New York: John Wiley & Sons.

Ross, S. (1978). Mutual fund separation in financial theory—the separating distributions. *Journal of Economic Theory*, *17*, 254–286.

Samorodnitsky, G., & Taqqu, M. S. (1994). *Stable non-Gaussian random processes: Stochastic models with infinite variance*. New York: CRC Press.

Sharpe, W. F. (1994). The Sharpe ratio. *Journal of Portfolio Management* (Fall), 45–58.

Sklar, A. (1959). Fonctions de Réparitition à n dimensions et Leurs Marges. *Publications de l'Institut de Statistique de Universitè de Paris*, *8*, 229–231.

Stoyanov, S., Rachev, S., & Fabozzi, F. (2007). Optimal financial portfolios. *Applied Mathematical Finance*, *14*(5), 401–436.

Sun, W., Rachev, S., Stoyanov, S., & Fabozzi, F. (2008). Multivariate skewed Student's *t* copula in analysis of nonlinear and asymmetric dependence in German equity market. To appear in *Studies in Nonlinear Dynamics & Econometrics*, *12*(2).

Tasche, D. (2000). Risk contributions and performance measurement. *Working Paper. TU München*. http://www-m4.mathematik.tu-muenchen.de/m4/pers/tasche/riskcon.pdf

Tversky, A., & Kahneman, D. (1992). Advances in prospect theory: Cumulative representation of uncertainty. *Journal of Risk and Uncertainty*, *5*, 297–323.

Zolatorev, V. M. (1986). One-dimensional stable distributions. *American Mathematical Society Translations*. Monographs 65. RI. Transl. of the original 1983 Russian, Providence.

6 Staying ahead on downside risk

Giuliano De Rossi

Executive Summary

Critics of the mean–variance approach have pointed out its shortcomings in countless academic papers. It is often argued that variance is a poor measure of risk because it treats symmetrically positive and negative returns. I argue that this feature is likely to have an adverse effect on the performance of mean–variance strategies when the distribution of asset returns changes rapidly over time. Nevertheless, a standard approach that can supersede mean–variance is yet to emerge.

This chapter discusses the advantages of an alternative measure recently advocated in the academic literature, expectile value at risk (EVaR). I illustrate a simple asset allocation procedure that incorporates a dynamic model of EVaR. The information available from time series of returns of raw assets is used to learn about the evolution of portfolio downside risk. Risk is then minimized by targeting the predicted EVaR. The new approach to asset allocation allows for changes in the overall distribution of asset returns while retaining a high degree of tractability.

6.1 Introduction

The aim of this chapter is provide a simple methodology to deal with downside risk at times when the distribution of asset returns experiences sudden and dramatic changes. The recent turmoil due to the credit crisis provides a remarkable example of such an environment.

Most of the existing asset allocation and portfolio construction procedures can be viewed as a way of finding the optimal balance between return and risk, given the portfolio manager's preferences. As a consequence, two intricately intertwined aspects of the procedure are fundamental to the performance of a trading strategy: how we measure risk and how we keep track of the changes in the risk characteristics of a portfolio.

Return volatility is arguably the most widespread risk measure in this context, partly due to the elegance and mathematical tractability of the mean–variance framework. Another popular risk measure, which requires a somewhat greater effort when used for asset allocation, is value at risk (VaR). Both these measures have been criticized in the recent academic research on risk measurement. In fact, Artzner, Delbaen, Eber, and Heath (1999) have drawn up a list of minimal

Doi:10.1016/B978-0-12-374952-9.00006-3.

requirements for what they call a *coherent* risk measure. VaR and volatility are known to violate the coherency assumptions. A popular risk measure that does meet all requirements is expected shortfall.

Intuitively, the practical consequences of the departure from the basic properties of risk measures are likely to be exacerbated by the need to track risk over time. For example, it is widely accepted that sudden increases in market volatility tend to be driven by the downside, rather than the upside, in the return distribution. In fact, it is well known that surges in market volatility coincide with economic downturns.[1] In a dynamic model, a symmetric measure like volatility may overstate or understate risk in crucial phases of the business cycle.

Another example is provided by the analysis of the comovements in asset returns when extreme events occur. The academic literature has often argued that asset returns are strongly dependent in the lower tail of the distribution, more than is implicitly assumed by a Gaussian model.[2] In other words, when assets experience heavy losses, the covariance matrix is of little help in determining the likely loss for the whole portfolio. The practical implications of these findings have been felt by portfolio managers in July 2007, when traditional quantitative factors seemed to stop working all at once and markets suddenly seemed to offer no place to hide.

Countless academic papers have highlighted the limitations of the Markowitz approach. Nevertheless, a standard approach that can supersede mean–variance is yet to emerge. Madan (2006) proposed a non-Gaussian factor model. The dynamics of the factors is tracked by independent component analysis and the portfolio setup by maximizing a utility function. Mencìa and Sentana (2008) and Jondeau and Rockinger (2006) are recent examples of portfolio allocation with higher moments. This approach faces formidable challenges because the multivariate density is difficult to model. Rockafellar and Uryasev (2000) developed a method to optimize a portfolio's CVaR. A similar approach, cast in a utility maximization framework, is taken by Basset, Koenker, and Kordas (2004). Finally, Meucci (2006) adopted a Bayesian approach. He extended the Black–Litterman model to a non-Gaussian context, deriving simulation-based methods to obtain the optimal allocation.

This chapter presents a dynamic model centered on a risk measure that has been recently advocated in the academic literature, expectile value at risk (EVaR). Intuitively, EVaR is closely related to VaR in that it can be interpreted as an optimal level of reserve capital requirement. Both minimize an asymmetric cost function, which strikes a balance between the expected opportunity cost of holding excessive reserves and the expected cost of losses that exceed the reserve capital. The advantages of measuring risk through EVaR are detailed in the next section. In short, EVaR satisfies all the basic properties of a coherent measure of risk, it is mathematically tractable, and lends itself to a simple interpretation in a utility maximization framework.

[1] See, among others, Schwert (1989) and Engle and Rangel (2008).

[2] See Patton (2004) and references therein.

In addition, several dynamic EVaR models have been developed in the recent econometric literature and will be briefly reviewed in the next section. My contribution is to suggest a simple asset allocation procedure that can be used to build the optimal portfolio by minimizing risk as measured by the predicted EVaR.

6.2 Measuring downside risk: VaR and EVaR

6.2.1 Definition and properties

EVaR is introduced in Kuan, Yeh, and Hsu (2009). The concept of ω-expectile, for $0 < \omega < 1$, can be found in Newey and Powell (1987). They defined $\mu(\omega)$, for any random variable Y with finite mean and cumulative distribution function $F(y)$, as the solution of the equation:

$$\mu(\omega) = E(Y) + \frac{2\omega - 1}{1 - \omega} \int_{\mu(\omega)}^{\infty} (y - \mu(\omega))\, dF(y) \tag{6.1}$$

The ω-EVaR of Kuan *et al.* (2009) is just $-\mu(\omega)$. Expression (6.1) can be rearranged to give:

$$\omega \int_{\mu(\omega)}^{\infty} (y - \mu(\omega))\, dF(y) = (1 - \omega) \int_{-\infty}^{\mu(\omega)} (\mu(\omega) - y)\, dF(y)$$

The differences that appear in the integrals are nonnegative. When $\omega = 0.5$, the solution is simply $\mu(\omega) = E(Y)$, while as ω varies between zero and 0.5, the expectile function $\mu(\omega)$ describes the lower tail of the distribution of Y. Intuitively, the quantities ω and $1 - \omega$ can be seen as asymmetric weights that multiply the integrals of the deviations from $\mu(\omega)$. If $\omega < 0.5$, then the weight on the outcomes y that lie below $\mu(\omega)$ (i.e., the weight on the integral on the right hand side) dominates.

It can be shown that a solution exists and it is unique. Furthermore, the expectiles of a linear transformation of Y, $Y^* = aY + b$ can be easily found since $\mu^*(\omega) = a\mu(\omega) + b$ for any real numbers a and b.

Suppose that we have obtained a sample of n observations Y_i. The sample equivalent of the population expectile can be found by solving:

$$\min_{\mu} \sum_{i=1}^{n} \rho_{\omega}(Y_i - \mu) \tag{6.2}$$

where

$$\rho_{\omega}(x) = |\omega - I(x < 0)| x^2$$

In other words, we minimize the weighted squared deviations from the observations. Weights are asymmetric: if $\omega < 0.5$ then negative residuals receive a larger weight than positive residuals. When $\omega = 0.5$, the problem boils down to minimizing the sum of squared errors, yielding the sample mean as the solution.

It is easy to show that VaR can be characterized as the value that minimizes asymmetric *absolute* deviations instead of square deviations. The function $\rho_\omega(x)$ is replaced by $\rho_\alpha(x) = |\alpha - I(x < 0)|\,|x|$, $0 < \alpha < 1$. As a consequence, the difference between VaR and EVaR can be stated in terms of the shape of the objective function in the optimization problem (6.2).

A fundamental property of expectiles, which follows from the first-order conditions of the minimization problem, is that the *weighted* sum of residuals is equal to zero:

$$\sum_{i=1}^{n} |\omega - I(Y_i - \mu)|(Y_i - \mu) = 0$$

EVaR has several advantages. First, when expectiles exist, they characterize the shape of the distribution just like quantiles. In fact, each ω-expectile corresponds to an α-quantile, although the relation between ω and α varies from distribution to distribution, as can be seen from Table 6.1. For example, a 5% EVaR when returns are normally distributed corresponds to a 12.6% VaR. However, for heavy tailed distributions like a $t(5)$, the 5% EVaR corresponds to a 10% VaR. If α is seen as an indication of how conservative the risk measure is, then it can be seen that EVaR is more conservative when extreme losses are more likely.

A second advantage of EVaR is that it is a coherent measure of risk.[3] It also admits an interpretation in terms of utility maximization. Manganelli (2007) shows that an expectile can be seen as the expected utility of an agent having asymmetric preferences on below target and above target returns. His argument is presented in more detail in the section on asset allocation. The resulting portfolio allocations are not dominated in the second-order stochastic dominance sense. In other words, a portfolio is second-order stochastic dominance efficient if and only if it is optimal for some investor who is nonsatiable (i.e., the higher the profit, the better) and risk averse. It is interesting to note that the mean–variance optimal portfolios do not satisfy this property: There may

Table 6.1 α corresponding to ω when ω-EVaR and α-VaR are equal

ω	Normal	$t(30)$	$t(5)$	$t(3)$
1%	4.3%	4.0%	3.0%	2.4%
5%	12.6%	12.3%	10.0%	8.5%
10%	19.5%	19.0%	16.6%	14.5%

[3] It can be shown that EVaR satisfies all the properties required by the definition of Artzner *et al.* (1999), including monotonicity and subadditivity.

exist portfolios that would be chosen by all nonsatiable risk averse agents over the mean–variance solution.

6.2.2 Modeling EVaR dynamically

Another notable advantage of expectiles is their mathematical tractability, as argued in Newey and Powell (1987). Dynamic expectile models have been derived in Taylor (2008), Kuan *et al.* (2009), and De Rossi and Harvey (2009). An early example appeared in Granger and Sin (2000). Computationally, simple univariate models compare favorably to GARCH or dynamic quantile models.

The analysis presented in this chapter is based on the dynamic model of De Rossi and Harvey (2009). Intuitively, their estimator produces a curve rather than a single value μ, so that it can adapt to changes in the distribution over time. The two parameters needed for the estimation are ω and q.

The former can be interpreted as a *prudentiality* level: the lower ω, the more risk aversion. Figure 6.1 shows a typical expectile plot for alternative values of ω. By decreasing ω, one focuses on values that are further out in the lower tail, i.e., more severe losses.

By increasing q, we can make the model more flexible in adapting to the observed data. The case $q = 0$ corresponds to the constant expectile (estimated by the sample expectile). As Figure 6.2 shows, larger values of q produce estimated curves that follow more and more closely the observations.

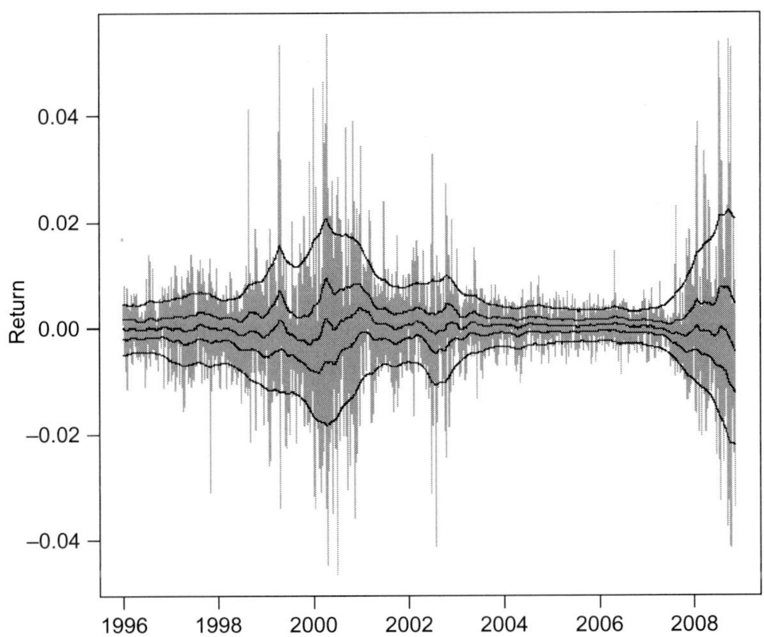

Figure 6.1 Time-varying expectiles. The solid black lines represent estimated dynamic expectiles for $\omega = 0.05, 0.25, 0.5$ (mean), 0.75, and 0.95.

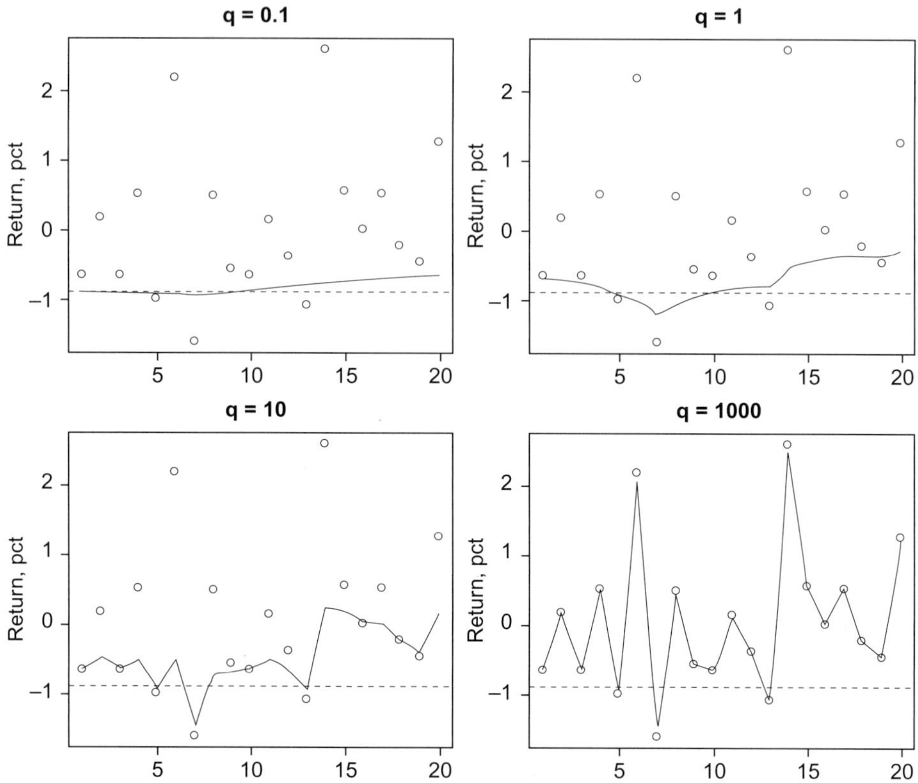

Figure 6.2 Time-varying expectiles for alternative values of q. The data is a simulated time series. The dotted line represents the sample 5% expectile, which corresponds to the case $q = 0$.

De Rossi and Harvey (2009) assume that a time series y_t, a risk tolerance parameter $0 < \omega < 1$, and a signal to noise ratio q are given. They then decompose each observation into its unobservable ω-expectile, $\mu_t(\omega)$, and an error term ε having ω-expectile equal to zero:

$$\begin{aligned} y_t &= \mu_t(\omega) + \varepsilon_t(\omega) \\ \mu_t(\omega) &= \mu_{t-1}(\omega) + \eta_t(\omega) \end{aligned}$$

The ω-expectile, $\mu_t(\omega)$, is assumed to change over time following a (slowly evolving) random walk that is driven by a normally distributed error term η_t having zero mean.[4] In the special case, $\omega = 0.5$, $\mu_t(0.5)$ is just the time-varying mean and therefore y_t is a random walk plus noise. The signal $\mu_t(0.5)$ can be estimated via the Kalman filter and smoother (KFS).

[4] The dynamics could be specified alternatively as following an autoregressive process or an integrated random walk.

Equivalently, the problem can be cast in a nonparametric framework. The goal is to find the optimal curve $f(t)$, plotted in Figure 6.1, that fits the observations. It is worth stressing that $f(t)$ is a continuous function, so here the argument t, with a slight abuse of notation, is allowed to be a positive real number such that $0 < t < T$. The solution minimizes:

$$\int_0^T [f'(t)]^2 dt + q\sum_{s=1}^{T} \rho_\omega(y_s - f(s)) \tag{6.3}$$

with respect to the whole function f, within the class of functions having square integrable first derivative. At any point $t = 1,...,T$, we then set $\mu_t(\omega) = f(t)$.

The first term in Equation (6.3) is a *roughness penalty*. Loosely speaking, the more the curve $f(t)$ wiggles, the higher the penalty. The second term is the objective function for the expectile, which as noted above gives asymmetric weights on positive and negative errors.

The constant q represents the relative importance of the expectile criterion. As q grows large, the objective function tends to become influenced less and less by the squared first derivative. In the limit, only the errors $y_t - \mu_t(\omega)$ matter and the solution becomes $\mu_t(\omega) = y_t$, i.e., the estimated expectile coincides with the corresponding observation. As q tends to zero, instead, the integral of the squared derivative is minimized by a straight line. As a result, the solution in the limit is to set all expectiles equal to the historical expectile. The role of q is illustrated in Figure 6.2.

It can be shown that the optimal curve is a piecewise linear spline. Computationally, finding the optimal spline boils down to solving a nonlinear system in $\boldsymbol{\mu}$, the vector of T expectiles. After some tedious algebra, the first-order conditions to minimize Equation (6.3) turn out to be:

$$[\boldsymbol{\Omega} + \mathbf{D}(\mathbf{y},\boldsymbol{\mu})]\boldsymbol{\mu} = \mathbf{D}(\mathbf{y},\boldsymbol{\mu})\mathbf{y}$$

where $\mathbf{D}(\mathbf{y}, \boldsymbol{\mu})$ is diagonal with element (t, t) given by

$$|\omega - I(y_t - \mu_t < 0)|$$

and

$$\boldsymbol{\Omega} = q^{-1}\begin{bmatrix} 1 & -1 & 0 & 0 & \dots & & 0 \\ -1 & 2 & -1 & 0 & \dots & & 0 \\ 0 & -1 & 2 & -1 & \dots & & 0 \\ \vdots & & & & & & \vdots \\ 0 & & & & \dots & 2 & -1 \\ 0 & & & & \dots & -1 & 1 \end{bmatrix}$$

Both $\boldsymbol{\Omega}$ and $\mathbf{D}$ are $T \times T$ matrices. Starting with an initial guess, $\boldsymbol{\mu}^{(1)}$, the optimal $\boldsymbol{\mu}$ is found by iterating the equation:

$$\boldsymbol{\mu}^{(i+1)} = [\boldsymbol{\Omega} + \mathbf{D}(\mathbf{y}, \boldsymbol{\mu}^{(i)})]^{-1}\mathbf{D}(\mathbf{y}, \boldsymbol{\mu}^{(i)})\mathbf{y}$$

until convergence. I define $\hat{\mathbf{D}}(\mathbf{y})$ the matrix $\mathbf{D}$ upon convergence of $\boldsymbol{\mu}^{(i)}$ to $\hat{\boldsymbol{\mu}}$. The repeated inversion of the $T \times T$ matrix in the formula can be efficiently carried out by using the KFS.[5] To this end, it is convenient to set up an auxiliary linear state space form at each iteration:

$$\begin{aligned} y_t &= \delta_t + u_t \\ \delta_t &= \delta_{t-1} + v_t \end{aligned} \qquad (6.4)$$

where

$$Var(u_t) = 1/|\omega - I(y_t - \mu_t^{(i)} < 0)|$$

and

$$Var(v_t) = q$$

The unobservable state δ_t replaces μ_t. The model in Equation (6.4) is just a convenient device that can be used to carry out the computation efficiently. It can be shown that the linear KFS applied to Equation (6.4) yields the optimal $\boldsymbol{\mu}$ characterized above.

The parameter q can be estimated from the data by cross-validation. The method is very intuitive: It consists of dropping one observation at a time from the sample and thus re-estimating the time-varying expectile with a missing observation T times. The T residuals are then used to compute the objective function:

$$CV_\omega(q) = \sum_{t=1}^{T} \rho_\omega(y_t - \tilde{\mu}_t^{(-t)})$$

where $\tilde{\mu}_t^{(-t)}$ is the estimated value at time t when y_t is dropped. $CV_\omega(q)$ depends on q through the estimator $\tilde{\mu}_t^{(-t)}$. De Rossi and Harvey (2006) devise a computationally efficient method to minimize $CV_\omega(q)$ with respect to q.

6.3 The asset allocation problem

This section illustrates a method for finding, given a set of basic assets, the portfolio with the lowest estimated risk—where risk is measured as EVaR. It is

[5] The connection between spline smoothing and signal extraction in linear Gaussian systems has been thoroughly investigated in the statistical literature. An early example is Wahba (1978).

straightforward to extend the analysis to the case in which we target a preassigned portfolio alpha given a set of asset alphas. The crucial point is that the method described here allows portfolio expectiles to change over time. In the mean–variance world, this corresponds to a minimum variance allocation coupled with a dynamic estimator of the covariance matrix.

How does one compute the optimal weights? Manganelli (2007) suggests a solution in the context of conditional autoregressive expectile models. A special case is the constant expectile. The problem can be expressed as a linear programming one, similar to the approach of Rockafellar and Uryasev (2000).

Assume that n assets are traded on the market and we have observed the history of returns $\mathbf{R} = [\mathbf{r}_1,\ldots,\mathbf{r}_T]'$, a $T \times n$ matrix that is taken to have full rank. The risk tolerance parameter $0 < \omega < 5$ and the signal noise ratio $q > 0$ are taken as given. Furthermore, I assume that an $m_1 \times n$ matrix $\mathbf{A}$ and an $m_2 \times n$ matrix $\mathbf{G}$ are given, both of full rank, for positive integers m_1, m_2. Finally, $\mathbf{b}$ and $\mathbf{g}$ are given $m_1 \times 1$ and $m_2 \times 1$ vectors. It is also assumed that $n > m_1$ and $T > n - m_1 - m_2 + 1$.

For any vector of n portfolio weights, $\boldsymbol{\lambda}$, historical portfolio returns are obtained as $\boldsymbol{\lambda}'\mathbf{R}$.

The matrix $\mathbf{A}$ is used to impose equality constraints, e.g., to force the sum of weights to be equal to one (fully invested portfolio). I obtain this result by optimizing on a vector of dimension $n - m_1$, $\boldsymbol{\lambda}^*$.

The matrix $\mathbf{G}$ is used to impose inequality constraints, e.g., long-only constraints or a minimum holding in any given asset.

The aim is to solve:

$$\max_{\boldsymbol{\lambda}^*} \hat{\mu}_{T+1}(\mathbf{y}(\boldsymbol{\lambda});\omega)$$

subject to

$$\boldsymbol{\lambda} = \mathbf{A}\boldsymbol{\lambda}^* + \mathbf{b}$$

and

$$\mathbf{G}_{i\bullet}\boldsymbol{\lambda}_i \geq g_i \; \forall i = 1,\cdots,n$$

where $\hat{\mu}_{T+1}$ is the predicted ω-expectile of $y_{T+1} = \boldsymbol{\lambda}'\mathbf{r}_{T+1}$, namely the portfolio return one step ahead.

It is worth stressing that the objective function depends on the parameter q, although I omitted the parameter from the formula in order to keep the notation simple. In fact, $\hat{\mu}_{T+1}$ is the prediction obtained from the dynamic model, which takes as inputs the (univariate) time series of portfolio returns $\mathbf{y}(\boldsymbol{\lambda})$ and the parameters ω and q.

The objective function $\hat{\mu}_{T+1}(\mathbf{y}(\boldsymbol{\lambda});\omega)$ is equal to the prediction from a random walk model of the unobserved expectile, which is in turn equal to the estimated expectile at the end of the sample period, $\hat{\mu}_T$. From the previous discussion, we know that it satisfies:

$$\hat{\mu}_T = [\mathbf{\Omega} + \mathbf{D}(\mathbf{R\lambda}, \hat{\boldsymbol{\mu}})]^{-1}_{T\bullet}\mathbf{D}(\mathbf{R\lambda}, \hat{\boldsymbol{\mu}})\mathbf{R\lambda}$$

where the subscript on the inverse matrix indicates the T-th row. Recall that each diagonal element of $\mathbf{D}(\mathbf{R\lambda}, \boldsymbol{\mu})$ can only take on two values, ω and $1 - \omega$. As a result, on any set $\Lambda \subset R^n$ such that $\hat{\mathbf{D}}(\mathbf{R\lambda})$ is constant $\forall\, \boldsymbol{\lambda} \in \Lambda$, the objective function is linear in $\boldsymbol{\lambda}$. De Rossi (2009) shows that the objective function $\hat{\mu}_{T+1}(\mathbf{y}(\boldsymbol{\lambda}); \omega)$, viewed as a function of $\boldsymbol{\lambda}^*$, is piecewise linear and convex. This implies that the solution to the maximization problem must be on a vertex. The simplex algorithm starts from an arbitrary vertex and moves along the edges of the "surface" until it reaches the maximum. At each step, the ω-expectiles of the portfolios corresponding to the new vertices are estimated and used for prediction one step ahead. The absence of local minima makes the optimization procedure robust and reliable.

It is straightforward to see that the procedure can be used to deal with objective functions that include alpha terms. Call $\boldsymbol{\alpha}$ a vector of expected asset returns. We can impose the additional inequality constraint $\boldsymbol{\lambda}'\boldsymbol{\alpha} \geq \bar{r}$, where $\bar{r}$ is a preassigned constant. The asset allocation procedure would then minimize risk subject to an expected return of at least $\bar{r}$.

The parameters q and ω are inputs to the asset allocation procedure. How should a portfolio manager choose their values?

First, q can be estimated from the data as argued above, e.g., from a time series of returns to the minimum variance portfolio. The parameter can be viewed as a signal to noise ratio: The higher q, the more information content can be found in the observations. In addition, q is related to turnover. The more responsive the model is to changes in expectiles over time, the larger the amount of trading. By looking at historical data, the value of q can be fine-tuned to match the desired turnover level.

As for the parameter ω, Manganelli (2007) highlights the fact that the maximum expectile objective can be interpreted in a utility maximization framework. Minimizing the EVaR is equivalent to maximizing:

$$E(U_\mu(Y))$$

where Y is the portfolio return and

$$U_\mu(x) = \begin{cases} x & \text{if } x \geq \mu \\ \mu - \dfrac{1-\omega}{\omega}(\mu - x) & \text{if } x < \mu \end{cases}$$

If we interpret $U(x)$ as a utility function, then it is easy to see that as $\omega \to 0.5$, the preferences approach risk neutrality: all the investor cares about is the expected return (which is the 50% expectile). For $\omega < 0.5$, however, lower levels of ω correspond to higher risk aversion.

In particular, this class of utility functions introduces loss aversion. Suppose $\omega = 0.1$. When we are *below the target level* μ, the decrease in utility caused by a 1% decrease in return is equal to 9. When we subtract 1% from the portfolio return *above the target*, utility drops by 1. Figure 6.3 illustrates the different utility functions corresponding to alternative levels of risk aversion.

6.4 Empirical Illustration

An example will help illustrate the proposed methodology. Assume that the only four available assets are the S&P 500, DAX, Nikkei and FTSE 100 indices. Domestic returns are used, i.e., all currency risk is assumed to be hedged. The data consists of 400 weekly returns, obtained from Bloomberg, covering the period April 2001–November 2008. The goal is to identify the optimal country allocation at the end of November 2008.

To abstract from the problem of estimating alphas, following Grinold and Kahn (2000), I generated a vector of predicted returns for December 2008 assuming a high information coefficient (0.5). The vector is displayed in the first row of Table 6.2.

The equality constraint used in the optimization forces the portfolio to be fully invested, i.e., $\Sigma_i \lambda_i = 1$. The inequality constraints ensure that the portfolio is long-only ($\lambda_i \geq 0 \forall i$) and that the expected monthly return based on the alphas in Table 6.2 is greater than or equal to 1% ($\boldsymbol{\lambda}'\boldsymbol{\alpha} \geq 0.01$).

I will first present the results obtained by using a mean–variance optimizer. I adopted the optimal shrinkage estimator devised by Ledoit and Wolf (2004) to

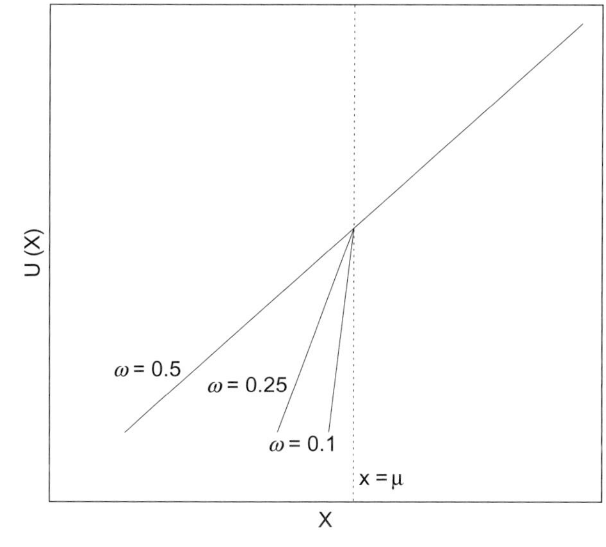

Figure 6.3 Shape of the loss aversion utility function for alternative values of ω.

Table 6.2 The first row shows a vector of randomly generated alphas. The remainder of the table displays volatilities and correlations estimated via the optimal shrinkage method of Ledoit and Wolf (2004)

	spx	dax	nky	ukx
Alpha	−1.88%	2.08%	4.76%	−0.15%
Ann volatility	17.73%	25.68%	22.72%	19.04%
spx	1.00	0.76	0.53	0.74
dax	0.76	1.00	0.61	0.79
nky	0.53	0.61	1.00	0.57
ukx	0.74	0.79	0.57	1.00

estimate the covariance matrix. Table 6.2 displays the estimated volatilities and correlations. The optimal portfolio is obtained by minimizing variance subject to the constraint on expected return. The S&P 500 receives a weight of 37.2%, the Nikkei 36.6%, and the FTSE 100 26.2%. The DAX index is excluded from the optimal portfolio.

The intuition behind this outcome is straightforward: To satisfy the alpha constraint, a portfolio must contain a long position in one of the indices with positive alphas, i.e., DAX or Nikkei. The latter is preferred because of its low volatility and lower correlation with S&P 500 and FTSE. Some exposure to S&P 500 and FTSE is required by the minimum variance objective because the two assets display the lowest volatilities.

Having characterized the optimal mean–variance portfolio, I will now focus on the results of the mean–EVaR optimization. Using the same set of alphas (first row of Table 6.2) and the same constraints, I have run the simplex optimizer for a range of values of the risk tolerance parameter ω. The signal to noise ratio parameter q was set to 10^{-3}. The resulting country allocation frontier is displayed in Figure 6.4, with the value of ω on the horizontal axis.

Setting $\omega = 0.15$ yields a portfolio that is very similar to the mean–variance solution. As ω decreases (and therefore risk aversion increases), the optimizer overweighs the UK index while reducing the exposure to S&P 500. This phenomenon can be explained by the differences in the left tail of the distribution of the returns to the two indices. In particular, the dynamic model that is built into the procedure generates a weighting pattern that tends to emphasize the recent information about downside extremes. Table 6.3 shows that during the last 6 months of the sample period, the S&P 500 has displayed fatter tails (as denoted by the higher kurtosis) than the FTSE 100.

As ω increases above 0.15, the portfolio manager is willing to tolerate the high downside risk carried by the DAX in order to capture some of its upside. As a result, the weight on the German index increases.

At $\omega = 0.5$, the risk neutrality case, the optimal portfolio is made up of just DAX and FTSE. Even though this case is irrelevant for practical purposes, it can help us understand the shape of the mean–EVaR frontier. It is useful to note

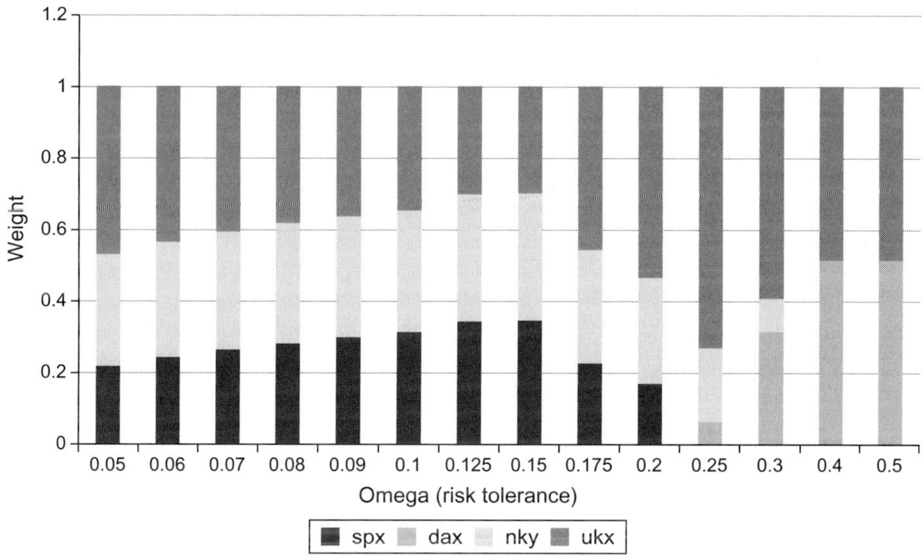

Figure 6.4 Country allocation frontier.

Table 6.3 Descriptive statistics for returns over the last 6 months in the sample period

	spx	ukx
Kurtosis	3.85	2.31
Worst return	−15.20%	−12.00%
Rank 2	−10.55%	−8.46%
Rank 3	−7.83%	−7.70%

that, in the limit as ω approaches 0.5, the dynamic model simplifies to a linear state space model where returns are assumed to follow a random walk plus noise. As a result, the estimated expectile in the limit (i.e., the estimated time-varying mean) is obtained from the KFS. While a random walk is arguably a good approximation to the dynamics of the expectiles in the tails, it is clearly an unrealistic model of the time-varying mean. Without constraints on the alphas, the optimizer would simply put 100% of the weight on the FTSE 100, because it turns out to have the highest estimated mean at the end of the sample if a random walk plus noise model is fitted to the data. In my example, some exposure to the DAX is needed in order to satisfy the constraint on the alphas. In other words, there is a conflict between the two estimates of the expected returns (i.e., the KFS and the simulated alphas) and the optimizer finds a compromise in maximizing expected returns at the portfolio level.

Table 6.4 further compares the mean–variance and the mean–EVaR portfolios. By using the covariance matrix estimated ex ante (Table 6.2), I computed ex ante estimates of the volatility of each optimal portfolio. The minimum is reached by the mean–variance solution, by construction, at 17.1%. However, for low risk tolerance levels, the EVaR portfolios yield very similar levels of ex ante volatility (ranging from 17.1% to 17.4% for ω between 0.05 and 0.2). In other words, the portfolios obtained by minimizing EVaR at low levels of risk tolerance are also low risk according to a volatility criterion. Similarly, the predicted 5% EVaR of the optimal mean–variance portfolio is similar to the one obtained by optimizing on EVaR directly.

The illustration so far has focused on one particular vector of simulated alphas. I carried out a Monte Carlo analysis by simulating 100 vectors with an information coefficient of 0.5. The three-step simulation procedure follows Grinold and Kahn (2000) and Ledoit and Wolf (2004). The benchmark weights are taken to reflect the capitalization of each market, in US dollars, at the end of November 2008: The US receives a weight of 63.9%, Germany 5.7%, Japan 18.7%, and the UK 11.7%. The goal in each optimization exercise was to obtain a portfolio alpha greater than or equal to the expected benchmark return plus 3% (annualized).[6]

Table 6.5 shows summary statistics for the simulated alphas. The first row displays the actual return of each asset for the month of December 2008. The means of the simulated values mirror the pattern found in the actual returns, while the accuracy of each prediction is negatively related to the volatility of the corresponding asset (Table 6.2).

The Monte Carlo results are displayed in Figure 6.5. Each plot refers to an individual asset and the horizontal axis (drawn on the same scale for all plots) measures the difference, in percentage points, between the asset's weight in the mean–variance portfolio and the weight in the mean–5% EVaR portfolio. If the distribution is concentrated to the right of zero, then the mean–EVaR

Table 6.4 Annualized ex ante volatility for alternative portfolios, based on the optimal shrinkage estimate of the covariance matrix displayed in Table 6.3

	Annualized ex ante vol
Mean–variance solution	17.1%
Mean, 5% EVaR	17.3%
Mean, 10% EVaR	17.1%
Mean, 20% EVaR	17.4%
Mean, 30% EVaR	19.7%
Mean, 40% EVaR	21.3%
50% EVaR (risk neutral)	21.3%

[6] In the event that for a particular set of alphas the constraints are inconsistent, the simulated alpha vector is discarded.

approach tends to underweight the asset compared to mean–variance. Once again, the main differences are found for the S&P and FTSE indices. By adopting a mean–EVaR criterion instead of mean–variance, we end up overweighting the UK index, for the reason explained above: The S&P 500 has experienced

Table 6.5 Actual returns observed for December 2008 and sample moments of the 100 simulated alpha vectors

	spx	dax	nky	ukx
Return in December 2008	−2.20%	1.51%	3.70%	1.54%
Simulated alphas:				
Mean	−0.74%	0.45%	1.08%	0.40%
Standard deviation	2.08%	2.84%	2.88%	1.97%

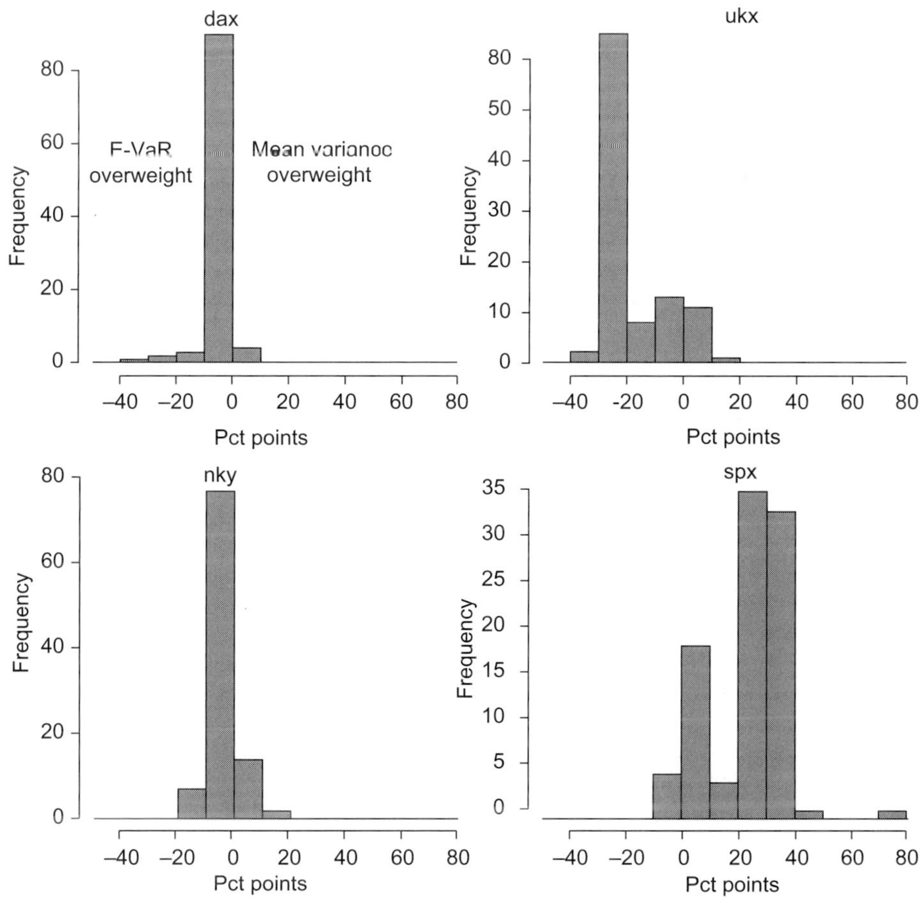

Figure 6.5 Difference between mean–variance and mean–EVaR allocation.

more severe extreme losses in the final part of the sample period. The results of the Monte Carlo analysis show that this conclusion is not limited to the particular set of alphas selected for the initial illustration. By averaging across randomly simulated alphas, I obtained a consistent pattern.

6.5 Conclusion

In this chapter, I have argued that the need for a coherent risk measure, i.e., one that satisfies the most intuitive principles of risk measurement, is particularly felt in a context where the risk characteristics of financial assets evolve rapidly. I have shown how this can be achieved by adopting EVaR, a measure recently proposed in the academic literature.

Furthermore, I have sketched a numerical algorithm that can be used for asset allocation or low-dimensional portfolio construction problems. The objective of the procedure is to minimize risk, possibly subject to a minimum target expected return. Risk is measured by the predicted EVaR of the portfolio, which adapts dynamically to the new information available from the time series of asset returns.

The disadvantage of the proposed methodology, compared to the existing approaches based on quantiles (VaR) or expected shortfall, is that its economic interpretation is less straightforward. The 1% VaR, for example, can be viewed as *the best of the 1% worst cases*.[7] No such intuitive characterization exists for EVaR. However, an advantage of the EVaR approach is that it lends itself naturally to a dynamic model of portfolio risk. In an environment where the risk characteristics of a portfolio evolve rapidly, it is crucial to be able to process the most recent information on downside events through a simple dynamic weighting scheme. This is exactly what the spline smoothing expectile estimator, which is at the heart of the methodology described in this chapter, achieves.

In my experience, the overall weighting pattern generated by a mean–EVaR approach tends to be similar to the solution to a traditional mean–variance problem. The main difference appears to be, as the empirical analysis in this chapter illustrates, that EVaR tends to underweight assets that have experienced extreme downside events, particularly if these have occurred in the recent past.

An attractive feature of the approach advocated in this chapter is its computational simplicity. Because the objective function is convex (and piecewise linear), the proposed optimizer is robust and reliable. Computation times are modest for typical asset allocation applications. It is likely that by adopting a more sophisticated optimization procedure, e.g., an interior point algorithm, it would be possible to improve the computational efficiency further.

Several questions remain open for future research. It would be interesting to backtest a simple trading strategy based on the mean–EVaR optimizer in order to assess whether it improves on the traditional mean–variance approach

[7] See Acerbi (2004).

when the same signal is used. An important question is whether it is possible to limit the downside without sacrificing the upside of portfolio returns. As for the theory, it would be possible to develop a model in which the objective function consists of a linear combination of several expectiles. This would allow the portfolio manager to incorporate his or her preferences on the shape of the distribution of returns in the asset allocation process.

References

Acerbi, C. (2004). Coherent representations of subjective risk aversion. In G. Szegö (Ed.), *Risk measures for the 21st century* (pp. 147–207). Chichester: Wiley.

Artzner, P., Delbaen, F., Eber, J.-M., & Heath, D. (1999). Coherent measures of risk. *Mathematical Finance*, *9*, 203–228.

Basset, G. W., Koenker, R., & Kordas, G. (2004). Pessimistic portfolio allocation and Choquet expected utility. *Journal of Financial Econometrics*, *2*(4), 477–492.

De Rossi, G. (2009). Time varying expectiles and dynamic asset allocation, Working paper.

De Rossi, G., & Harvey, A.C. (2006). Time-varying quantiles, Cambridge Working Papers in Economics, 0619.

De Rossi, G., & Harvey, A.C. (2009). Quantiles, expectiles and splines. *Journal of Econometrics*, Forthcoming.

Engle, R., & Rangel, J. G. (2008). The spline GARCH model for low frequency volatility and its global macroeconomic causes. *Review of Financial Studies*, *21*, 1187–1222.

Granger, C. W. J., & Sin, C. Y. (2000). Modelling the absolute returns of different stock indices: Exploring the forecastability of an alternative measure of risk. *Journal of Forecasting*, *19*, 277–298.

Grinold, R. C., & Kahn, R. N. (2000). *Active portfolio management*. New York: MGraw-Hill.

Jondeau, E., & Rockinger, M. (2006). Optimal portfolio allocation under higher moments. *European Financial Management*, *12*, 29–55.

Kuan, C.-M., Yeh, J.H., & Hsu, Y.-C. (2009). Assessing value at risk with CARE, the Conditional AutoRegressive Expectile Models. *Journal of Econometrics*, Volume 150(2), pp. 261–270.

Ledoit, O., & Wolf, M. (2004). Honey, I shrunk the sample covariance matrix. *Journal of Portfolio Management*, *30*, 110–119.

Madan, D. B. (2006). Equilibrium asset pricing: With non-Gaussian factors and exponential utilities. *Quantitative Finance*, *6*(6), 455–463.

Manganelli, S. (2007). Asset allocation by penalized least squares. ECB Working paper 723.

Mencìa, J,. & Sentana, E. (2008). Multivariate location-scale mixtures of normals and mean–variance–skewness portfolio allocation. *Journal of Econometrics*, Forthcoming.

Meucci, A. (2006). Beyond Black–Litterman: Views on non-normal markets. *Risk Magazine*, *19*(2), 96–102.

Newey, W. K., & Powell, J. L. (1987). Asymmetric least squares estimation and testing. *Econometrica*, *55*, 819–847.

Patton, A. J. (2004). On the out-of-sample importance of skewness and asymmetric dependence for asset allocation. *Journal of Financial Econometrics*, 2, 130–168.

Rockafellar, R., & Uryasev, S. (2000). Optimization of conditional VaR. *Journal of Risk*, 2, 21–41.

Schwert, G. (1989). Why does stock market volatility change over time? *Journal of Finance*, *44*, 1115–1153.

Taylor, J. (2008). Estimating value at risk and expected shortfall using expectiles. *Journal of Financial Econometrics*, 6, 231–252.

Wahba, G. (1978). Improper priors, spline smoothing and the problem of guarding against model errors in regression. *Journal of the Royal Statistical Society B*, 40, 364–372.

7 Optimization and portfolio selection

Hal Forsey and Frank Sortino

Executive Summary

Dr. Hal Forsey and Dr. Frank Sortino present a new Forsey–Sortino Optimizer that generates a mean–downside risk efficient frontier. Part 2 develops a secondary optimizer that finds the best combination of active managers, to add value, and passive indexes, to lower costs.

7.1 Introduction

It is important for the reader to understand the assumptions that apply to all optimizers before I discuss the development of SIA's portfolio selection routines. Utility Theory[1] provides a backdrop for discussing the limitations of mathematics with respect to finding an optimal solution to portfolio selection. The underlying assumption of most people who use optimizers is that the probability distribution is known. Well, in portfolio management it is not known. It can only be estimated, which means that the portfolios on the so-called efficient frontier

[1] From Utility Theory:

(a) If the outcome from a choice of action is known with certainty, then the optimal choice of action is the one with outcome of greatest utility.
(b) If the probability distribution of the outcomes from a choice of action is known, then the optimal choice of action is the one with the greatest expected utility of its outcomes.
(c) If the probability distribution of the outcomes from a choice of action is only approximate, then the action with the greatest expected utility may or may not be a reasonable choice of action.

Utility Theory applied to choosing an optimal portfolio:

(a) If the return from each possible investment is known with certainty, then the optimal investment portfolio is a 100% allocation to the investment with the greatest return. This only assumes that the utility function is an increasing function of return.
(b) If the joint probability of returns for the set of possible investments is known and if the utility function for portfolios with a given variance increases as the expected return increases, then the optimal portfolio is on the mean-variance efficient frontier.
(c) If the joint probability of returns is only approximate, then portfolios on the efficient frontier may or may not be reasonable choices.

Doi:10.1016/B978-0-12-374952-9.00007-5.

may or may not be reasonable (see end notes). Early assumptions were that distributions were bell shaped, i.e., followed a standard normal distribution. Common sense says otherwise. Since one cannot lose an infinite amount of money, even if you are a hedge fund, the distribution must be truncated on the downside and it must be positively skewed in the long run. Only to the extent that the estimate of the joint distribution of returns of a portfolio is reasonable should one put any faith in the veracity of an efficient frontier. We have expended considerable effort to obtain reasonable estimates of the joint distributions of portfolios, which leads us to the following conclusions:

- Optimal solutions to a mathematical problem are often on the boundary of possible solutions. This causes problems in applying mathematics to real-world situation as the mathematical model used to describe the situation is often only an approximation. So the mathematical solutions to the model will often be extreme and since the model is only approximate the solution to the model may be far from optimal. Think of the case in which the returns of the possible investments are thought to be known with certainty. The optimal solution might be a portfolio of 100% in alternative investments like oil futures. People who behave as if they know or their model knows with certainty what is going to happen are unknowingly taking a dangerous amount of risk.
- Even when it is only assumed that the joint probability of returns is known, limiting solutions to some efficient frontier will give extreme portfolios that may be far from optimal if the probability model does not fit reality. For example, many probability models have thin tails that may lead to underestimating probabilities of large losses. This may, for example, lead to portfolios with too much weight in equities.
- To the extent we can accurately describe the joint distribution of returns, we should get reasonably reliable estimates of efficient portfolios. If that statement is true, the pioneering work of the innovators discussed elsewhere should prove beneficial. However, if the input to the optimizer is seriously flawed so will be the output (GIGO).

7.2 Part 1: The Forsey–Sortino Optimizer

This is a model we built after Brian Rom terminated his relationship with the Pension Research Institute. It has never been marketed because we have no interest in becoming a software provider. Neither do we want to keep our research efforts a secret. Therefore, we will provide an executable version and the source code for the Forsey–Sortino Optimizer on the Elsevier web site (http://booksite.elsevier.com/Sortino/ - Password: SAT7TQ46SH27). Be advised, we have no intention of supporting the software in any way. Our intention is to provide a starting point from which other researchers around the world can make improvements and in that way make a contribution to the state of the art.

7.2.1 Basic assumptions

We begin with the assumption that the user wants to maximize the geometric average rate of return in a multiperiod framework. Therefore, the three-parameter lognormal distribution suggested by Atchison and Brown should provide a better estimate of the shape of the joint distribution than assuming a bell shape (standard normal distribution). We recognize that this shape should change when the market is undervalued in that it should be more positively skewed than normal, and that it should be more negatively skewed when the market is overvalued.

However, when the market has been undervalued in the past, the subsequent result is seldom, if ever, the same. This is where we call on Bradley Efron's bootstrap procedure to generate a distribution of returns that could have happened from all those periods in the past when the market was undervalued. The same is done for all those periods in the past when the market was overvalued. This saves us from saying foolish things like "The future is going to be like 1932 or 1990, etc." What we are assuming is that the distribution of returns will be more closely approximated by bootstrapping all of the monthly returns from all past periods when the market proved to be undervalued.

A representation of that distribution is given in Figure 7.1.

Next (see Figure 7.2) we allow the user to identify which part of the world he or she is operating from. That will determine which currency the indexes will be denominated in and which indexes to use.

Next we allow the user to select combinations of scenarios from the three buckets of returns. If the user does not wish to make such a decision, the choice should be "unknown," in which case the returns from all three buckets are used.

Let us assume the choice is US and the user chooses "Unknown" for the scenario. A distribution for each index is then presented. Figure 7.3 displays the distribution for the MSCI Japan index with a mean of 15%.

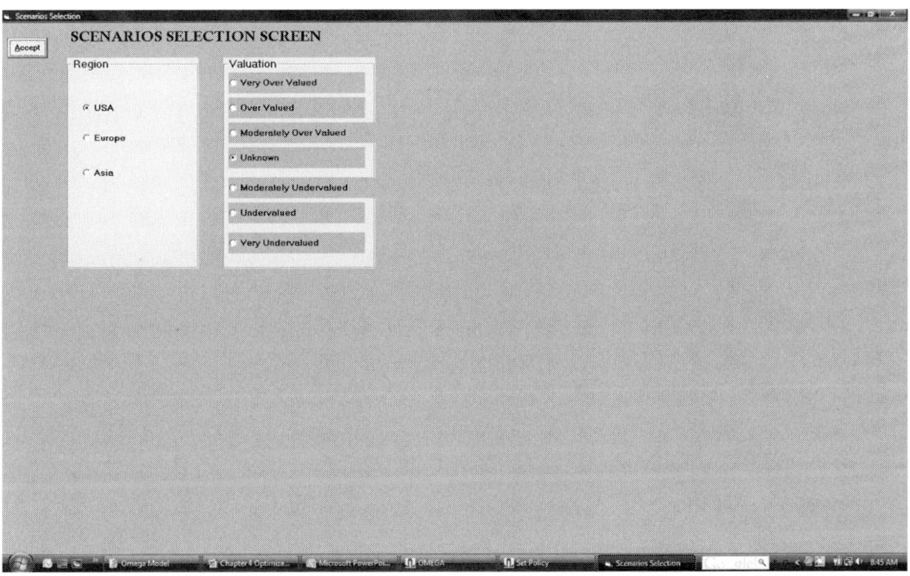

Figure 7.1 Scenario selection screen.

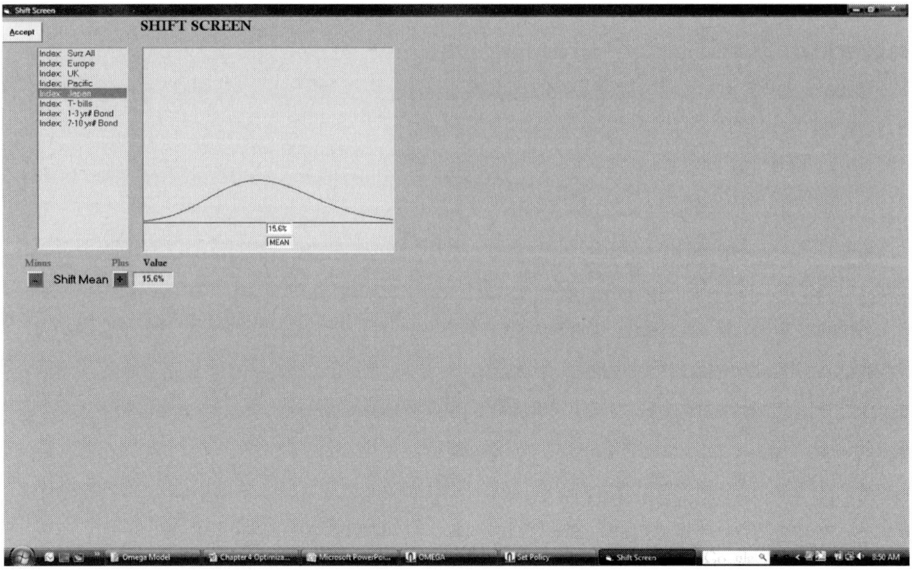

Figure 7.2 Shift Screen before.

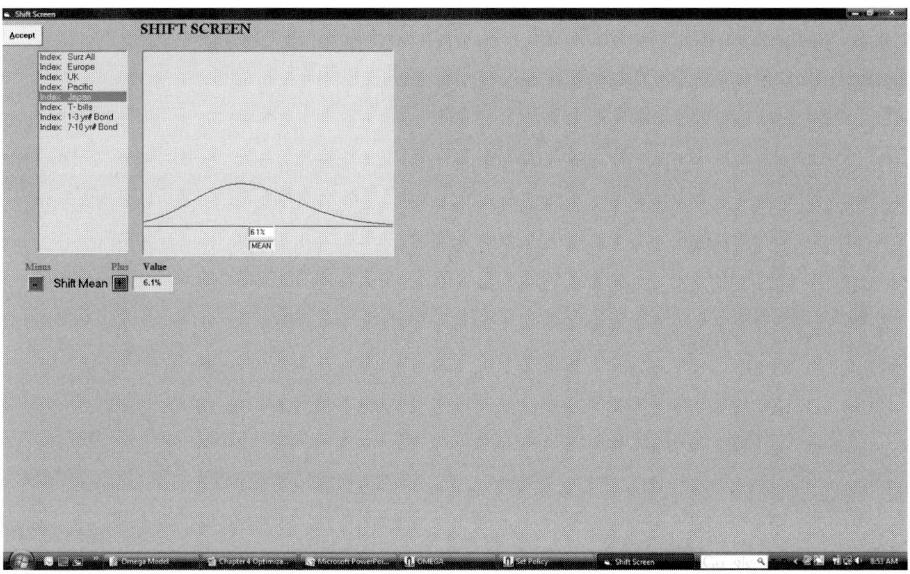

Figure 7.3 Shift screen after.

The user now has the option to change the location point of the distribution by shifting the mean to the left or to the right (Figure 7.3).

In Figure 7.3, the mean has been shifted to 6.1% and the distribution has become more positively skewed.

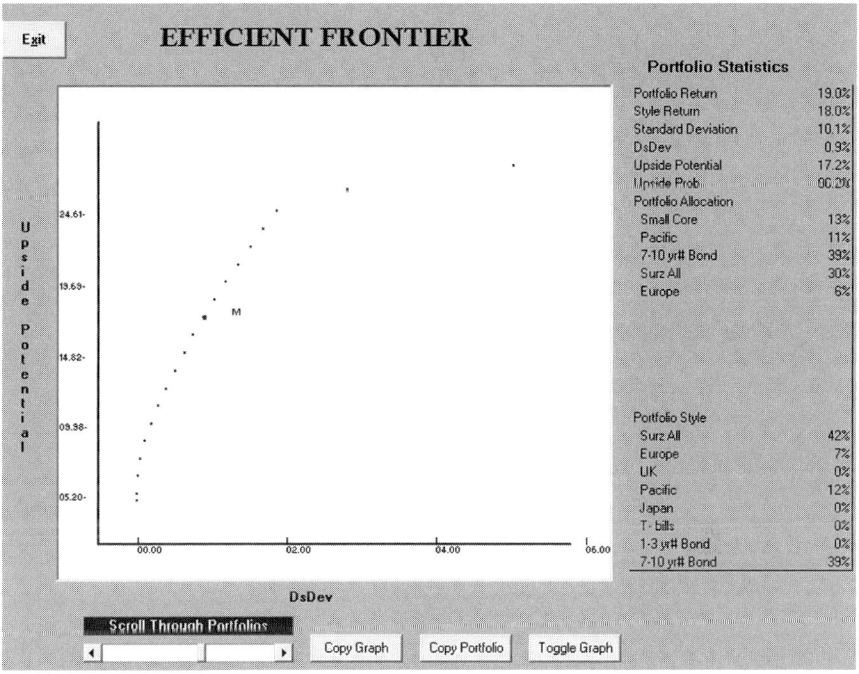

Figure 7.4 Efficient Frontier.

7.2.2 Optimize or measure performance

The model now allows the user to use the optimizer in a mean–variance mode or a mean–downside risk mode, or to view performance measurement in several modes. Figure 7.5 shows the screen for the asset allocation choice for a defined contribution plan and an 8% DTR™* portfolio (Moderate). We allow a mean–variance as well as a mean–downside risk choice because we wanted to be able to compare our mean–variance result with the output from a standard mean–variance optimizer. If we did not get very similar results, it would call into question our methodology for generating joint distributions.

Figure 7.5 shows the efficient frontier in upside potential to downside risk space. The letter M is approximately where the moderate portfolio is located. The user can scroll to that approximate location and see the results to the right of the efficient frontier. The asset allocation is shown in the lower statistics table titled "Portfolio Style." Forty-two percent is allocated to the nine US equity styles identified by Ron Surz, president of PPCS-inc. Nothing is allocated to Japan because we shifted the mean down from 15% to 6% without making any adjustments to the other indexes. This would not prevent a mean–variance optimizer from assigning some weight to Japan if it was less than perfectly positively correlated with the other indexes.

* DTR is a trademark of Sortino Investment Advisors.

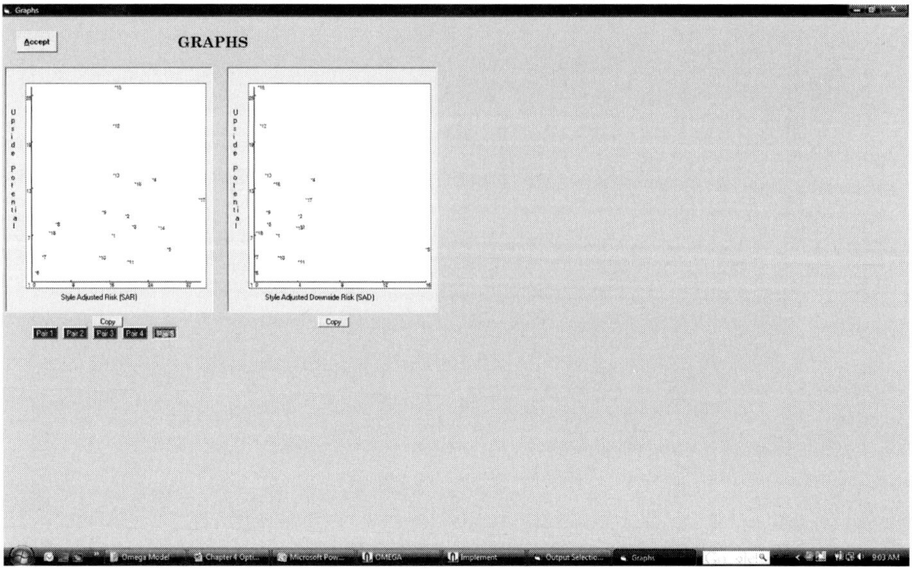

Figure 7.5 Efficient frontier for the upside potential ratio.

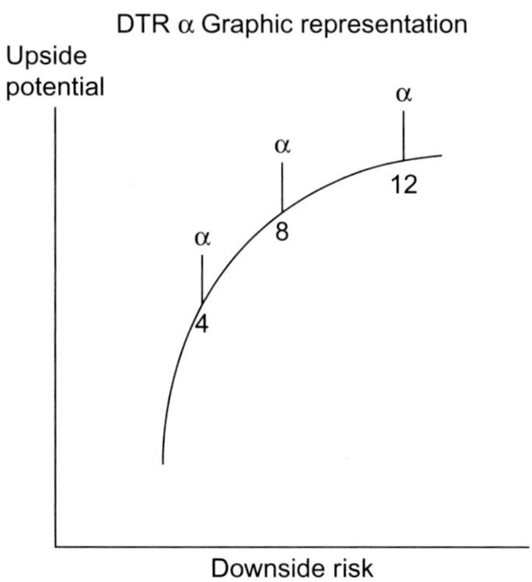

Figure 7.6 Two views of performance measurement.

Figure 7.6 shows one of five possible performance descriptions.

Notice the pattern of the risk–return framework. On the left, risk is measure as standard deviation. On the right, it is downside risk. In each case, it is adjusted for a term we call the style beta (see Appendix).

Name	R^2	U-P ratio	DTRα	LrgVal	LrgCor	LrgGro	MidVal	MidCor	MidGro	MinVal	MinCor	MinGro
LrgValu	1	1.25	0%	100%	0%	0%	0%	0%	0%	0%	0%	0%
LrgCore	1	0.84	0%	0%	100%	0%	0%	0%	0%	0%	0%	0%
LrgGrow	1	0.61	0%	0%	0%	100%	0%	0%	0%	0%	0%	0%
MidValu	1	1.53	0%	0%	0%	0%	100%	0%	0%	0%	0%	0%
MidCore	1	1.1	0%	0%	0%	0%	0%	100%	0%	0%	0%	0%
MidGrow	1	0.66	0%	0%	0%	0%	0%	0%	100%	0%	0%	0%
MinValu	1	1.91	0%	0%	0%	0%	0%	0%	0%	100%	0%	0%
MinCore	1	0.75	0%	0%	0%	0%	0%	0%	0%	0%	100%	0%
MinGrow	1	0.34	0%	0%	0%	0%	0%	0%	0%	0%	0%	100%
Mellon Capital	0.99	1.03	–1%	43%	19%	19%	11%	8%	0%	0%	0%	0%
HW Large Value	0.82	1.53	–8%	1%	0%	0%	99%	0%	0%	0%	0%	0%

Figure 7.7 DTR α efficient frontier representation.

7.3 Part 2: The DTR optimizer

No matter what kind of optimizer one uses to generate an efficient frontier of passive indexes, it is important to know if there may be some combination of active managers and passive indexes that would lie above the efficient frontier as shown in Figure 7.6.

Figure 7.6 depicts a hypothetical representation of what this efficient frontier would look like. The numbers 4, 8, and 12 represent points on the efficient frontier for portfolios of passive indexes that have the highest upside potential ratio for a given level of downside risk. The points correspond to DTRs of 4%, 8%, and 12%.

To accomplish this, we begin with a quadratic optimizer developed by Bill Sharpe. Using the Surz indexes, we can quickly identify the style blend of thousands of managers. For illustrative purposes, a partial list is shown in Figure 7.7. Three years of monthly returns were used. M Capital is a fund that attempts to track the S&P 500 while SW claims to be a large value manager.

The S&P 500 is often used as a surrogate for the market. For details on the error of this assumption, see Chapter 2 by Ron Surz. All a manager would have to do is change the weights to claim an enhanced index. In this case, large cap is overweighted. If an S&P 500 index is employed, it should be evaluated as an active manager competing with the passive indexes that more accurately reflect the complete market of 6,000 plus stocks. HW loads almost entirely on the midcap value index yet only has an R^2 of 0.82. In other words, 18% of the difference in the variance in returns between the midcap index plus 1% large value index is not being explained. Something is missing. We asked a friend to run the data on a standard style analyzer and it came up with 54% large cap value. The reason? He unknowingly was using quarterly data. There were only 12 observations for 9 variables. That is totally unreliable. When he used 5 years of quarterly data, his answer was more similar to ours. The point we wish to make is that all quantitative models that attempt to do the same thing are not the same and even those that are the same are influenced by the periodicity of the data.

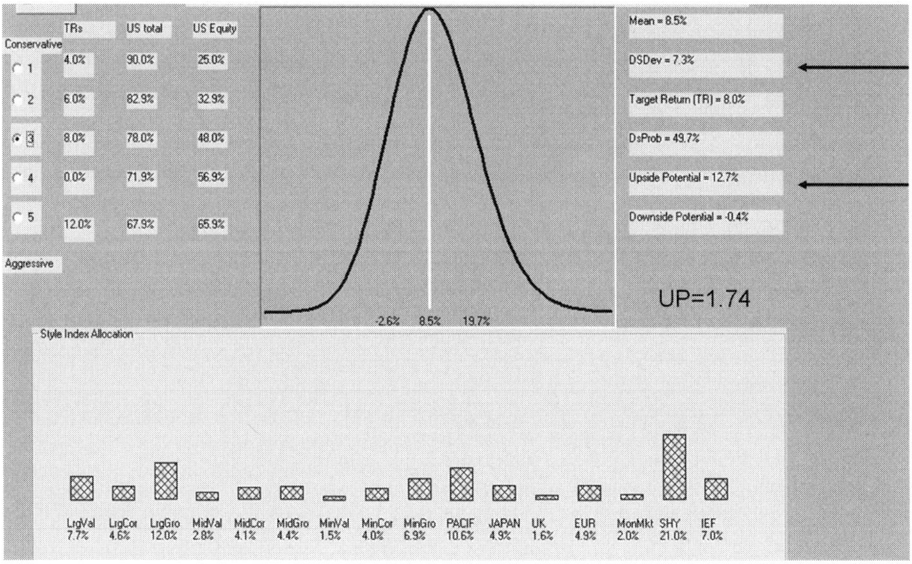

Figure 7.8 Style analyzer.

Following the style analysis, the DTR optimizer solves for that combination of active and passive managers that fulfills the desired asset allocation shown at the bottom of Figure 7.9 and provides the highest overall added value as measured by DTR α.

Portfolio 3 shown in Figure 7.9 is for the DTR 8 portfolio. Seventy-eight percent of the portfolio is in US securities and 48% of the portfolio is in US equity. That leaves 30% in US fixed income (30 + 48 = 78). The mean of this joint distribution is 8.5%, which is approximately equal to the DTR of 8%. If one were making decisions on the basis of probabilities, there is about the same chance of returns being higher than the mean as there is of being lower. However, this portfolio has the potential for returning 12.7% in any given year and the downside risk of achieving that is only 7.3%. Dividing the upside potential by the downside risk (see arrows) yields an upside potential ratio of 1.74, indicating 74% more upside potential than downside risk. Expected Utility Theory, explained in the Appendix, would claim that risk averse investors should make decisions on the basis of the upside potential ratio instead of mere probabilities. We further claim that the proper focus is the DTR and not the mean.

What is actually taking place is a shift in the joint distribution of the portfolio. Figure 7.9 depicts a 500 basis point shift from left to right for a DTR α of 5%.

Figure 7.10 shows the final output of the SIA optimizer. At the bottom is the asset allocation that was stipulated as input to the optimizer. At the top is the combination of active managers and passive indexes that fulfills the asset allocation and adds value in terms of DTR α while reducing costs by using indexes. To reiterate, the first requirement is to ensure that the asset allocation that was

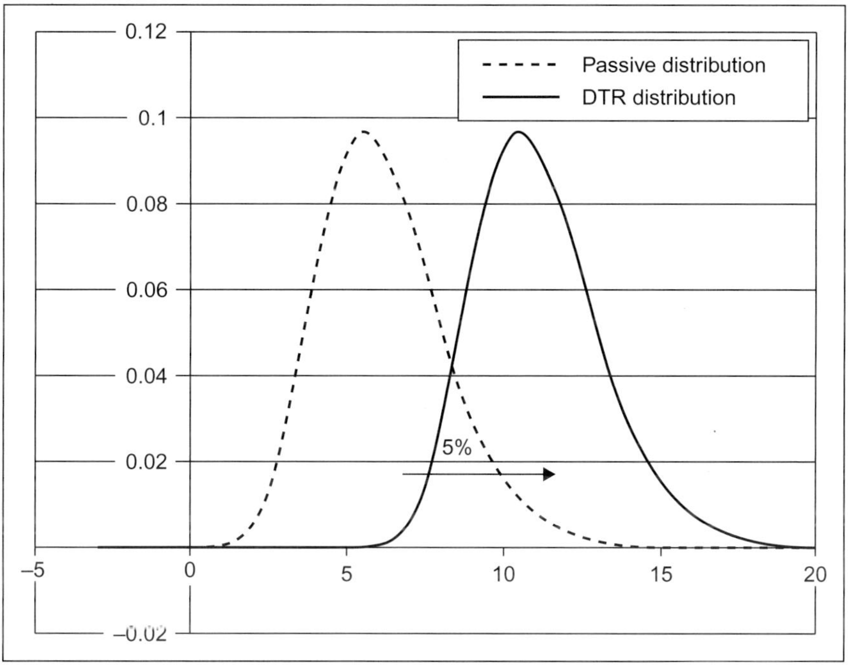

Figure 7.9 SIA optimizer.

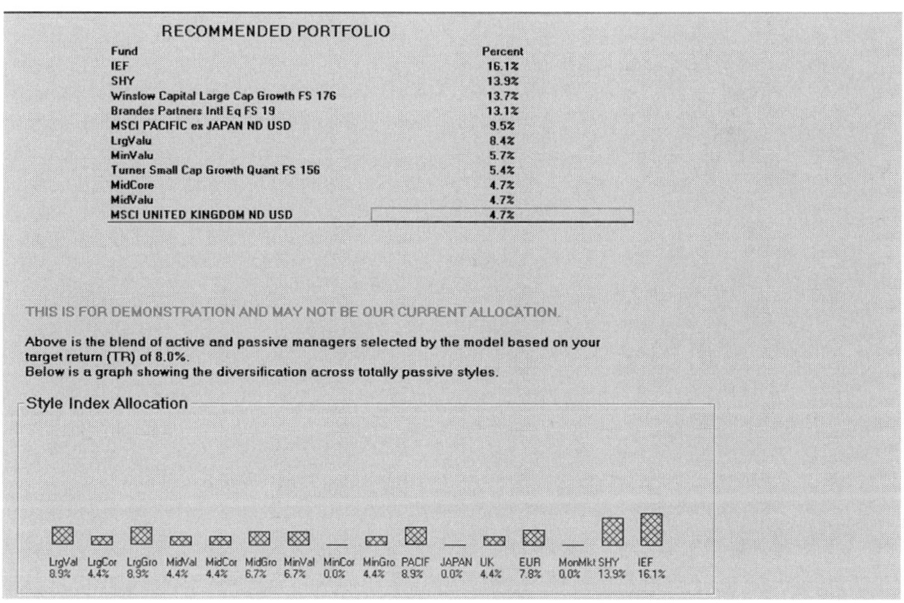

Figure 7.10 DTR α added value representation.

specified in the beginning is the asset allocation that one ends up with. This is the most important decision made in that it accounts for 80% or more of the future return of the portfolio. The second consideration is to hire those managers who can add the greatest value to the portfolio. This must take into consideration each manager's style blend in order to not corrupt the desired asset allocation. Third, use passive indexes to fill out the asset allocation in areas where the active managers do not add value.

For a manager to come into solution, the R^2 must be 0.70 or greater, the upside potential ratio must be among the highest, and the DTR α must be in the top three. In the portfolios case, three active managers added value to the portfolio and passive indexes were used to fulfill the asset allocation. This is a quantitative way of reducing costs while adding value as measured by a manager's ability to beat his or her style blend.

Appendix: Formal definitions and procedures

Overview

Our portfolio selection is based on several criteria. Some subjective, but most based on the results of an analysis of return data. Each fund has a **style blend** described by a portfolio of passive indexes obtained by a quadratic fit of return data. As a result, we are able to compare the fund with its style blend. We summarize this comparison with a difference in risk-adjusted returns called the DTR α. The DTR α is a measure of added value and is key to our final selection process. However, before making this calculation, possible funds are carefully screened. One part of the screening process is based on an analysis of risk and return statistics defined in terms of a DTR. The purpose of this appendix is to give formal definitions to these statistics and explain how they are estimated.

The DTR

The DTR is the annualized rate of return required from investments to be able to support expenditures during retirement. This rate of return depends on many factors, including life expectancy, retirement contributions, and retirement income from sources other than investments. As many of these factors are not known, this rate can only be approximated. This is especially true if retirement is many years in the future. We have developed a calculator to help an investor determine a reasonable combination of retirement contribution level, retirement age, and DTR. This calculator is available on the Sortino Investment Advisors web site www.sortinoia.com.

About Example 3

The calculations for Example 3 are based on assuming a discrete probability function of returns determined by the sample of returns. Here are the formal definitions in this case.

The basic statistics—discrete version

Let $p(x)$ be the discrete probability function of returns and T = DTR.

$$\text{Upside Probability} = \sum\nolimits_{x>T} p(x)$$

$$\text{Upside Potential} = \sum\nolimits_{x>T} (x - T) \cdot p(x)$$

$$\text{Downside Deviation} = \sqrt{\sum\nolimits_{x \le T} (T - x)^2 p(x)}$$

$$\text{Upside Potential Ratio} = \text{Upside Potential/Downside Deviation}$$

$$= \frac{\sum_{x>T}(x - T) \cdot p(x)}{\sqrt{\sum_{x \le T}(T - x)^2 p(x)}}$$

The basic statistics—continuous version

Rather than using a discrete probability function and following a procedure similar to that of Example 3 to estimate our statistics, we fit a 3+ parameter lognormal distribution by using a bootstrap procedure on historical data. First, we give the general formulas for a probability density function and then we will consider the special case of the lognormal. A more expansive justification and further details for using the lognormal is given in Chapter 4 of *Managing Downside Risk in Financial Markets*.

Let $f(x)$ be the probability density function of returns and T = DTR. As explained in the next section, we use a 3+ parameter lognormal density function, and in that case there are analytic expressions for each of the following integrals.

$$\text{Upside Probability} = \int_{x>T} f(x)dx$$

$$\text{Upside Potential} = \int_{x>T} (x - T)f(x)dx$$

$$\text{Downside Deviation} = \sqrt{\int_{x<T} (T - x)^2 f(x)dx}$$

$$\text{Upside Potential Ratio} = \text{Upside Potential/Downside Deviation}$$

Estimating the probability of returns with a 3+ parameter lognormal

The formulas for the lognormal are not pretty. They are generally described in terms of the parameters for the underlying normal. So, with the logarithmic translations, they can get a bit involved. The resulting basic formulas are collected together below. So for now, we will not concern ourselves with these technical details. What we will do is describe the essentials.

First, what are the three parameters? There is some choice in selecting the parameters. We made these selections so that the meanings of the parameters would be easily understood in terms of the annual returns. The parameters are the mean, the standard deviation, and the extreme value of annual returns. You are probably already familiar with the mean and standard deviation. The mean is a measure of the central tendency and the standard deviation a measure of the spread of the curve. These two parameters are enough to describe a normal and the standard lognormal. But the three-parameter version of the lognormal uses another parameter. A lognormal curve has either a largest value or a smallest value. This third parameter, the extreme value, allows us to shift and flip the distribution.

Now we must find a way of estimating these parameters from a sample. We choose to solve this problem by using the sample mean and sample standard deviation to estimate the mean and standard deviation of the underlying lognormal. Our estimate of the extreme value was selected on the basis of simulations. These simulations showed that only a rough estimate of the extreme value is required to obtain a reasonable lognormal fit. We estimate it as follows: First, calculate the minimum and the maximum of the sample and take the one closest to the mean. The extreme value is obtained from this value by moving it four standard deviations further from the mean. For example, if the mean, standard deviation, minimum, and maximum of a sample are 12%, 8%, −15%, and 70%, respectively, then the extreme value is (−15%)—(4)(8%) − 47%, since the minimum is closer to the mean than the maximum.

Consider the following table with statistics computed from a lognormal fit:

Parameters and statistics	Example 1	Example 2	Example 3
Mean	12%	12%	12%
Standard deviation	22%	22%	22%
Extreme value	−50%	74%	−50%
DTR	7.5%	7.5%	13.5%
Upside probability	51.9%	74%	40.4%
Downside deviation	10.5%	15.5%	14.3%
Upside potential	10.6%	11.2%	7.8%
Upside potential ratio	1.01	0.73	0.55

The first two examples differ only in extreme values. One is a maximum and the other a minimum. Please notice how different these two examples are even though they have the same mean and standard deviation. These differences cannot be captured by the normal curve. The third example is similar to the first but with a higher DTR.

The three basic parameters are estimated from a sample obtained from a bootstrap procedure on historical returns.

Mean = sample mean
SD = sample standard deviation
Tau = τ = extreme value computed as described above

Some auxiliary parameters

$$\text{Dif} = |Mean - Tau|$$

$$\sigma = \ln\left(\left(\frac{SD}{Dif}\right)^2 + 1\right)$$

$$\mu = \ln(Dif) - \frac{\sigma}{2}$$

$$\alpha = \frac{1}{(\sqrt{2\pi} \cdot \sigma)}$$

$$\beta = -\frac{1}{(2\sigma^2)}$$

Formula for the lognormal curve f(x)

If the extreme value is a minimum and x is greater than the extreme value, then:

$$f(x) = \frac{\alpha}{x - \tau} \cdot \exp(\beta \cdot (\ln(x - \tau) - \mu)^2)$$

If the extreme value is a maximum and x is less than the extreme value, then:

$$f(x) = \frac{\alpha}{\tau - x} \cdot \exp(\beta \cdot (\ln(\tau - x) - \mu)^2)$$

Formula for the lognormal cumulative distribution function F(x)

If the extreme value is a minimum and x is greater than the extreme value, then:

$$F(x) = 1 - \frac{erfc(\ln(x - \tau) - \mu)}{2\sqrt{2} \cdot \sigma}$$

If the extreme value is a maximum and x is less than the extreme value, then:

$$F(x) = 1 - \frac{erfc(\ln(\tau - x) - \mu)}{2\sqrt{2} \cdot \sigma}$$

Note: erfc is the complementary error function (Ref. 3).

Analytic expression for the downside and upside statistics

Using the above formula for the lognormal density function, it is possible to obtain analytic expressions for the integrals defining the basic statistics. Following are fragments of Visual Basic code in the case Tau in a minimum for analytic expressions of the basic statistics for a lognormal distribution (as above, T is the DTR).

Upside probability or U can be written in terms of $F(x)$, the distribution function as:

$$U = 1 - F(x)$$

Upside potential or UP

```
If T > tau Then
        b = T-tau
        c = sqrt2 * sigma
      UP = 0.5 * Exp(mu + 0.5 * sigma ^ 2) * (2-erfc((mu +
     sigma ^ 2-Log(b)) /c))
Else
      UP = tau + Exp(mu + sigma * sigma/2)
EndIf
```

Downside Variance or DV (this is the square of downside deviation)

```
If T > tau Then
        b = T-tau
        c = sqrt2 * sigma
        a = (Log(b) -mu) /c
     DV = 0.5 * Exp(2 * mu + 2 * sigma ^ 2) * (2-erfc(a-c))
  -b * (2-erfc(a-c/2)) *Exp(mu+ (sigma ^ 2) /2) +0.5 * b ^ 2 * (2-erfc(a))
Else
     DV = 0
End If
```

And, of course:
Upside Potential Ratio = Upside Potential/Downside Deviation

Style analysis—determining the style blend of a fund

The costs of investing in an actively managed fund are often much higher than owning a portfolio of index funds. So how can we determine if these extra costs are warranted? One approach is through returns-based style analysis (Sharpe, 1988) to determine a portfolio of index funds that most closely matches the investment style of the fund.

Here, briefly, are the procedures for determining this portfolio, called the style blend.

A small (less than 20) set of indexes is chosen to divide, without overlap, the space of potential investments. This selection should be made so as to reduce the problem of colinearity but still allow a good fit as measured by R^2. The selection of the time interval of data to fit is also important. The interval must be long enough for a good fit but short enough to capture just the current style of the management of the fund. We call this interval the style interval.

Weights to form a passive portfolio of indexes are chosen so that the variance of the difference between the fund's returns and the passive portfolio's returns over the style interval is minimized. This choice is accomplished with a quadratic optimizer.

A style statistic

Now we have our comparison portfolio—the style blend of passive indexes. How do we make the comparison? First, we do not even try unless the style fit, as measured by R^2, is high enough, say 0.75 or higher. Then, the obvious thing is to look at the difference of annual returns over the style interval. This is not bad, but an improvement would be to adjust for difference in risk. One way to do this is to use a style Fishburn utility function. Here is the definition for the statistic we use, called DTR α.

$$\begin{aligned}\text{DTR}\ \alpha = {} & \text{Risk Adjusted Return of the Fund} \\ & - \text{Risk Adjusted Return of the Style Blend}\end{aligned}$$

$$\text{Risk Adjusted Return} = \text{Annual Return} - \lambda^{*}\,\text{Downside Variance}$$

Downside Variance is the square of the Downside Deviation and λ is a constant we usually take to be 3.

So we will consider using a fund in our portfolio only if it has a positive DTR α.

There is more about this selection process in the previous chapter on optimization.

References

Aitchison, J., & Brown, J. A. C. (1957). *The lognormal distribution.* Cambridge: Cambridge University Press.

Malkiel, B. (1993). *A random walk down Wall Street.* New York: W.W. Norton.

Press, W., et al. (1992). *Numerical recipes in* C. Cambridge University Press, 220.

8 Computing optimal mean/downside risk frontiers: the role of ellipticity

Tony Hall and Stephen E. Satchell

Executive Summary

The purpose of this chapter is to analyze and calculate optimal mean/downside risk frontiers for financial portfolios. Focusing on the two important cases of mean/value at risk and mean/semivariance, we compute analytic expressions for the optimal frontier in the two asset case, where the returns follow an arbitrary (nonnormal) distribution.

Our analysis highlights the role of the normality/ellipticity assumption in this area of research. Formulae for mean/variance, mean/expected loss, and mean/semi–standard deviation frontiers are presented under normality/ellipticity. Computational issues are discussed and two propositions that facilitate computation are provided.

Finally, the methodology is extended to nonelliptical distributions where simulation procedures are introduced. These can be presented jointly with our analytical approach to give portfolio managers deeper insights into the properties of optimal portfolios.

8.1 Introduction

There continues to be considerable interest in the trade-off between portfolio return and portfolio downside risk. This has arisen because of inadequacies of the mean–variance framework and regulatory requirements to calculate value at risk and related measures by banks and other financial institutions.

Much of the literature focuses on the case where asset and portfolio returns are assumed normally or elliptically distributed, see Alexander and Bapista (2001), Campbell, Huisman, and Koedijk (2002), Kaynar, Ilker Birbil, and Frenk (2007), and Kring, Rachev, Höchstötter, Fabozzi, and Bianchi (2009). In most cases, the problem of a mean/value at risk frontier differs only slightly from a classic mean–variance analysis. However, the nature of the optimal mean–risk frontiers under normality is not discussed in any detail and we

[1] We thank John Knight and Peter Buchen for many helpful comments, particularly on Proposition 4.1.
[2] Correspondence to Steve.Satchell@econ.cam.ac.uk

Doi:10.1016/B978-0-12-374952-9.00008-7.

present some results in Section 8.2. These are a consequence of two results we present, Proposition 8.1 and a generalization, Proposition 8.2, which prove that the set of minimum risk portfolios are essentially the same under ellipticity for a wide class of risk measures. In addition to these results, we present three extensions. The extensions we propose in this chapter are threefold. First, we consider extensions for portfolio simulation of those advocated for value at risk simulation by Bensalah (2002). Second, under normality we compute explicit expressions for mean/value at risk, mean/expected loss, and mean/semivariance frontiers in the two asset case and in the general N asset case, complementing the results for mean/value at risk under normality provided by Alexander and Bapista (2001). Finally, our framework allows us to consider fairly arbitrary risk measures in the two asset case with arbitrary return distributions, and in particular some explorations under bivariate lognormality are considered. In Section 8.6, we present issues to do with the simulation of portfolios, pointing out some of the limitations of our proposed methodology. These methodologies are applied to general downside risk frontiers for general distributions. Conclusions follow in Section 8.7.

8.2 Main proposition

It is worth noting that although it is well known that normality implies mean–variance analysis for an arbitrary utility function (see, for example, Sargent (1979, p. 149)), it is not clear what happens to the mean/downside risk frontier under normality. What is known is that the (μ_p, θ_p) frontier should be concave under appropriate assumptions for θ_p and that the set of minimum downside risk portfolios should be the same as those that make up the set of minimum variance portfolios. We note that Wang (2000) provides examples that show that when returns are not elliptical, the two sets do not coincide. A proof of this second assertion is provided below in Proposition 8.1. We shall initially assume that the $(N \times 1)$ vector of returns is distributed as $N_N(\mu, \Sigma)$ where μ is an $(N \times 1)$ vector of expected returns and Σ is an $(N \times N)$ positive definite covariance matrix. We define the scalars $\alpha = \mu'\Sigma^{-1}\mu$, $\beta = \mu'\Sigma^{-1}e$, and $\gamma = e'\Sigma^{-1}e$, where e is an $(N \times 1)$ vector of ones.

Proposition 8.1: For any risk measure $\phi_p = \phi\left(\mu_p, \sigma_p^2\right)$, where $\mu_p = \mu'x$, $\sigma_p^2 = x'\Sigma x$, $\phi_1 = \partial\phi/\partial\mu_P$, $\phi_2 = (\partial\phi/\partial\sigma_p^2)$, and ϕ_2 is assumed non-zero, the mean minimum risk frontier (μ_p, ϕ_p) is spanned by the same set of vectors for all ϕ, namely, $x = \Sigma^{-1}E(E'\Sigma^{-1}E)^{-1}\ \Psi_p$, where $\Psi_p{}' = (\mu_p, 1)$ and $E = (\mu, e)$.

Proof: Our optimization problem is to minimize ϕ_p subject to $E'x = \Psi_p$. That is,

$$\min_x \quad \phi\left(\mu_p,\ \sigma_p^2\right) - \lambda'(E'x - \Psi_p) \tag{8.1}$$

where λ is a (2×1) vector of Lagrangians. The first-order conditions are:

$$\frac{\partial \phi}{\partial \mu_p}\frac{\partial \mu_p}{\partial x} + \frac{\partial \phi}{\partial \sigma_p^2}\frac{\partial \sigma_p^2}{\partial x} - E\lambda = 0 \tag{8.2}$$

and

$$E'x - \Psi_p = 0 \tag{8.3}$$

Here,

$$\frac{\partial \mu_p}{\partial x} = \mu \tag{8.4}$$

and

$$\frac{\partial \sigma_p^2}{\partial x} = 2\Sigma x \tag{8.5}$$

This implies that:

$$\phi_1 \mu + 2\phi_2 \Sigma x = E\lambda \tag{8.6}$$

or

$$\phi_1 \Sigma^{-1}\mu + 2\phi_2 x = \Sigma^{-1}E\lambda \tag{8.7}$$

or

$$\lambda = (E'\Sigma^{-1}E)^{-1}(\phi_1 E'\Sigma^{-1}\mu + 2\phi_2 \Psi_p) \tag{8.8}$$

Thus, the general "solution" x satisfies:

$$\phi_1 \Sigma^{-1}\mu + 2\phi_2 x = \Sigma^{-1}E(E'\Sigma^{-1}E)^{-1}(\phi_1 E'\Sigma^{-1}\mu + 2\phi_2 \Psi_p) \tag{8.9}$$

We now show that:

$$x = \Sigma^{-1}E(E'\Sigma^{-1}E)^{-1}\Psi_p \tag{8.10}$$

satisfies Equation (8.9) for any ϕ_1 and ϕ_2. This is because the right-hand side of Equation (8.9) can be written as:

$$\phi_1\Sigma^{-1}E(E'\Sigma^{-1}E)^{-1}E'\Sigma^{-1}\mu + 2\phi_2\Sigma^{-1}E(E'\Sigma^{-1}E)^{-1}\Psi_p \tag{8.11}$$

But,

$$(E'\Sigma^{-1}E)^{-1}E'\Sigma^{-1}\mu = \begin{pmatrix}1\\0\end{pmatrix} \tag{8.12}$$

so that

$$\phi_1\Sigma^{-1}E(E'\Sigma^{-1}E)^{-1}E'\Sigma^{-1}\mu = \phi_1\Sigma^{-1}\mu \tag{8.13}$$

and Equation (8.9) then simplifies to:

$$2\phi_2 x = 2\phi_2\Sigma^{-1}E(E'\Sigma^{-1}E)^{-1}\Psi_p \tag{8.14}$$

or

$$x = \Sigma^{-1}E(E'\Sigma^{-1}E)^{-1}\Psi_p \tag{8.15}$$

Corollary 8.1: The above applies to a large family of risk measures for a range of distributions that reduce to the "mean–variance" analysis, namely the elliptical class as outlined in Ingersoll (1987).

Ingersoll (1987, p. 104) defines a vector of N random variables to be elliptically distributed if its density (pdf) can be written as:

$$pdf(y) = |\Omega|^{-1/2}\, g\left((y-\mu)'\Omega^{-1}(y-\mu)\right) \tag{8.16}$$

If means exist, then $E[y] = \mu$ and if variances exist, then the covariance matrix cov(y) is proportional to Ω. The characteristic function of y is:

$$\varphi_N(t) = E\left[\exp(it'y)\right] = \exp(it'\mu)\psi(t'\Omega t) \tag{8.17}$$

for some function ψ that does not depend on N. It is apparent from Equation (8.17) that if $\omega'y = z$ is a portfolio of elliptical variables, then

$$E\left[\exp(isz)\right] = E\left[\exp\left(is\omega'y\right)\right] = \exp(is\omega'\mu)\psi(s^2\omega'\Omega\omega) \tag{8.18}$$

and all portfolios from the joint *pdf* given by Equation (8.16) will have the same marginal distribution, which can be obtained by inverting Equation (8.17). Furthermore, the distribution is location scale, in the sense that all portfolios differ only in terms of $\omega'\mu$ and $\omega'\Omega\omega$ and it is for this reason that mean–variance analysis holds for these families.

Corollary 8.2: Our result includes as a special case the value at risk calculations of Alexander and Bapista (2001) since $\phi\left(\mu_p, \sigma_p^2\right) = t\sigma_p - \mu_p$ for $t > 0$.

We note that Alexander and Bapista (2001) are more concerned with the efficient set than the minimum risk set. The distinction between the efficient set and the minimum risk set is addressed by finding the minimum point on the mean minimum risk frontier. Nevertheless, their Proposition 2 (page 1,168) that implicitly relates normality goes to some length to show that the mean–value at risk efficient set of portfolios is the same as the mean minimum variance portfolios. This follows as a consequence of our Proposition 8.2 and its corollaries.

We now turn to the question as to whether the mean minimum risk frontier is concave (i.e., $(\partial^2 v/\partial\mu^2) > 0$). Suppose that returns are elliptical so that our risk measure v can be expressed as:

$$\nu = \varphi\left(\mu, \sigma_p^2\right) \text{ and } \mu = \mu_p, \tag{8.19}$$

where

$$\sigma_p^2 = \frac{\mu^2\gamma - 2\beta\mu + \alpha}{\Delta} \tag{8.20}$$

and

$$\Delta = \alpha\gamma - \beta^2, \tag{8.21}$$

so that

$$\frac{\partial\nu}{\partial\mu} = \varphi_1 + \frac{\varphi_2}{\sigma_p}\frac{(\mu\gamma - \beta)}{\Delta} \tag{8.22}$$

so that $\partial v/\partial\mu \geq = \leq 0$ when

$$\varphi_1 + \frac{\varphi_2}{\sigma_p}\frac{(\mu\gamma - \beta)}{\Delta} \geq = \leq 0 \tag{8.23}$$

We shall assume the existence of a unique minimum risk portfolio; if it exists, the minimum risk portfolio occurs when ($\partial v/\partial\mu = 0$) or when

$$\mu^* = \varphi_1 + \frac{\beta}{\gamma} - \frac{\sigma_p\varphi_1\Delta}{\gamma\varphi_2} \tag{8.24}$$

We note that β/γ is the expected return of the global minimum variance portfolio, so the global minimum "v" portfolio is to the right or to the left depending on the signs of ϕ_1 and ϕ_2.

For value at risk, $\varphi_2 = t > 0$ and $\varphi_1 = -1$, whereas for variance, $\varphi_2 = 2\sigma_p$ and $\varphi_1 = 0$. In general, one may wish to impose the restriction that $\varphi_2 > 0$ and $\varphi_1 \leq 0$, then $\mu^* \geq (\beta/\gamma)$ if $\sigma_p\Delta \geq \varphi_2\gamma$. Other cases can be elaborated. Multiple solutions may be possible, but we shall ignore these. Considering now the second derivatives,

$$\frac{\partial^2 v}{\partial\mu^2} = \varphi_{11} + \frac{2\varphi_{12}}{\sigma_p}\frac{(\mu\gamma - \beta)}{\Delta} + \frac{\varphi_{22}}{\sigma_p^2}\frac{(\mu\gamma - \beta)^2}{\Delta^2} + \frac{\varphi_2}{\sigma_p}\frac{\gamma}{\Delta} \tag{8.25}$$

For $v^2 v/\partial\mu^2 > 0$ as required, we need the matrix of second derivatives $\begin{pmatrix}\varphi_{11} & \varphi_{12}\\ \varphi_{21} & \varphi_{22}\end{pmatrix}$ to be positive definite and $\varphi_2 > 0$. This condition is satisfied for variance and for value at risk as in this case $\varphi_{12} = 0$ whilst $\varphi_2 = t$.

8.3 The case of two assets

We now consider the nature of the mean/semivariance frontier. When $N = 2$, some general expressions can be calculated for the frontier. In particular, for any distribution, $r_p = \omega r_1 + (1 - \omega)r_2$, and

$$\mu_p = \omega\mu_1 + (1 - \omega)\mu_2 \tag{8.26}$$

Two special cases arise: if $\mu_1 = \mu_2 = \mu$, then μ_p always equals μ, and the $\left(\mu_p, \theta_p^2\right)$ frontier is degenerate consisting of a single point. Otherwise, assume that $\mu_1 \neq \mu_2$ when Equation (8.26) can be solved for ω^*, so that

$$\omega^* = \frac{\mu_p - \mu_2}{\mu_1 - \mu_2} \tag{8.27}$$

If we assume, without loss of generality that $\mu_1 > \mu_2$, when there is no short selling allowed, then $\mu_1 \geq \mu_p \geq \mu_2$ for $0 \leq \omega^* \leq 1$. We shall concentrate on this part of the frontier, but could consider extensions for $\omega^* > 1$ or $\omega^* < 0$.

Now define $\theta_p^2(\tau)$ as the lower partial moment of degree 2 with truncation point τ. Then,

$$\theta_p^2(\tau) = \int_{-\infty}^{\tau} (\tau - r_p)^2 pdf(r_p) dr_p \tag{8.28}$$

However, an alternative representation in terms of the joint *pdf* of r_1 and r_2 is available. Namely,

$$\theta_p^2(\tau) = \int_{\Re} (\tau - r_p)^2 pdf(r_1, r_2) dr_1 dr_2 \tag{8.29}$$

where $\Re = \left\{ (r_1, r_2) \mid r_1\omega^* + r_2(1 - \omega^*) \leq \tau \right\}$.

We can now change variables from (r_1, r_2) to (r_1, r_p) by the (linear) transformations

$$r_p = \omega^* r_1 + (1 - \omega^*) r_2 \quad \text{and} \quad r_1 = r_1 \tag{8.30}$$

Therefore,

$$dr_1 dr_2 = \frac{1}{(1 - \omega^*)} dr_1 dr_p \quad \text{if} \quad 0 \leq \omega^* < 1 \tag{8.31}$$

If $\omega^* = 1$, then the transformation is $(r_1, r_2) \mapsto (r_2, r_p)$.

Now,

$$\theta_p^2(\tau) = \int_{-\infty}^{\infty} \int_{-\infty}^{\tau} \frac{(\tau - r_p)^2}{(1 - \omega^*)} pdf\left(r_1, \frac{r_p - \omega^* r_1}{1 - \omega^*} \right) dr_1 dr_p \tag{8.32}$$

This equation gives us the mean/semivariance locus for **any** joint *pdf*, i.e., *pdf*(r_1, r_2). As μ_p changes, ω^* changes and so does $\theta_p^2(\tau)$. In certain cases, i.e., ellipticity, we can compute explicitly *pdf*(r_p) and we can directly use Equation (8.28). In general, however, we have to resort to Equation (8.32) for our calculations.

In what follows, we present two results: First, for $N = 2$ under normality, where Equation (8.28) can be applied directly and a closed form solution derived. Second, assuming joint lognormality, where we use either Equation (8.28) or (8.32) and numerical methods.

We see, by manipulating the above, that if we can compute the marginal probability density function of r_p, the results will simplify considerably. For the case of normality, and for general elliptical distributions, the $pdf(r_p)$ is known. We proceed to compute the $\left(\mu_p, \theta_p^2\right)$ frontier under normality.

Proposition 8.2: Assuming that

$$\begin{pmatrix} r_1 \\ r_2 \end{pmatrix} \sim N\left[\begin{pmatrix} \mu_1 \\ \mu_2 \end{pmatrix}, \begin{pmatrix} \sigma_1^2 & \sigma_{12} \\ \sigma_{12} & \sigma_2^2 \end{pmatrix}\right] \tag{8.33}$$

the mean/semivariance frontier can be written as:

$$\theta_p^2 = \left(\sigma_p^2 + (t - \mu_p)^2\right) \quad \Phi\left(\frac{t - \mu_p}{\sigma_p}\right) + (t - \mu_p)\sigma_p \quad \phi\left(\frac{t - \mu_p}{\sigma_p}\right) \tag{8.34}$$

where $\phi(\cdot)$ and $\Phi(\cdot)$ are the standard normal density and distribution functions respectively, μ_p is given by Equation (8.26), and $\sigma_p^2 = \omega^2\sigma_1^2 + 2\omega(1 - \omega)\sigma_{12} + (1 - \omega)^2\sigma_2^2$. Moreover, if r_1 and r_2 are any two mean–variance efficient portfolios, the above result will hold for any $N > 2$.

Proof: Consider the integral (letting $\theta_p^2(t) = I(t)$),

$$I(t) = \int_{-\infty}^{t} (t - r_p)^2 pdf(r_p) dr_p \tag{8.35}$$

where $r_p \sim N\left(\mu_p, \sigma_p^2\right)$, so that

$$pdf(r_p) = \frac{1}{\sigma_p\sqrt{2\pi}} \exp\left(-\frac{(r_p - \mu_p)^2}{2\sigma_p^2}\right) \tag{8.36}$$

Transform $r_p \mapsto y = t - r_p \Rightarrow rp = t - y$ and $\left|dr_p\right| = |dy|$

$$I(t) = \int_0^{\infty} y^2 \frac{1}{\sigma_p\sqrt{2\pi}} \exp\left(-\frac{(y + t - \mu_p)^2}{2\sigma_p^2}\right) dy \tag{8.37}$$

$$I(t) = \frac{\partial^2}{\partial q^2}\left[\int_0^{\infty} \exp(qy) \frac{1}{\sigma_p\sqrt{2\pi}} \exp\left(-\frac{(y + t - \mu_p)^2}{2\sigma_p^2}\right) dy\right]_{q=0} \tag{8.38}$$

So, examine the integral in brackets:

$$J = \int_0^\infty \exp(qy) \frac{1}{\sigma_p \sqrt{2\pi}} \exp\left(-\frac{(y + t - \mu_p)^2}{2\sigma_p^2} \right) dy \tag{8.39}$$

and let $t - \mu_p = -\mu_t$, so

$$J = \int_0^\infty \frac{1}{\sigma_p \sqrt{2\pi}} \exp\left(-\frac{\left(y^2 - 2y\mu_t - 2\sigma_p^2 qy + \mu_t^2\right)}{2\sigma_p^2} \right) dy \tag{8.40}$$

$$J = \exp\left(-\frac{\mu_t^2}{2\sigma_p^2} \right) \exp\left(\frac{\left(\mu_t^2 + q\sigma_p^2\right)^2}{2\sigma_p^2} \right)$$
$$\int_0^\infty \frac{1}{\sigma_p \sqrt{2\pi}} \exp\left(-\frac{\left(y - \left(q\sigma_p^2 - \mu_t\right)\right)^2}{2\sigma_p^2} \right) dy \tag{8.41}$$

Transform $y \mapsto z = \dfrac{\left(y - \left(q\sigma_p^2 - \mu_t\right)\right)}{\sigma_p} \Rightarrow y = z\sigma_p + q\sigma_p^2 - \mu_t$, so

$$J = \exp\left(\frac{-2q\mu_t + q^2\sigma_p^2}{2} \right) \int_{-\left(q\sigma_p^2 - \mu_t\right)/\sigma_p}^\infty \frac{1}{\sqrt{2\pi}} \exp\left(-\frac{z^2}{2} \right) dz \tag{8.42}$$

$$J = \exp\left(\frac{-2q\mu_t + q^2\sigma_p^2}{2} \right) \left(1 - \Phi\left(-\frac{q\sigma_p^2 - \mu_t}{\sigma_p} \right) \right) \tag{8.43}$$

That is,

$$I(t) = \frac{\partial^2}{\partial q^2} \left[\exp\left(\frac{-2q\mu_t + q^2\sigma_p^2}{2} \right) \Phi\left(\frac{q\sigma_p^2 - \mu_t}{\sigma_p} \right) \right]_{q=0} \tag{8.44}$$

$$I(t) = \frac{\partial}{\partial q} \left[\left(q\sigma_p^2 - \mu_t\right) \exp\left(\frac{-2q\mu_t + q^2\sigma_p^2}{2} \right) \Phi\left(\frac{q\sigma_p^2 - \mu_t}{\sigma_p} \right) \right.$$
$$\left. + \sigma_p \exp\left(\frac{-2q\mu_t + q^2\sigma_p^2}{2} \right) \Phi'\left(\frac{q\sigma_p^2 - \mu_t}{\sigma_p} \right) \right]_{q=0} \tag{8.45}$$

$$I(t) = \frac{\partial}{\partial q}\left[\sigma_p^2 \exp\left(\frac{-2q\mu_t + q^2\sigma_p^2}{2}\right)\Phi\left(\frac{q\sigma_p^2 - \mu_t}{\sigma_p}\right)\right.$$
$$+ \left(q\sigma_p^2 - \mu_t\right)^2 \exp\left(\frac{-2q\mu_t + q^2\sigma_p^2}{2}\right)\Phi\left(\frac{q\sigma_p^2 - \mu_t}{\sigma_p}\right)$$
$$+ 2\sigma_p\left(q\sigma_p^2 - \mu_t\right)\exp\left(\frac{-2q\mu_t + q^2\sigma_p^2}{2}\right)\Phi'\left(\frac{q\sigma_p^2 - \mu_t}{\sigma_p}\right)$$
$$\left.+ \sigma_p^2 \exp\left(\frac{-2q\mu_t + q^2\sigma_p^2}{2}\right)\Phi''\left(\frac{q\sigma_p^2 - \mu_t}{\sigma_p}\right)\right]\Bigg|_{q=0} \tag{8.46}$$

$$I(t) = \sigma_p^2\Phi\left(-\frac{\mu_t}{\sigma_p}\right) + \mu_t^2\Phi\left(-\frac{\mu_t}{\sigma_p}\right) - 2\mu_t\sigma_p\Phi'\left(-\frac{\mu_t}{\sigma_p}\right) + \sigma_p^2\Phi''\left(-\frac{\mu_t}{\sigma_p}\right) \tag{8.47}$$

Now,

$$\Phi(x) = \int_{-\infty}^{x} \frac{1}{\sqrt{2\pi}} \exp\left(-\frac{z^2}{2}\right) dz \tag{8.48}$$

$$\Phi'(x) = \frac{1}{\sqrt{2\pi}} \exp\left(-\frac{x^2}{2}\right) = \phi(x) \tag{8.49}$$

and

$$\Phi''(x) = -x\frac{1}{\sqrt{2\pi}} \exp\left(-\frac{x^2}{2}\right) = -x\phi(x) \tag{8.50}$$

Thus,

$$I(t) = \left(\sigma_p^2 + \mu_t^2\right)\Phi\left(-\frac{\mu_t}{\sigma_p}\right) - 2\mu_t\sigma_p\phi\left(\frac{\mu_t}{\sigma_p}\right) + \sigma_p^2\frac{\mu_t}{\sigma_p}\phi\left(\frac{\mu_t}{\sigma_p}\right) \tag{8.51}$$

or

$$I(t) = \left(\sigma_p^2 + \mu_t^2\right)\Phi\left(-\frac{\mu_t}{\sigma_p}\right) - \mu_t\sigma_p\phi\left(\frac{\mu_t}{\sigma_p}\right) \tag{8.52}$$

Finally, if r_1 and r_2 are any two mean–variance efficient portfolios, then they span the set of minimum risk portfolios as described in Proposition 8.1, and thus the result follows.

Corollary 8.3: If we wish to consider expected loss under normality, which we denote by L where $L = E[r_p | r_p < t]$, then

$$L = (t - \mu_p)\Phi\left(\frac{t - \mu_p}{\sigma_p}\right) - \sigma_p\phi\left(\frac{t - \mu_p}{\sigma_p}\right) \tag{8.53}$$

Proof: The same argument as before.

The above results can be generalized to any elliptical distribution with a finite second moment since in all cases we know, at least in principle, the marginal distribution of any portfolio. Let $f(\cdot)$ and $F(\cdot)$ be the *pdf* and *cdf* of any portfolio return r_p, and let μ_p and σ_p be the relevant mean and scale parameters. Then, it is a consequence of ellipticity that a density (if it exists) $f(z)$ has the following property. For any (μ_p, σ_p) and $r_p = \mu_p + \sigma_p z$,

$$f(z) - \sigma_p pdf\left(\frac{r_p - \mu_p}{\sigma_p}\right) \tag{8.54}$$

Furthermore, there are incomplete moment distributions:

$$F^k(x) = \int_{\infty}^{x} z^k f(z)dz, F^0(x) = F(x) \tag{8.55}$$

for k a positive integer (existence of the kth moment is required for $F^k(x)$ to exist). So,

$$L = (t - \mu_p)F\left(\frac{t - \mu_p}{\sigma_p}\right) - \sigma_p F^1\left(\frac{t - \mu_p}{\sigma_p}\right) \tag{8.56}$$

and letting $\omega = (t - \mu_p)/\sigma_p$,

$$\begin{aligned} \text{Semivariance} = sv = \; & (t - \mu_p)^2 F(\omega) - 2(t - \mu_p)\sigma_p F^1(\omega) \\ & + \sigma_p^2 F^2(\omega) \end{aligned} \tag{8.57}$$

To illustrate the problems that arise when the *pdf* of r_p is not available in closed form, we consider the case of bivariate lognormality; as we see below, the previous simplifications no longer occur. Suppose that:

$$r_1 = \left(\frac{P_1}{P_0} - 1\right) \tag{8.58}$$

thus,

$$(1 + r_p) = \left(\frac{P_1}{P_0}\right) = \exp(y_1) \tag{8.59}$$

So that

$$(1 + r_p) = 1 + \sum \omega_i r_i = \omega \exp(y_1) + (1 - \omega)\exp(y_2) \tag{8.60}$$

and

$$r_p = \omega \exp(y_1) + (1 - \omega)\exp(y_2) - 1 \tag{8.61}$$

Therefore,

$$sv(r_p) = \int_{-\infty}^{t} (t - r_p)^2 pdf(r_p)dr_p \tag{8.62}$$

If

$$r_p < t, \text{ then } \omega \exp(y_1) + (1 - \omega)\exp(y_2) < 1 + t \tag{8.63}$$

The above transforms to a region $\Re$ in (y_1, y_2) space. Hence,

$$sv(r_p) = \int_{\Re} (1 + t - \omega \exp(y_1) - (1 - \omega)\exp(y_2))^2 pdf(y_1, y_2)dy_1 dy_2 \tag{8.64}$$

or

$$\begin{aligned} sv(r_p) = {} & \int_{\Re} (1 + t)^2 pdf(y_1, y_2)dy_1 dy_2 + c_1 \int_{\Re} \exp(y_1) pdf(y_1, y_2)dy_1 dy_2 \\ & + c_1 \int_{\Re} (\exp(y_1 + y_2))^2 pdf(y_1, y_2)dy_1 dy_2 \end{aligned} \tag{8.65}$$

where c_1 and c_2 are some constants. None of the above integrals can be computed in closed form, although they can be calculated by numerical methods.

8.4 Conic results

Using our definition of value at risk as $VaR_p = t\sigma_p - \mu_p$ with $t > 0$, and noting from Proposition 8.1 that σ_p^2 must lie on the minimum variance frontier so that

$$\sigma_p^2 = \frac{(\mu_p^2 \gamma - 2\beta\mu_p + \alpha)}{(\alpha\gamma - \beta^2)} \tag{8.66}$$

we see that

$$(VaR_p + \mu_p)^2 = \frac{(t^2)(\mu_p^2\gamma - 2\beta\mu_p + \alpha)}{(\alpha\gamma - \beta^2)} \tag{8.67}$$

Equation (8.67) is a general quadratic (conic) in μ_p and VaR_p and we can apply the methods of analytical geometry to understand what locus it describes.

The conic for v and u is:

$$v^2 + 2vu + u^2(1 - \gamma\theta) + 2\beta\theta u - \alpha\theta = 0 \tag{8.68}$$

where $\theta = t^2/(\alpha\gamma - \beta^2)$, $u = \mu_p$, $v = VaR_p$, and α, γ, and θ are always positive.

Following standard arguments, we can show that this conic must always be a hyperbola since $\gamma\theta > 0$ (see Brown and Manson (1959, p. 284)). Furthermore, the center of the conic in (u, v) space is $(\beta/\gamma, -\beta/\gamma)$ so that the center corresponds to the same expected return as the global minimum variance portfolio. The center of the hyperbola divides the (u, v) space into two regions.

We now consider implicit differentiation of Equation (8.68) for the region where $\mu \geq \beta/\gamma$, which corresponds to the relevant region for computing our frontier.

$$2v\frac{\partial v}{\partial u} + 2\frac{\partial v}{\partial u}u + 2v + 2u(1 - \gamma\theta) + 2\beta\theta = 0 \tag{8.69}$$

So,

$$\frac{\partial v}{\partial u} = \frac{u(\gamma\theta - 1) - v - \beta\theta}{u + v} \tag{8.70}$$

Thus,

$$v = u(\gamma\theta - 1) - \beta\theta \quad \text{when} \quad \frac{\partial v}{\partial u} = 0 \tag{8.71}$$

or,

$$u = \frac{(v + \beta\theta)}{(\gamma\theta - 1)} \tag{8.72}$$

Substituting into Equation (8.68),

$$v^2 + 2v\frac{(v + \beta\theta)}{(\gamma\theta - 1)} + \frac{(v + \beta\theta)^2}{(\gamma\theta - 1)^2}(1 - \gamma\theta) + 2\beta\theta\frac{(v + \beta\theta)}{(\gamma\theta - 1)} - \alpha\theta = 0 \tag{8.73}$$

Simplifying, and letting $\Delta = \alpha\gamma - \beta^2 > 0$,

$$\begin{aligned} & v^2(\gamma\theta - 1) + 2v^2 + 2v\beta\theta - v^2 - 2\beta\theta v - \beta^2\theta^2 + 2\beta\theta v \\ & \quad + 2\beta^2\theta^2 - \alpha\theta(\gamma\theta - 1) = 0 \end{aligned} \tag{8.74}$$

or,

$$v^2\gamma + 2v\beta + (\alpha - \Delta\theta) = 0 \tag{8.75}$$

So that

$$v = -\frac{\beta}{\gamma} \pm \frac{\sqrt{\beta^2 - \alpha\gamma + \Delta\theta\gamma}}{\gamma} \tag{8.76}$$

Since,

$$\beta^2 - \alpha\gamma + \Delta\theta\gamma \geq 0 \tag{8.77}$$

or

$$-\Delta + \Delta\theta\gamma \geq 0 \tag{8.78}$$

or

$$\Delta(\theta\gamma - 1) \geq 0 \tag{8.79}$$

so that

$$\theta\gamma \geq 1 \tag{8.80}$$

The solution for the upper part of the hyperbola is where $v > -\dfrac{\beta}{\gamma}$ and this corresponds to:

$$v = -\frac{\beta}{\gamma} + \frac{\sqrt{\beta^2 - \alpha\gamma + \Delta\theta\gamma}}{\gamma} \tag{8.81}$$

To check that this is a minimum, we compute:

$$u = \frac{\left(-\frac{\beta}{\gamma} + \frac{1}{\gamma}\sqrt{\Delta(\theta\gamma - 1)}\right) + \beta\theta}{(\gamma\theta - 1)} \tag{8.82}$$

$$u = \frac{\left(-\beta + \sqrt{\Delta(\theta\gamma - 1)}\right) + \beta\theta\gamma}{\gamma(\gamma\theta - 1)} \tag{8.83}$$

$$u = \frac{(\beta(\theta\gamma - 1) + \sqrt{\Delta(\theta\gamma - 1)})}{(\gamma\theta - 1)} \tag{8.84}$$

$$u = \frac{\beta}{\gamma} + \frac{1}{\gamma}\frac{\sqrt{\Delta(\theta\gamma - 1)}}{(\theta\gamma - 1)} \tag{8.85}$$

$$u = \frac{\beta}{\gamma} + \frac{1}{\gamma}\sqrt{\frac{\Delta}{(\theta\gamma - 1)}} \tag{8.86}$$

$$u > \frac{\beta}{\gamma} \tag{8.87}$$

Considering now the second-order conditions, first:

$$u + v = \frac{\beta}{\gamma} + \frac{1}{\gamma}\sqrt{\frac{\Delta}{(\theta\gamma - 1)}} - \frac{\beta}{\gamma} + \frac{1}{\gamma}\sqrt{\Delta(\theta\gamma - 1)} \tag{8.88}$$

$$u + v = \frac{1}{\gamma}\left[\sqrt{\frac{\Delta}{(\theta\gamma - 1)}} + \sqrt{\Delta(\theta\gamma - 1)}\right] > 0 \tag{8.89}$$

Differentiating Equation (8.69),

$$\left(\frac{\partial v}{\partial u}\right)^2 + v\frac{\partial^2 v}{\partial u^2} + u + v\frac{\partial^2 v}{\partial u^2} + \frac{\partial v}{\partial u} + \frac{\partial v}{\partial u} + (1 - \gamma\theta) = 0 \tag{8.90}$$

Since $\partial v/\partial u = 0$ at the minimum,

$$\frac{\partial^2 v}{\partial u^2} = \frac{(\gamma\theta - 1)}{(u + v)} \tag{8.91}$$

and since $\gamma\theta > 1$ and $(u + v) > 0$ at (u^*, v^*) from Equation (8.89), the minimum point is established.

We note that the condition $\gamma\theta > 1$ corresponds to the condition given in Proposition 1 of Alexander and Bapista (2001). By substituting Equation (8.69) back into Equation (8.73), we can recover the mean–variance portfolio that is the minimum value at risk portfolio as described by their Equation (10).

8.5 Simulation methodology

We consider simulation of portfolios ω of length ω_2^* subject to symmetric linear constraints, i.e., $\sum \omega_i = 1$ and $a \leq \omega_i \leq b$. We assume ω_1 is distributed as *uniform* $[a,b]$ and let ω_1^* be the sampled value. Then, ω_2 is also sampled from a *uniform* $[a,b]$ distribution with sampled value ω_2^*. The procedure is repeated sequentially as long as $\sum_{j=1}^{m} \omega_j^* \leq (N - m)a$. If a value ω_{m+1}^* is chosen such that $\sum_{j=1}^{m+1} \omega_j^* \geq (N - m - 1)a$, we set ω_{m+1}^* so that $\sum_{j=1}^{m+1} \omega_j^* + (N - m - 1)a = 1$. This is always feasible. Because the sequential sampling tends to make the weights selected early in the process larger, for any feasible ω we consider all $N!$ permutations of ω^*. From the symmetric constraints, these will also be feasible portfolios. With $N = 8$, $N! = 40{,}320$, so it may be feasible, but at $N = 12$, $N! = 479{,}001{,}600$, which may no longer be a feasible approach. We have to rely on random sampling in the first instance.

If we have a history of N stock returns for T periods, then for any vector of portfolio weights, we can calculate the portfolio returns for the T periods. These values can then be used to compute an estimate of the expected return μ_i, and the risk measure φ_i, which can be mapped into points on a diagram as in Figure 8.1. If we believe the data to be elliptically generated, then, following Proposition 8.2, we can save the weights of the set of minimum risk portfolios.

We amend an algorithm suggested by Bensalah (2000). If the risk–return value of individual portfolios can be computed, then an intuitive procedure is to draw the surface of many possible portfolios in a risk–return framework and then identify the optimal portfolio in a mean minimum risk sense. In the case of no short selling, an algorithm for approximating any frontier portfolios of N assets each with a history of T returns can be described as follows:

Step 1: Define the number of portfolios to be simulated as M.

Step 2: Randomize the order of the assets in the portfolio.

Step 3: Randomly generate the weight of the first asset $\omega_1 \sim U[0,1]$ from a Uniform distribution, $\omega_2 \sim U[0,1 - \omega_1]$, $\omega_3 \sim U[0,1 - \omega_1 - \omega_2]$, $\omega_N = 1 - \omega_1 - \omega_2 - \ldots - \omega_{(N-1)}$

Step 4: Generate a history of T returns for this portfolio, compute the average return and risk measure.

Step 5: Repeat Step 2 to Step 4, M times.

Step 6: From the M sets of risk–return measures, rank the returns in ascending order and allocate the returns and allocate each pair to B buckets equally spaced from the smallest to the largest return. Within each bucket, determine the portfolio with the minimum (or maximum as required) risk measure. Across the B buckets, these portfolios define the approximate risk–return frontier.

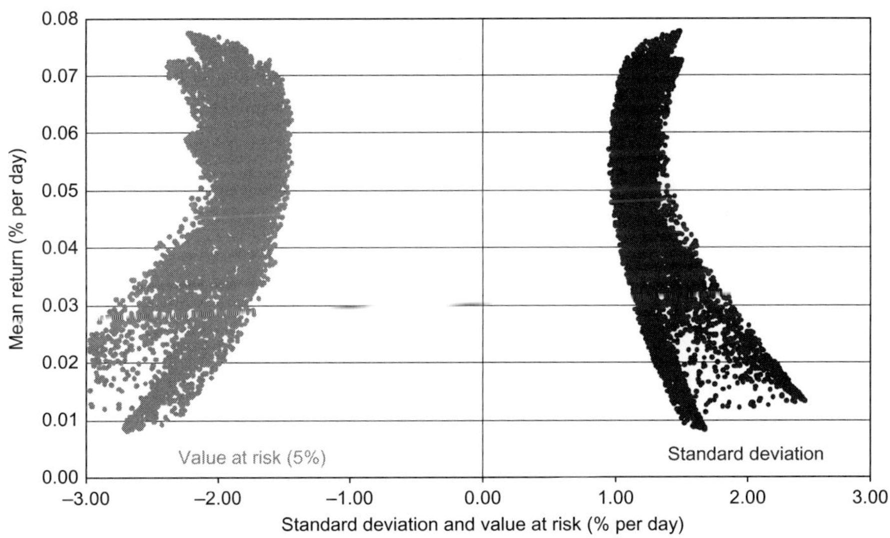

Figure 8.1 Illustration of mean/standard deviation and mean value at risk surfaces using 8,000 portfolios.

We illustrate the feasibility and accuracy of this algorithm using data from the Australian Stock Exchange. Daily returns were obtained for the trading days from 1 June 1995 to 30 August 2002 on the eight largest capitalization stocks giving a history of 1,836 returns. Summary statistics for the daily percentage return on these stocks are reported in Table 8.1. They display typical properties of stock returns, in particular, with all stocks displaying significant excess kurtosis. 128,000 random portfolios were generated and for each portfolio the average return, the sample standard deviation, the sample semi–standard deviation, the 5% value at risk, and the sample expected loss below zero were computed. Figure 8.1 illustrates the surface of the mean/standard deviation and mean/value at risk pairs obtained from 8,000 random portfolios. The frontiers for these risk measures are readily identified. From the 128,000 random portfolios, the approximated risk frontiers are presented in Figure 8.2 for

Table 8.1 Summary statistics for eight stocks

Rank	Stock	Average	St Dev	Min	Max	Skewness	Kurtosis
1	NAB	0.058	1.407	−13.871	4.999	−0.823	9.688
2	CBA	0.066	1.254	−7.131	7.435	−0.188	5.022
3	BHP	0.008	1.675	−7.617	7.843	0.107	4.205
4	ANZ	0.073	1.503	−7.064	9.195	−0.122	4.617
5	WBC	0.059	1.357	−6.397	5.123	−0.181	3.985
6	NCP	0.013	2.432	−14.891	24.573	0.564	11.321
7	RIO	0.035	1.659	−12.002	7.663	0.069	5.360
8	WOW	0.078	1.482	−8.392	11.483	0.025	6.846

Note: Stocks are ranked by market capitalization on 30 August 2002. The statistics are based on 1,836 daily returns, and expressed as percentage per day. Average is the sample mean; St Dev is the sample standard deviation; Min is the minimum observed return; Max is the maximum observed return; Skewness is the sample skewnees; and Kurtosis is the sample kurtosis.

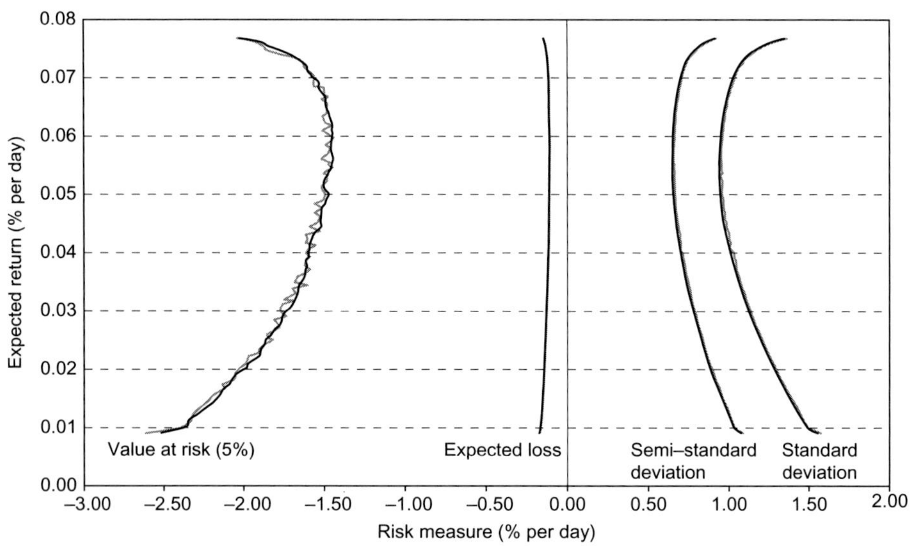

Figure 8.2 Risk frontiers obtained using quadratic programming and 128,000 simulated portfolios. Black line: frontiers using weights from solving minimum standard deviation frontiers with no short selling. Grey line: frontiers obtained using the simulated portfolios where portfolio weights are derived by minimizing standard deviation.

the mean/standard deviation, the mean/semi–standard deviation, the mean/value at risk (5%), and the mean/expected loss. The black lines represent the frontiers if it is assumed that the returns are elliptically distributed. In this case, a quadratic programming algorithm has solved for the optimal mean/standard deviation portfolio with no short selling for a number of target returns. The portfolio weights were then used to compute a history of returns to obtain the

Table 8.2 Summary statistics: distance between frontiers, eight assets, 128,000 simulated portfolios

		Standard deviation	Semi–standard deviation	Value at risk	Expected loss
			Panel A		
St Dev	Average	0.0105	0.0072	0.0175	0.0001
	St Dev	0.0049	0.0036	0.0284	0.0008
	Min	0.0004	0.0003	−0.0508	−0.0015
	Max	0.0242	0.0158	0.0955	0.0020
			Panel B		
Semi–St Dev	Average	0.0111	0.0069	0.0185	0.0000
	St Dev	0.0055	0.0035	0.0295	0.0008
	Min	0.0004	0.0003	−0.0508	−0.0015
	Max	0.0270	0.0166	0.0981	0.0023
			Panel C		
VaR	Average	0.0218	0.0148	−0.0106	0.0008
	St Dev	0.0123	0.0093	0.0187	0.0012
	Min	0.0001	0.0001	−0.0508	−0.0013
	Max	0.0614	0.0481	0.0817	0.0048
			Panel D		
Expected loss	Average	0.0125	0.0082	0.0122	−0.0003
	St Dev	0.0064	0.0050	0.0258	0.0007
	Min	0.0004	0.0003	−0.0480	−0.0015
	Max	0.0391	0.0276	0.0955	0.0014

Note: The units of measurement are percent per day. For standard deviation and semi–standard deviation, we measure the deviation as the simulated portfolio value minus the quadratic programming value. For value at risk and expected loss, we measure the quadratic programming value minus the simulated portfolio value. The statistics reported here are based on 100 points equally spaced along the respective frontiers from the minimum to the maximum observed sample portfolio returns.

other three risk measures. The grey lines represent the frontiers identified by the algorithm. In this case, we use the portfolios weights for the optimal portfolios in each bucket found by minimizing the sample standard deviation. The identified frontiers for the mean/standard deviation and mean/semi–standard deviation are very close to that identified by the quadratic programming algorithm. The two expected loss frontiers are superimposed and cannot be distinguished on this diagram. By contrast, the deviations between the two value at risk frontiers are relatively larger. The quality of the approximations for the 128,000 random portfolios is summarized in Table 8.2, where 100 buckets of portfolios have been used. In Panel A, we chose the optimal portfolio in each bucket as that portfolio with the smallest standard deviation. The metric here is returns measured in daily percentages. For the mean/standard deviation frontier, we measure the error as the distance of the approximation from the quadratic programming solution, and across 100 points on the frontier, the average error is 1.05 basis points with a standard deviation of 0.49 basis points and the largest error is 2.42 basis points. For the mean/semi–standard deviation frontier, the average error is 0.72 basis points with a standard deviation of 0.36 basis points. The expected loss frontiers almost coincide, with the average error 0.01 basis points with a standard deviation of 0.08 basis points. The value at risk frontiers show the largest discrepancies with an average error of 1.75 basis points, a standard deviation of 2.84 basis points and a range in values from −5.08 up to 9.55 basis points. Panels B, C, and D report similar measures when the optimal portfolios in each bucket are chosen by minimizing semi–standard deviation, maximizing value at risk (5%), and maximizing the associated expected loss. The results are qualitatively the same as in Panel A, but it is notable that the portfolios chosen using value at risk result in the largest deviations from the quadratic programming solutions.

This example does illustrate that it is feasible to approximate the frontiers for a variety of risk measures using this intuitive simulation methodology, but that in this case there is little to be gained over the frontiers identified from assuming that the returns are elliptically distributed.

8.6 Conclusion

In this chapter, we have presented analytical results that allow us to understand better what optimal mean–risk frontiers look like. For elliptical returns, these simplify to explicit formulae and we present closed form expressions for mean/value at risk frontiers under ellipticity and mean/expected loss and mean/semivariance frontiers under normality. For nonelliptical distributions, a simulation methodology is presented that can be applied easily to historical data. We do not consider the case of a riskless asset since this only has relevance when index-linked bonds are available. However, our results could be extended to this case.

References

Alexander, G. J., & Bapista, A. M. (2001). Economic implications of using a mean–VaR model for portfolio selection: A comparison with mean–variance analysis. *Journal of Economic Dynamics and Control*, *26*, 1159–1193.

Bensalah, Y. *Asset allocation using extreme value theory*. Bank of Canada Working Paper 2002-2.

Bradley, B. O., & Jaggia, M. S. (2002). Financial risk and heavy tails. In F. Comroy & S. T. Rader (Eds.), *Heavy-tailed distributions in finance*. North Holland.

Brown, J. T., & Manson, C. W. M. (1959). *The elements of analytical geometry*. New York: Macmillan.

Campbell, R., Huisman, R., & Koedijk, K. (2002). Optimal portfolio selection in a value at risk framework. *Journal of Banking and Finance*, *25*, 1789–1804.

Embrecht, P., McNeil, A. J., & Straumann, D. (2002). Correlation and dependence in risk management: Properties and pitfalls. In M. Dempster & H. K. Moffatt (Eds.), *Risk management: Value at risk and beyond*. Cambridge University Press.

Ingersoll, J. E. (1987). *Theory of financial decision making*. Totowa: Rowman & Littlefield.

Kaynar, B., Ilker Birbil, D., & Frenk, J. B. G. (2007). *Application of a general risk management model to portfolio optimization problems with elliptical distributed returns for risk neutral and risk averse decision makers*. Econometrics Institute report. Department of Economics, Erasmus University Rotterdam.

Kring, S., Rachev, S. T., Höchstötter, M., Fabozzi, F. J., & Bianchi, M. L. (2009). Multi-tail generalized elliptical distributions for asset returns. *Econometrics Journal*, 1–20.

Sargent, T. J. (1979). *Macroeconomic theory*. New York: Academic Press.

Wang, J. (2000). Mean–variance–VaR based portfolio optimization. Mimeo, Valdosta State University.

9 Portfolio optimization with "Threshold Accepting": a practical guide

Manfred Gilli and Enrico Schumann

Executive Summary

Recent years have seen a proliferation of new risk and performance measures in investment management. These measures take into account stylized facts of financial time series like fat tails or asymmetric return distributions. In practice, these measures are mostly used for ex post performance evaluation, only rarely for explicit portfolio optimization. One reason is that, other than in the case of classical mean–variance portfolio selection, the optimization under these new risk measures is more difficult since the resulting problems are often not convex and can thus not be solved with standard methods. We describe a simple but effective optimization technique called "Threshold Accepting (TA)," which is versatile enough to be applied to different objective functions and constraints, essentially without restrictions on their functional form. This technique is capable of optimizing portfolios under various recently proposed performance or (downside) risk measures, like value at risk, drawdown, Expected Shortfall, the Sortino ratio, or Omega, while not requiring any parametric assumptions for the data, i.e., the technique works directly on the empirical distribution function of portfolio returns. This chapter gives an introduction to TA and details how to move from a general description of the algorithm to a practical implementation for portfolio selection problems.

9.1 Introduction

The aim of portfolio selection is to determine combinations of assets like bonds or stocks that are "optimal" with respect to performance scores based, for instance, on capital gains, volatility, or drawdowns. For any decision rule, if it is supposed to be practical, a compromise needs to be found between the model's financial aspects, and statistical and (particularly) computational considerations. Thus, the question "what do I want to optimize?" has to be traded off against "what can I estimate?" and "what can I compute?"

The workhorse model for portfolio selection is mean–variance optimization (Markowitz, 1952). To a considerable extent, this specification is owed to

Doi:10.1016/B978-0-12-374952-9.00009-9.

computational restrictions. In fact, already in the 1950s Markowitz pondered using downside semivariance as a measure for risk—but rejected it on mainly computational grounds. Today still, standard optimization methods like linear or quadratic programming allow to solve portfolio selection problems only under simplifying assumptions. So altogether, the third question is actually a very strong limit to the implementation and testing of alternative, possibly superior models.

To give just one example (which we do not claim to be a superior model), assume we wish to minimize the value at risk (VaR) of an equity portfolio. Figure 9.1 shows the search space of such a problem, which is a direct mapping from different weight combinations to the portfolio's VaR.

As can be seen, there are many local minima. Hence, standard methods, which usually use gradient information, cannot be applied here, since they will stop at the first local minimum.

There are different approaches to deal with such problems. Sometimes we can reformulate the model until it can be solved with standard methods, in particular linear programs; for example see Rockafellar and Uryasev (2000), Chekhlov, Uryasev, and Zabarankin (2005), or Mausser, Saunders, and Seco (2006). In such cases, powerful solvers are available that can efficiently handle even large-scale instances of given problems. But there is also a cost, namely that the formulation is very problem-specific and often relies on simplifications or specific assumptions. Passow (2005) for instance shows how to optimize

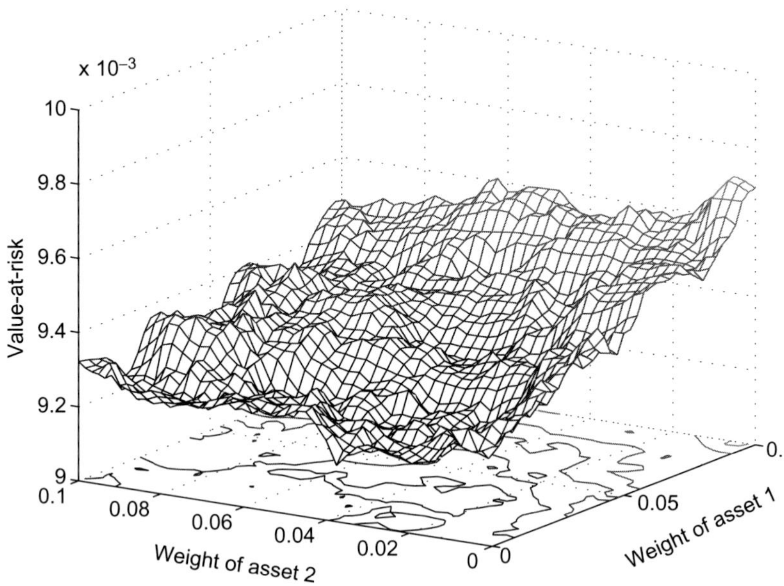

Figure 9.1 Search space for VaR.

portfolios with the Omega ratio (Keating & Shadwick, 2002), but this is only possible by imposing a specific distributional assumption on the data.

In this chapter, we will outline an alternative approach, namely optimization with a heuristic technique. The term "heuristic" is used in different scientific disciplines, with different but related meanings. Mathematicians use it to describe an explanation that is strictly speaking not correct, but leads to the correct conclusion nonetheless; in the language of psychologists, heuristics are "rules of thumb" for decision making that, though sometimes seemingly crude, work robustly in many circumstances (Gigerenzer, 2004, 2008). In optimization theory, the term is characterized by Winker and Maringer (2007), Barr, Golden, Kelly, Resende, and Stewart (1995) via several criteria:

- The method should produce "good" stochastic approximations of the true optimum, where "good" is measured in terms of solution quality and computing time.
- The method should be robust in case of comparatively small changes to the given problem (e.g., when a constraint is added). Furthermore, obtained results should not vary too much for changes in the parameter settings of the heuristic.
- The technique should be easy to implement.
- Implementation and application of the technique should not require subjective elements.

It must be stressed that heuristics, following this definition, are not *ad hoc* procedures that are tailored to specific problems. For many techniques like Genetic Algorithms (Holland, 1992) or Simulated Annealing (Kirkpatrick, Gelatt, & Vecchi, 1983), a considerable theoretical background is available, including convergence proofs. Heuristics have been shown to work well for problems that are completely infeasible for classical optimization approaches (Michalewicz & Fogel, 2004). Conceptually, they are often very simple; implementing them rarely requires high levels of mathematical sophistication or programming skills. Heuristics are flexible as adding, removing, or changing constraints or exchanging objective functions can be accomplished very easily.

These advantages come at a cost as well, as the obtained solution is only a stochastic approximation, a random variable. But still, such an approximation may be better than a poor deterministic solution (which, even worse, may not even become recognized as such) or no solution at all when other methods cannot be applied.

In this chapter, we will outline how to implement such a heuristic technique, Threshold Accepting (TA), for a portfolio optimization problem. TA was introduced in the 1980s (Dueck & Scheuer, 1990) and was, incidentally, the first heuristic method to be applied to non-mean–variance portfolio selection (Dueck & Winker, 1992). The reader we have in mind is primarily the analyst or engineer who wishes to implement the algorithms, hence all relevant procedures are given in pseudocode. But this chapter should also be of interest to the portfolio manager who uses the resulting implementation, since he or she should be aware of the possibilities and limitations of the method. The chapter is structured as follows: Section 9.2 will briefly outline the portfolio selection problem.

Section 9.3 will detail the optimization technique. Two aspects relevant for practical applications are the stochastic nature of the solutions obtained from TA, and of course, some diagnostics for the algorithm. These issues are discussed in Sections 9.4 and 9.5, respectively. Section 9.6 concludes.

9.2 Portfolio optimization problems

We assume that there are n_A risky assets available, with current prices collected in a vector p_0. We are endowed with an initial wealth v_0, and intend to select a portfolio x of the given assets. We can thus write down a budget constraint:

$$v_0 = x' p_0 \tag{9.1}$$

The vector x stores the number of shares or contracts, i.e., integer numbers. If we need portfolio weights, we divide both sides of Equation (9.1) by v_0.

The chosen portfolio is held for one period, from now (time 0) to time T. End-of-period wealth is given by:

$$v_T = x' p_T$$

where the vector p_T holds the asset prices at T. Since these prices are not known at the time when the portfolio is formed, v_T will be a random variable, following some unknown distribution. It is often convenient to rescale v_T to a return r_T, i.e.:

$$r_T = \frac{v_T}{v_0} - 1$$

9.2.1 *Risk and reward*

The first step in an optimization model is to characterize a given portfolio in terms of its desirability, thus we want to map a given portfolio into a real number. This mapping is called an objective function, or selection criterion. The great insight of Markowitz was that such a function cannot consist of just the expected profits. A common approach is hence to define two properties, "reward" and "risk," that describe the portfolio, and to trade them off against each other. Markowitz identified these properties with the mean and variance of returns, respectively.

With our optimization technique, we are not bound to this choice. In our exposition here, our objective functions will be general ratios of risk and reward to be minimized. (We could equivalently maximize reward–risk ratios.) Ratios have the advantage of being easy to communicate and interpret (Stoyanov, Rachev, & Fabozzi, 2007). Even though, numerically, linear combinations are often more stable and thus preferable, working with ratios practically never caused problems in our studies. See also Section 9.5.3.

Many special cases for such ratios have been proposed in the literature, the best known certainly being the Sharpe ratio (Sharpe, 1966). For this ratio, reward is the mean portfolio return and risk is the standard deviation of portfolio returns, so it corresponds closely to mean–variance optimization. Rachev, Fabozzi, and Menn (2005) give an overview of various alternative ratios that have been proposed in the academic literature; further possible specifications come from financial advisors, in particular from the hedge fund and commodity trading advisor field (Bacon, 2008, Chapter 4).

In general, these ratios can be decomposed into "building blocks"; we will discuss some examples now.

Partial moments

For any portfolio return r_t, the equation

$$r_t = \underbrace{r_d}_{\text{desired return}} + \underbrace{\max(r_t - r_d, 0)}_{\text{upside}} - \underbrace{\max(r_d - r_t, 0)}_{\text{downside}} \tag{9.2}$$

always holds for any desired-return threshold r_d (Scherer, 2004). Partial moments are a convenient way to distinguish between returns above and below r_d, i.e., the "upside" and "downside" terms in Equation (9.2), and thus to capture potential asymmetry around this threshold. For a sample of portfolio returns $r = [r_1\ r_2\ \ldots\ r_{n_S}]'$ with n_S observations, the partial moments $\mathcal{P}_\gamma^{(\cdot)}(r_d)$ can be estimated as:

$$\mathcal{P}_\gamma^+(r_d) = \frac{1}{n_S} \sum_{r>r_d} (r - r_d)^\gamma \tag{9.3a}$$

$$\mathcal{P}_\gamma^-(r_d) = \frac{1}{n_S} \sum_{r<r_d} (r_d - r)^\gamma \tag{9.3b}$$

The superscripts + and − indicate the tail (i.e., upside and downside). Partial moments take two more parameters: an exponent γ and the threshold r_d. The expression "$r > r_d$" indicates that we sum only over those returns that are greater than r_d.

Conditional moments

Conditional moments can be estimated by:

$$\mathcal{C}_\gamma^+(r_d) = \frac{1}{\#\{r > r_d\}} \sum_{r>r_d} (r - r_d)^\gamma \tag{9.4a}$$

$$\mathcal{C}_\gamma^-(r_d) = \frac{1}{\#\{r < r_d\}} \sum_{r<r_d} (r_d - r)^\gamma \tag{9.4b}$$

where again + and − indicate the tail, and "$\#\{r > r_d\}$" is a counter for the number of returns higher than r_d.

Conditional and partial moments are closely related. For a fixed threshold r_d, the lower partial moment of order γ equals the lower tail's conditional moment of the same order times the lower partial moment of order 0. That is,

$$\mathcal{P}_\gamma^+(r_d) = \mathcal{C}_\gamma^+(r_d)\mathcal{P}_0^+(r_d)$$

$$\mathcal{P}_\gamma^-(r_d) = \mathcal{C}_\gamma^-(r_d)\mathcal{P}_0^-(r_d)$$

The partial moment of order 0 is simply the probability of obtaining a return beyond r_d. Still, for a given r_d, both conditional and partial moments convey different information, since both the probability and the conditional moment need to be estimated from the data to obtain a partial moment. In other words, the conditional moment measures the magnitude of returns around r_d, while the partial moment also takes into account the probability of such returns.

Both partial and conditional moments, in our definitions, are centered around r_d. In the finance literature, r_d is often regarded as exogenously fixed, for instance at the risk-free rate or at some minimal acceptable return. In this case, we can directly work with Equations (9.3) and (9.4). An alternative way to set r_d, often chosen in risk management applications, is to equate r_d to some quantile of the return distribution. This is the usual convention for conditional moments like Expected Shortfall, as they can then be compared with the corresponding VaR values. But now we must not optimize with equations like (9.4) anymore: Fixing a quantile says nothing about the location of the return distribution that our optimization algorithm selects, and we may end up with a dominated distribution (see Figure 9.2).

The simplest remedy is not to center around r_d, and modify Equation (9.4) to:

$$\mathcal{C}_\gamma^+(r_d) = \frac{1}{\#\{r > r_d\}} \sum_{r>r_d} r^\gamma \tag{9.5a}$$

$$\mathcal{C}_\gamma^-(r_d) = \frac{1}{\#\{r < r_d\}} \sum_{r<r_d} r^\gamma \tag{9.5b}$$

But now without centering, we have no more guaranty for the sign of r. Hence, for $\gamma \neq 1$, we replace r by $\max(r, 0)$ in Equation (9.5a), and by $\min(r, 0)$ in Equation (9.5b).

When r_d is fixed

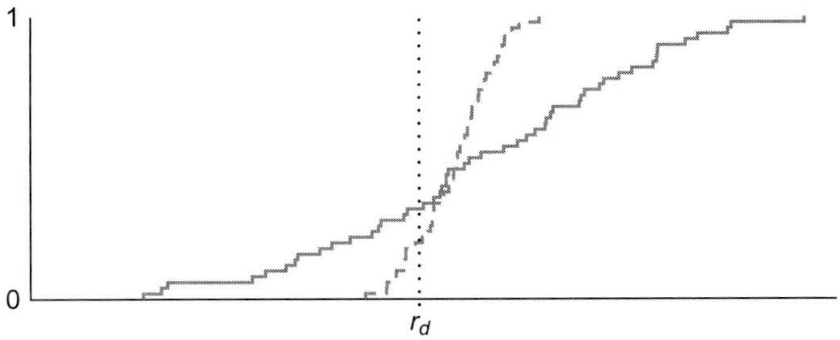

The illustration pictures the return distributions of two portfolios. If we aim to find a portfolio that minimizes a function of the returns below r_d, there should be little debate about which portfolio to choose.

When r_d equals some quantile

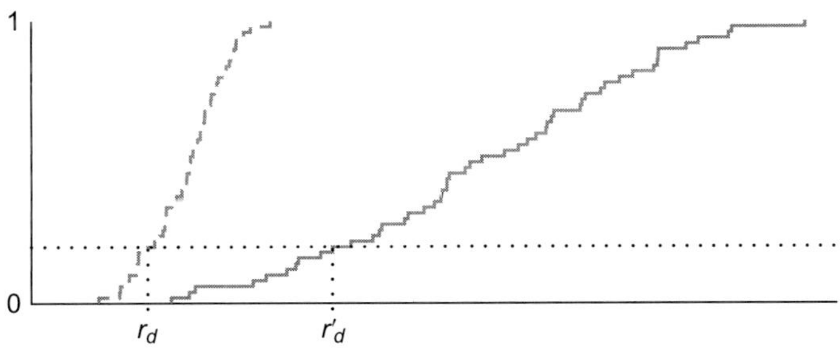

We would certainly prefer the more variable distribution on the right; a decision rule based on a moment centered around r_d would, however, choose the dominated distribution on the left.

Figure 9.2 Setting the desired return r_d.

Quantiles

A quantile of the cumulative distribution function (CDF) of a sample r is defined as:

$$\mathcal{Q}_q = \text{CDF}^{-1}(q) = \min\{r \mid \text{CDF}(r) \geq q\}$$

where q may range from 0% to 100% (we drop the % sign in subscripts). So, in words, the qth quantile is a number $\mathcal{Q}_q$ such that q of the observations are smaller, and $(100\% - q)$ are larger than $\mathcal{Q}_q$. When we estimate quantiles, several numbers may satisfy this definition (Hyndman & Fan, 1996). A simple estimator is the order statistics of portfolio returns $[r_{[1]}\ r_{[2]} \cdots r_{[n_S]}]'$, i.e., a step function. If k is the smallest integer not less than qn_S, then the qth quantile is the $\max(k, 1)$th order statistic. This is consistent with the convention in many statistical packages like Matlab or R that $\mathcal{Q}_0$ is the minimum of the sample (an estimator for the worst-case return), and $\mathcal{Q}_{100}$ is the maximum. Of course, for quantiles in the tails this requires that we have a reasonably large number of observations.

VaR is the loss (negative return) only to be exceeded with a given, usually small, probability at the end of the investment period. Thus, VaR is a quantile of the distribution of returns for this period. In our notation, VaR for a probability of 1% can be written as $\mathcal{Q}_1$. Quantiles may also be used as reward measures; we could, for example, maximize a higher quantile (e.g., the 90th).

Drawdown

The functions described so far are functions of the distribution of final wealth, but we may also observe the time path of portfolio wealth. Let v be a time series of portfolio values, with observations at $t = 0, 1, 2, \ldots, T$. Then, the drawdown $\mathcal{D}_t$ of this series at time t is defined as:

$$\mathcal{D}_t = v_t^{\max} - v_t \tag{9.6}$$

where $v_t^{\max}$ is the running maximum, i.e., $v_t^{\max} = \max\{v_s \mid s \in [0, t]\}$.

The symbol $\mathcal{D}$ stands for the whole vector of length $T + 1$; subscripts indicate a scalar value, for instance the drawdown's maximum, or the drawdown at a particular point in time. Other functions may be computed to capture the information in the drawdown vector, e.g., the mean time underwater (i.e., the average time elapsed between two consecutive values in $\mathcal{D}$ that are sufficiently close to 0), or the correlation between a portfolio's drawdown and the drawdown of an alternative asset like an index.

Typical examples used in risk–reward ratios may be the drawdown's mean, its maximum, or its standard deviation, which can be computed by:

$$\begin{aligned}
\mathcal{D}_{\text{mean}} &= \frac{1}{T}\sum_{t=1}^{t=T}\mathcal{D}_t \\
\mathcal{D}_{\max} &= \max(\mathcal{D}) \\
\mathcal{D}_{\text{std}} &= \sqrt{\frac{1}{T-1}\sum_{t=1}^{t=T}(\mathcal{D}_t - \mathcal{D}_{\text{mean}})^2}
\end{aligned}$$

respectively. The definition in Equation (9.6) gives $\mathcal{D}$ in currency terms (i.e., the absolute drawdown). Usually, a relative drawdown is preferred, obtained by using the logarithm of v or by dividing every $\mathcal{D}_t$ by $v_t^{\max}$.

9.2.2 The problem summarized

Our optimization problem can be summarized as follows: let $x = [x_1\ x_2\ \dots\ x_{n_A}]'$ be the holdings of the individual assets, and $\mathcal{J}$ be the set of assets in the portfolio, i.e.:

$$\mathcal{J} - \{i | x_i \neq 0\}$$

The objective function Φ will be a ratio of reward and risk. If we just want to minimize some risk function, we replace the reward by a constant; for solely maximizing reward, we set the risk to a constant. The following table gives some examples.

Reward	Risk	
Constant	$\mathcal{C}_1^-(\mathcal{Q}_q)$	Minimize Expected Shortfall for qth quantile
Constant	$-\mathcal{Q}_0$	Minimize maximum loss
$\mathcal{P}_1^+(r_d)$	$\sqrt{\mathcal{P}_2^-(r_d)}$	Upside potential ratio (Sortino, van der Meer, & Plantinga, 1999)
$\mathcal{P}_1^+(r_d)$	$\mathcal{P}_1^-(r_d)$	Omega (Keating & Shadwick, 2002) for threshold r_d
$\frac{1}{n_S}\sum r$	$\mathcal{D}_{\max}$	Calmar ratio (Young, 1991)
$C_\gamma^+(\mathcal{Q}_p)$	$C_\delta^-(\mathcal{Q}_q)$	Rachev generalized ratio for exponents γ and δ (Biglova, Ortobelli, Rachev, & Stoyanov, 2004)

Now the problem, including constraints, can be written as:

$$\min_x \Phi$$

$$x_j^{\inf} \leq x_j \leq x_j^{\sup} \quad \text{for } j \in \mathcal{J}$$

$$K_{\inf} \leq \#\{\mathcal{J}\} \leq K_{\sup}$$

where $x_j^{\inf}$ and $x_j^{\sup}$ are minimum and maximum holding sizes, respectively, for those assets included in the portfolio (i.e., those in $\mathcal{J}$). $K_{\inf}$ and $K_{\sup}$ are cardinality constraints that set a minimum and maximum number of assets in $\mathcal{J}$. We do not include minimum return constraints. Note that we will always minimize, which is not restrictive since it is equivalent to maximizing $-\Phi$.

9.3 Threshold accepting

9.3.1 The algorithm

TA is a local search method. A local search starts with a random feasible solution x^c (i.e., a random portfolio), which we call the "current solution," as it represents the best we have so far. Then again randomly, a new solution x^n close to x^c is chosen. What "close to x^c" means will be discussed shortly. This new solution is called a neighbor solution. If x^n is better than x^c, the new solution is accepted (i.e., the neighbor becomes the current solution), if not, it is rejected, and another neighbor is selected. For a given objective function, a local search is completely described by how it chooses a neighbor solution and by its stopping criterion (i.e., the rule that states when the search is finished). Algorithm 1 gives this procedure in pseudocode. The stopping criterion here is a preset number of iterations n_{Steps}; the final x^c becomes our overall solution x^{sol}. The symbol $\mathcal{X}$ stands for the set of all feasible portfolios.

Algorithm 1 Local Search.

initialize n_{Steps}
randomly generate current solution $x^c \in \mathcal{X}$
for $i = 1 : n_{\text{Steps}}$ **do**
    generate $x^n \in \mathcal{N}(x^c)$ and compute $\Delta = \Phi(x^n) - \Phi(x^c)$
    if $\Delta < 0$ **then** $x^c = x^n$
end for
$x^{\text{sol}} = x^c$

In a convex problem, a local search will, given an appropriate neighborhood function and enough iterations, succeed in finding the global minimum, even though it is certainly not the most efficient method. But we are compensated for this inefficiency: all the method requires is that the objective function can be evaluated for a given portfolio x; there is no need for the objective function to be continuous or differentiable. Unfortunately, for problems with many local minima, a local search will stop at the first local optimum it finds.

TA now builds on this approach, but it adds a simple strategy to escape such local minima: It will not only accept a new solution that improves on the current one, but it will also allow "uphill moves," as long as the deterioration of Φ does not exceed a fixed threshold (hence the method's name). Over time, this threshold decreases until eventually TA turns into a classical local search. The whole procedure is summarized in Algorithm 2.

Algorithm 2 Threshold Accepting.

1: initialize n_{Steps} and n_{Rounds}
2: compute threshold sequence τ
3: randomly generate current solution $x^{c} \in \mathcal{X}$
4: **for** $r = 1 : n_{\text{Rounds}}$ **do**
5:     **for** $i = 1 : n_{\text{Steps}}$ **do**
6:         generate $x^{n} \in \mathcal{N}(x^{c})$ and compute $\Delta = \Phi(x^{n}) - \Phi(x^{c})$
7:         **if** $\Delta < \tau_r$ **then** $x^{c} = x^{n}$
8:     **end for**
9: **end for**
10: $x^{\text{sol}} = x^{c}$

Compared with local search, the changes are actually small. We add an outer loop that controls the thresholds τ. In Statement 7, the acceptance criterion is changed from "$\Delta < 0$" (i.e., improvement) to "$\Delta < \tau_r$" (i.e., change not worse than τ_r). For an actual implementation, we need to discuss the objective function Φ, the neighborhood function $\mathcal{N}$, the thresholds τ, and the stopping criterion in more detail.

9.3.2 Implementation

The objective function

Conceptually, the objective function Φ is given by the problem at hand. In our case, Φ will be a ratio of a risk and a reward function, for instance the ratio of downside semivariance to upside semivariance (i.e., $\mathcal{P}_2^- / \mathcal{P}_2^+$), to be minimized.

In the standard mean–variance model, we estimate the properties of single assets (means, variances, and correlations), and then aggregate them on a portfolio level. Thus, we capture the relevant information about the distribution of assets independently from the specific portfolio chosen. This approach does often not generalize to other specifications of risk and reward, or only in ways that are computationally very costly (see for instance Jondeau, Poon, and Rockinger (2007, chapter 9) for how to set up skewness or kurtosis matrices). Fortunately, our optimization procedure does not require such an aggregation, and we will instead always work with a return sample $r = [r_1\ r_2\ \dots\ r_{n_s}]'$ for a specific portfolio. We will not assume a parametric distribution for the returns and rather work with the CDF of this sample. Thus, we will conduct a scenario optimization. The simplest case is to regard every historical return observation as one scenario. We often found, however, that modeling the data (i.e., creating "artifical" scenarios) improved the performance of portfolios in the out-of-sample period (Gilli & Schumann, 2009a; Gilli, Schumann, di Tollo, & Cabej, 2008). Since we discuss the optimization here, we will assume that we have a scenario set, be it historical data, expert forecasts, or observations obtained from some resampling procedure.

For each of our n_A assets, we assume n_S return scenarios, all collected in a matrix R of size $n_S \times n_A$. One row of this matrix thus stores the returns for one state of nature. We can equivalently work with price scenarios, computed as:

$$P = (1 + R) \times \text{diag}(p_0)$$

where 1 is a matrix of ones of size $n_S \times n_A$ and "diag" is an operator that transforms a vector into a diagonal matrix. Note that the columns of P here are not time series. Every row of P holds the prices for one possible future scenario that might occur, given initial prices of p_0. In fact, for many objective functions (like partial moments), it is not relevant whether the scenarios are sorted in time, since such criteria only capture the cross section of returns. The portfolio values in these scenarios can be obtained by $v = Px$. Equivalently, we can compute returns $r = Rw$.

For selection criteria that need a path of portfolio wealth (like drawdowns), we need to work with time series. Resampling is still possible: We may, for instance, estimate a model that captures serial dependencies (like a GARCH model) and then resample from its residuals, or use a block bootstrap method. For simplicity, assume we work with historical data, and arrange the prices in a matrix P^{hist} of size $(T + 1) \times n_A$, where each column holds the historical prices of one asset. The portfolio values over time are then computed by $v = P^{\text{hist}}x$.

Let us stress the difference between Px and $P^{\text{hist}}x$ here: Px gives a sample of portfolio values over the cross section of scenarios, while $P^{\text{hist}}x$ gives one path of the portfolio value from time 0 to time T. For both scenarios and paths, given a vector v, any objective function constructed from the building blocks in Section 9.2 can easily be computed.

Like many other heuristics, TA is computationally intensive, with the main part of running time spent on evaluating the objective function. It therefore often pays (in terms of reduced computing time) to analyze and profile the objective function extensively. In some cases, the objective function can be updated, so that certain results from prior iterations can be reused. Let us give a concrete example. Assume we work with the matrix P of price scenarios (the same holds for P^{hist}). In practice, this matrix is often fairly large, storing thousands of scenarios for hundreds or thousands of assets. The algorithm started with an initial random portfolio x^{c} and now has to evaluate x^{n}. This means that the product Px^{c} has already been computed. As will be discussed below, a new portfolio will be created by a small perturbation of the original portfolio, hence:

$$x^{\mathrm{n}} = x^{\mathrm{c}} + x^{\Delta}$$

where x^{Δ} is a vector with few nonzero elements (usually only two). Then,

$$Px^{\mathrm{n}} = P(x^{\mathrm{c}} + x^{\Delta}) = \underbrace{Px^{\mathrm{c}}}_{\text{known}} + Px^{\Delta}$$

Since many elements of x^{Δ} are zero, the part of P relevant for the matrix multiplication consists of only a few columns. Thus, by creating a matrix P_* that only holds the columns where x^{Δ} is nonzero, and a vector x_*^{Δ} that comprises only the nonzero elements of x^{Δ}, we can often considerably speed up the matrix computation Px^{Δ} by replacing it by $P_* x_*^{\Delta}$.

The neighborhood function

To move from one solution to the next, we need to define a neighborhood $\mathcal{N}$ from which new candidate solutions are chosen. For portfolio selection problems, there exists a very natural way to create neighbor solutions: Pick one asset in the portfolio randomly, "sell" a small quantity of this asset, and "invest" the amount obtained in another asset. If short positions are allowed, the chosen asset to be sold does not have to be in the portfolio. The "small quantity" may either be a random number or a small fixed fraction (e.g., 0.1%). Experiments suggest that, for practical purposes, both methods give similar results.

Note that the neighborhood definition stays unchanged over the course of the optimization. A variant is to decrease the portfolio changes over time, thus to make smaller steps. In our experience, this is not helpful for portfolio selection problems. It is true that we often want the algorithm to become more select over time, but this can be achieved by setting the threshold sequence appropriately.

The threshold sequence

The threshold sequence is an ordered vector of positive numbers that decrease to zero or at least become very small. The neighborhood definition and the thresholds are strongly connected. Larger neighborhoods, which imply larger changes from one candidate portfolio to the next, need generally be accompanied by larger initial threshold values, and vice versa.

Winker and Fang (1997) were the first to suggest a data-driven method to obtain the thresholds: generate a large number of random solutions and select a neighbor for every solution. All these solutions are then evaluated according to the objective function, so for every pair (random solution–neighbor solution), we obtain a difference in the objective function value. The thresholds are then a number of decreasing quantiles of these changes. The procedure is summarized in Algorithm 3.

Algorithm 3 Computing the threshold sequence—Variant 1.

initialize n_{Rounds} (# of thresholds), n_{Deltas} (# of random solutions)
for $i = 1$ to n_{Deltas} **do**
    randomly generate current solution $x^{c} \in \mathcal{X}$
    generate $x^{n} \in \mathcal{N}(x^{c})$
    compute $\Delta_i = |\Phi(x^{n}) - \Phi(x^{c})|$
end for
sort $\Delta_1 \leq \Delta_2 \leq \cdots \leq \Delta_{n_{\text{Deltas}}}$
set $\tau = \Delta_{n_{\text{Rounds}}}, \ldots, \Delta_1$

The number of thresholds n_{Rounds} with this approach is usually large, hence the number of steps n_{Steps} per threshold (in the inner loop of Algorithm 2) can be low; often it is only one step per threshold.

A variation of this method, described in Gilli, Këllezi, and Hysi (2006) and stated in Algorithm 4, is to take a random walk through the data where the steps are made according to the neighborhood definition. At every iteration, the changes in the objective function value are recorded. The thresholds are then a number of decreasing quantiles of these changes.

Algorithm 4 Computing the threshold sequence—Variant 2.

initialize n_{Rounds} (# of thresholds), n_{Deltas} (# of random steps)
randomly generate current solution $x^{c} \in \mathcal{X}$
for $i = 1 : n_{\text{Deltas}}$ **do**
    generate $x^{n} \in \mathcal{N}(x^{c})$ and compute $\Delta_i = |\Phi(x^{n}) - \Phi(x^{c})|$
    $x^{c} = x^{n}$
end for
compute empirical distribution CDF of Δ_i, $i = 1, \ldots, n_{\text{Deltas}}$
compute threshold sequence $\tau_r = \text{CDF}^{-1}\left(\frac{n_{\text{Rounds}} - r}{n_{\text{Rounds}}}\right)$, $r = 1, \ldots, n_{\text{Rounds}}$

Many variations are possible. For example, Algorithm 4 uses equidistant quantiles, so for $n_{\text{Rounds}} = 5$, the 80th, 60th, 40th, 20th, and 0th quantiles are used. The convention in software packages like Matlab or *R* is to set the 0th quantile equal to the minimum of the sample, hence the last threshold is not necessarily 0. There is some evidence that the efficiency of the algorithm can be increased by starting with a lower quantile (e.g., the 50th), but generally, TA is very robust to different settings of these parameters.

For this second variant, Gilli and Schumann (2008) study the quality of solutions obtained in portfolio optimization problems for different numbers of thresholds for a fixed $n_{\text{Rounds}} \times n_{\text{Steps}}$. They find that the performance of the algorithm deteriorates for very small numbers of thresholds (e.g., 2 or 3), but stays roughly the same beyond 10 thresholds.

Constraints

There are several generic approaches to include constraints into the optimization. A first one is to create new solutions such that they conform with the given constraints. The budget constraint, for example, is automatically enforced by the specification of the neighborhood. Cardinality constraints can be implemented in this way as well. An alternative technique is to implement restrictions by penalty terms. If a constraint is violated, the objective function is impaired by an amount that increases with the magnitude of the violation. The penalty term often also increases over time, so to allow the algorithm to move relatively freely initially.

When penalizing the objective function to enforce constraints, the computational architecture needs hardly be changed, since we only add a scalar to the objective function. A further advantage is that the approach works very well if the search space is not convex, or even disconnected. Finally, penalties also allow to incorporate soft constraints, which can sometimes be more appropriate than hard constraints.

The stopping criterion

A stopping criterion is introduced by setting a fixed number of iterations, i.e., by the product $n_{\text{Rounds}} \times n_{\text{Steps}}$. An emergency brake may be included that stops the search if there has been no improvement in the current portfolio for a large number of iterations.

9.4 Stochastics

For a convex problem, repeatedly running a deterministic method like a linear program will always result in the same solution. But now assume we have a problem like minimizing a portfolio's VaR (see Figure 9.1), so Φ is given by $\mathcal{Q}_q/1$ (note that the reward in this ratio is a constant, hence we just minimize risk). This problem has many local minima, and if we vary the starting point (for instance, choose it randomly), we are not guaranteed any more to end up in the same solution: our solution becomes stochastic, despite having used a deterministic method.

To escape local minima, many heuristic optimization methods make deliberate use of randomness. In TA, for example, new neighbor portfolios are selected randomly. Thus, even for identical starting points (but different seeds for the random number generator that is used), repeated runs of TA will produce different solutions—in fact, even in convex problems. In the following discussion, we will characterize a solution by its associated objective function value, though sometimes it may also be appropriate to look at the parameter values. For a given problem, these solutions can be regarded as realizations of a random variable with an unknown distribution $\mathcal{F}$ (Gilli & Winker, 2009).

The distribution $\mathcal{F}$ is not symmetric, but bounded to the left (since we minimize) and degenerates for increasing computational resources to a single point, namely the global minimum. A proof of this convergence for TA is given in Althöfer and Koschnick (1991). (Of course, if our problem is not bounded, then neither is $\mathcal{F}$.) The particular shape of $\mathcal{F}$ will depend on the chosen algorithm, and on the amount of computational resources spent on the optimization. In a serial computing environment, this amount of computational resources is usually roughly proportional to computing time (see Gilli and Schumann (2008) for the case of parallel computing).

For a single optimization run, computational resources can be measured by the number of objective function evaluations, or the total number of iterations of the algorithm. For TA, the number of iterations is given by the number of thresholds times the number of steps per threshold, i.e., $n_{\text{Rounds}} \times n_{\text{Steps}}$ (note that we use "steps" always for "steps per threshold," whereas "iterations" refer to the product rounds times steps). Thus, one way to increase computational resources is to give more iterations.

There is a second way to increase computational resources: just rerun the optimization, i.e., conduct so-called restarts. To give an analogy: assume you have a die, and you wish to roll a 6. The die here is the optimization algorithm, and the "6" represents a good solution. Now, one approach is to manipulate the die (that would be cheating, but here it is only an analogy), so that the probability of rolling a 6 increases. This is equivalent to increasing the number of iterations. You could also roll the die several times (all you need is one 6, i.e., one good solution), i.e., you could conduct restarts.

Let $\mathcal{I}$ denote available computational resources, then we can split these resources between restarts, numbers of thresholds, and numbers of steps per threshold, i.e.:

$$\mathcal{I} = n_{\text{Restarts}} \times \underbrace{n_{\text{Rounds}} \times n_{\text{Steps}}}_{\#\text{ of iterations}}$$

The shape of $\mathcal{F}$ will depend on the number of iterations, but also on how this number is subdivided into rounds and steps. Every restart is a draw from $\mathcal{F}$.

Increasing $\mathcal{I}$ should lead to better solutions. In particular, with an increasing number of iterations, the mean solution per restart should decrease while the solutions' variance should also decrease and eventually go to zero as the number of iterations goes to infinity. Thus, for every restart, the probability of obtaining a good solution increases.

Unfortunately, there is no general rule like "2,000 iterations will give a good solution." Fortunately, investigating the convergence behavior is straightforward: run a reasonably large number of optimizations (i.e., conduct a large number of restarts) for a given fixed number of iterations. For each of these runs, store the obtained objective function value. Every restart is a draw from $\mathcal{F}^i$, where the superscript indicates the chosen number of iterations. Hence, the empirical CDF of these collected values is an estimator of $\mathcal{F}^i$. To give an example, in Gilli and Schumann (2008), we ran such tests for a portfolio problem where we minimized the Omega function (Keating & Shadwick, 2002). Figure 9.3 illustrates the empirical distribution functions of the solutions for 100,000, 50,000, 35,000, 20,000, and 10,000 iterations.

We can clearly see how the distribution gets steeper (i.e., variance decreases), and moves to the left (i.e., average solution decreases). With many iterations, the remaining variability becomes very small, though it is unlikely to disappear altogether. For all practical purposes, however, when working with real (noisy) data, this remaining variability becomes negligible (Gilli & Schumann, 2009b). When implementing TA, such experiments can help to decide how many iterations to use.

There is a second issue to be investigated: Assume we have a fixed amount of computational resources at our disposal, then it is also important to know how to allocate these resources among restarts, rounds, and steps. Intuitively, we may use fewer iterations and thus have a "worse" distribution of solutions, but still by restarting several times produce better results than from just running one optimization with many iterations. Formal studies on this question generally find that giving more iterations and using fewer restarts works better, see Winker (2001, pp. 129–134). A caveat is in order, though: these studies assume that TA is properly implemented and works correctly for the given problem. In a testing phase, several restarts are definitely advisable.

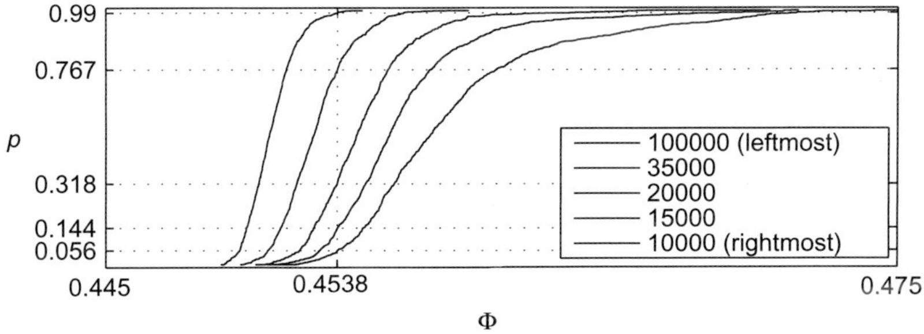

Figure 9.3 Distribution of solutions for increasing number of iterations.

9.5 Diagnostics

In this section, we will discuss several potential problems that may occur when implementing portfolio selection models and some "rules of thumb" that we found helpful in spotting trouble.

It is difficult to clearly characterize where TA works well and where it does not. Generally, TA is a local search method, so when it comes to the exploration of very large search spaces, there may be more appropriate methods (e.g., Genetic Algorithms). The algorithm may also run into trouble if the objective function is very noisy, or very "flat" overall.

9.5.1 Benchmarking the algorithm

When implementing the algorithm afresh, it is good practice to work with a well-known problem, one which can also be solved with another method—the obvious candidate is to conduct a mean–variance optimization with a quadratic programming solver and compare the results with those obtained from TA. This does not only help to spot errors in the implementation, but also gives an intuition of how closely TA approximates the exact solution.

If, for a convex problem like mean–variance, the solutions from TA differ widely across different optimization runs (restarts), this indicates insufficient computational resources, i.e., too few iterations.

9.5.2 Arbitrage opportunities

A serious problem in scenario optimization is the existence of arbitrage opportunities in the scenario set. This is a problem for any optimization algorithm, not just for TA. There exist formal tests to detect arbitrage (see for instance Ingersoll (1987, chapter 2)), but they rarely help to remove them. Furthermore, these tests will not find spurious "good deals" in the data (i.e., situations "close" to arbitrage). The resulting overfit is particularly pronounced if short positions are allowed, because then the algorithm will finance seemingly advantageous positions by short-selling less-attractive assets. This becomes clearest if we consider portfolios that also include options. If some underlyer never drops more than 10%, say, in our scenarios, it will always appear a good idea to write a put on this stock at 90% of spot. In fact, an unconstrained algorithm will sell as many puts as possible. Nonetheless, such problems are not limited to long–short portfolios; long-only portfolios also overfit in such cases, even though the effect is less pronounced (Gilli et al., 2008).

A practical solution for the equity-only case is to increase the number of observations in relation to the number of assets. When working with historical data, we may either use a longer historical time horizon or reduce the number of selectable assets. If we work with "artifical" scenarios, e.g., obtained from resampling, we can simply increase the number of replications.

Including restrictions like maximum holding sizes is practically always advisable, even though it just reflects the sad fact that we cannot model asset prices properly. Such constraints may not just limit position sizes, but if applicable, we can also include constraints on aggregate Greeks like Delta or Gamma, or a minimum required performance in added artificial crash-scenarios.

9.5.3 Degenerate objective functions

Numerically, ratios have some unfavorable properties when compared with linear combinations: If the numerator becomes zero, the ratio becomes zero and the search becomes unguided; if the denominator is zero, we get an error (inf). If the sign of numerator or denominator changes over the course of the search, the ratio often becomes uninterpretable; an example is the Sharpe ratio for negative mean excess return.

So we generally need safeguards. To avoid sign problems, we can use centered quantities: Lower partial moments, for example, are computed as $r_d - r$ for returns lower than r_{d}, thus the numbers will always be nonnegative. An alternative is to use operations like $\max(\cdot, 0)$ or $\min(\cdot, 0)$ to assure the sign of some quantity.

There is, however, also a valuable aspect in these instabilities, for they are not only of a numerical nature. In fact, problems when computing a ratio may indicate problems with the model or the data, which would go unnoticed with a linear combination. For instance, if a risk–reward ratio turns zero, this means we have found a portfolio with no risk at all, which, unfortunately, is often an indication of a peculiar data sample rather than a really excellent portfolio.

9.5.4 The neighborhood and the thresholds

It is helpful to have some insight into the local structure (the "surface") of the objective function in the search space. For low-dimensional problems, we can plot the objective function as in Figure 9.1; with more dimensions, we can take random walks through the data (like in Algorithm 4). Thus, we start with a random porfolio and move through the search space according to our neighborhood function, accepting any new portfolio. The changes in the objective function accompanying every step should be visually inspected, for instance with histograms or CDF-plots. A large number of zero changes indicates a flat surface, i.e., in such regions of the search space, the algorithm will get no guidance from the objective function. Another sign of potential trouble is the clustering of changes, i.e., if we have a large number of very small changes and a large number of very large changes, which may indicate a bad scaling of the problem.

The observed changes should be of a roughly similar magnitude as the thresholds—which is why such random walks are often used to inform the threshold setting, as was described in Algorithms 3 and 4. If the thresholds are too large compared with average changes, our optimization algorithm will become a random search, since any new portfolio will be accepted. If the thresholds are too small, the algorithm will be too restrictive, and become stuck in local minima.

During the actual optimization run, it is good practice to store the accepted changes in the objective function values (i.e., the accepted Δ-values in Algorithm 2).

As a rule of thumb, the standard deviation of these accepted changes should be of the same order of magnitude as the standard deviation of the changes in the objective function values recorded from the random walk (i.e., the Δ_i-values in Algorithm 3 or 4).

Figure 9.4 shows the thresholds for a given problem (with $\Phi = \mathcal{D}_{\text{mean}}/1$) as computed from the procedure given in Algorithm 4. The lower panel shows the value of the objective function over time during an actual optimization run. A vertical line indicates a switch to a lower threshold. Initially, the algorithm moves freely, moving often to portfolios that map to higher (i.e., worse) objective function values. Over time, as the thresholds decrease, the objective function descends more smoothly. In the last few rounds, fewer portfolios are accepted (as can be seen from the distance between the vertical lines), as the algorithm becomes much more select.

9.6 Conclusion

In this chapter, we have outlined the use of a heuristic optimization method, TA, for portfolio optimization problems. Heuristics are usually simple conceptually, but in our experience there is often quite a distance between a general description of an algorithm and an actual implementation. We hope that this chapter offers some helpful advice in this respect.

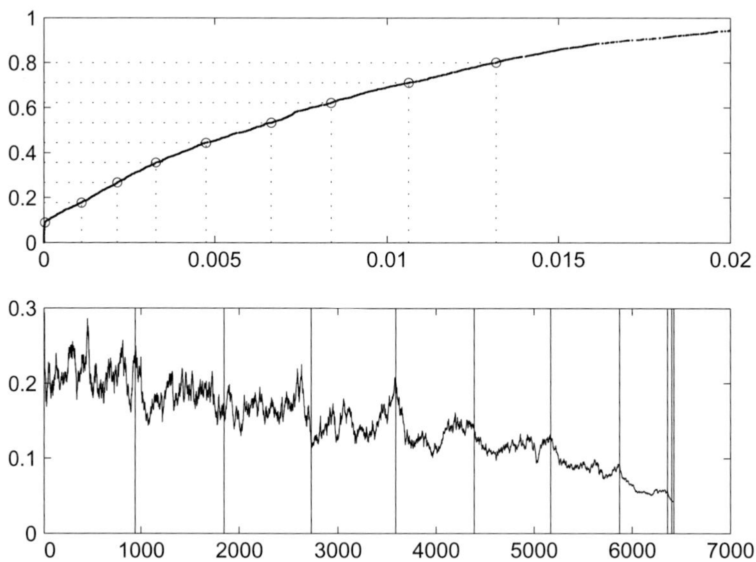

Figure 9.4 A threshold sequence (upper panel) and the current value of the objective function (lower panel).

Throughout, we always assumed that we have a model, that we have the data (i.e., scenarios), so that we could concentrate on the optimization procedure. But setting up an optimization algorithm can only be the first step. The main problem in portfolio optimization, no matter what technique is used, is how to tell noise from information, i.e., how to reduce overfitting. (An unfortunate rule of thumb in financial optimization: if something seems too good to be true, then it is not true.) To improve in this direction, more effort needs to be put into data modeling (i.e., the scenario generation process) and toward the testing of the empirical effectiveness of different portfolio selection criteria.

Both issues are, in our view, strongly "under-researched" in the academic world, at least when it comes to actual empirical performance. Note that this strongly relates to the first two of our initial questions: "what do I want to optimize?" and "what can I estimate?" Much research in portfolio optimization relates to in-sample properties of different methods: given a data set, we can now minimize drawdown, or ratios of losses to gains. But what we actually want is to minimize future drawdown, or the ratio of future losses to future gains. There exists comparatively little academic research into how these objectives relate to the quantities that we actually optimize.

The advantage of heuristic optimization methods in this respect is that when we formulate a model, we are quite unconstrained with regard to its tractability, so the third question ("what can I compute?") becomes much less of an obstacle to progress. Optimization is a tool, and like with any tool, it is its application that matters.

Acknowledgment

Both authors gratefully acknowledge financial support from the EU Commission through MRTN-CT-2006-034270 COMISEF.

References

Althöfer, I., & Koschnick, K.-U. (1991). On the convergence of "threshold accepting." *Applied Mathematics and Optimization*, *24*(1), 183–195.

Bacon, C. A. (2008). *Practical portfolio performance measurement and attribution* (2nd ed.). New York: John Wiley & Sons.

Barr, R. S., Golden, B. L., Kelly, J. P., Resende, M. G. C., & Stewart, W. R. (1995). Designing and reporting on computational experiments with heuristic methods. *Journal of Heuristics*, *1*(1), 9–32.

Biglova, A., Ortobelli, S., Rachev, S., & Stoyanov, S. (2004). Different approaches to risk estimation in portfolio theory. *Journal of Portfolio Management*, *31*(1), 103–112.

Chekhlov, A., Uryasev, S., & Zabarankin, M. (2005). Drawdown measure in portfolio optimization. *International Journal of Theoretical and Applied Finance*, *8*(1), 13–58.

Dueck, G., & Scheuer, T. (1990). Threshold accepting. A general purpose optimization algorithm superior to simulated annealing. *Journal of Computational Physics*, *90*(1), 161–175.

Dueck, G., & Winker, P. (1992). New concepts and algorithms for portfolio choice. *Applied Stochastic Models and Data Analysis*, *8*(3), 159–178.

Gigerenzer, G. (2004). Fast and frugal heuristics: The tools of bounded rationality. In D. J. Koehler & N. Harvey (Eds.), *Blackwell handbook of judgment and decision making* (pp. 62–88). Blackwell Publishing.

Gigerenzer, G. (2008). Why heuristics work. *Perspectives on Psychological Science*, *3*(1), 20–29.

Gilli, M., & Schumann, E. (2008). *Distributed optimisation of a portfolio's omega*. Swiss Finance Institute Research Paper No. 08-17.

Gilli, M., & Schumann, E. (2009a). *An empirical analysis of alternative portfolio selection criteria*. Swiss Finance Institute Research Paper No. 09-06.

Gilli, M., & Schumann, E. (2009b). *Optimal enough*? COMISEF Working Paper Series No. 10.

Gilli, M., & Winker, P. (2009). Heuristic optimization methods in econometrics. In D. A. Belsley & E. J. Kontoghiorghes (Eds.), Handbook of Computational Econometrics, New York: Wiley.

Gilli, M., Këllezi, E., & Hysi, H. (2006). A data-driven optimization heuristic for downside risk minimization. *Journal of Risk*, *8*(3), 1–18.

Gilli, M. & Schumann, E., di Tollo, G., & Cabej, G. (2008). *Constructing long/short portfolios with the omega ratio*. Swiss Finance Institute Research Paper No. 08-34.

Holland, J. H. (1992). *Adaptation in natural and artificial systems: An introductory analysis with applications to biology, control and artificial intelligence*. Cambridge: MIT Press.

Hyndman, R. J., & Fan, Y. (1996). Sample quantiles in statistical packages. *The American Statistician*, *50*(4), 361–365.

Ingersoll, J. E. (1987). *Theory of financial decision making*. Savage, MD: Rowman & Littlefield.

Jondeau, E., Poon, S.-H., & Rockinger, M. (2007). *Financial modeling under non-Gaussian distributions*. Berlin: Springer.

Keating, C., & Shadwick, B. (2002). An introduction to omega. April. *AIMA Newsletter*.

Kirkpatrick, S., Gelatt, C. D., & Vecchi, M. P. (1983). Optimization by simulated annealing. *Science*, *220*(4598), 671–680.

Markowitz, H. M. (1952). Portfolio selection. *Journal of Finance*, *7*(1), 77–91.

Mausser, H., Saunders, D., & Seco, L. (2006). Optimising omega. November. *Risk Magazine*.

Michalewicz, Z., & Fogel, D. B. (2004). *How to solve it: Modern heuristics*. Berlin: Springer.

Passow, A. (2005). Omega portfolio construction with Johnson distributions. April. *Risk Magazine*.

Rachev, S. T., Fabozzi, F. J., & Menn, C. (2005). Fat-tailed and skewed asset return distributions: Implications for risk management, portfolio selection, and option pricing. In *Risk measures in portfolio optimization and performance measures* (pp. 199–212). New York: John Wiley & Sons.

Rockafellar, R. T., & Uryasev, S. (2000). Optimization of conditional value-at-risk. *Journal of Risk*, *2*(3), 21–41.

Scherer, B. (2004). *Portfolio construction and risk budgeting* (2nd ed.). London: Risk Books.

Sharpe, W. F. (1966). Mutual fund performance. *Journal of Business*, *39*(1), 119–138.

Sortino, F., van der Meer, R., & Plantinga, A. (1999). The Dutch triangle. *Journal of Portfolio Management*, *26*(1), 50–58.

Stoyanov, S. V., Rachev, S. T., & Fabozzi, F. J. (2007). Optimal financial portfolios. *Applied Mathematical Finance*, *14*(5), 401–436.

Winker, P. (2001). *Optimization heuristics in econometrics: Applications of threshold accepting*. New York: John Wiley & Sons.

Winker, P., & Fang, K.-T. (1997). Application of threshold accepting to the evaluation of the discrepancy of a set of points. *SIAM Journal on Numerical Analysis*, *34*(5), 2028–2042.

Winker, P., & Maringer, D. (2007). The threshold accepting optimisation algorithm in economics and statistics. In E. J. Kontoghiorghes & C. Gatu (Eds.), *Optimisation, econometric and financial analysis. Vol. 9 of Advances in Computational Management Science* (pp. 107–125). Berlin: Springer.

Young, T. W. (1991). Calmar ratio: A smoother tool. October. *Futures Magazine*.

10 Some properties of averaging simulated optimization methods

John Knight and Stephen E. Satchell

Executive Summary

We consider problems in optimization that are addressed through the use of Monte Carlo simulated averages. To do this, we revisit the problem of calculating the exact distribution of optimal investments in a mean–variance world under multivariate normality. These results help us understand when Monte Carlo–based averaging methods will work well, and when they will fail. The results derived allow some exact and numerical analysis, which we use to illustrate the magnitude of biases that could occur in practical situations. For stock selection optimization, these can be very large.

10.1 Section 1

The major motivation for this research has been to try and understand the magnitude of estimation error; i.e., the extent to which the outcome of the portfolio decision is influenced by parameter uncertainty. The last decade has seen an overlap in academic and practitioner research in this area. The practitioner research has been exemplified by the work of Michaud. In response to parameter uncertainty, Michaud (1998) has proposed an averaging simulated optimization procedure, the outcome of which can only be really understood by an analysis of the exact properties of optimal portfolios. His innovation was followed by other commercial products based on similar methods. Michaud's procedure purports to solve some of the problems of portfolio optimization that arise from estimation error. Other authors have criticized Michaud's approach, see Scherer (2002) and Harvey, Leichty, Leichty, and Muller (2004). We investigate the merits of the averaging simulated Mote Carlo approach using exact distribution theory.

The application of exact distribution theory to mean–variance (MV) analysis has been undertaken by a number of authors, see Jobson and Korkie (1989), Jobson (1991), Britten-Jones (1999), Stein (2002), Hillier and Satchell (2003), Okhrin and Schmid (2006), Bodnar and Schmid (2006), and Frahm (2007). The usual assumptions are that returns are i.i.d. multivariate normal and that there may or may not be a riskless asset. However, in all listed cases, the analysis is in

Doi:10.1016/B978-0-12-374952-9.00010-5.

terms of absolute, i.e., unbenchmarked, portfolios. These portfolios are typically unconstrained; the only constraint they satisfy is that the weights add up to 1. This is a limitation since most institutional risk analysis is based on MV analysis using returns relative to a benchmark, and involves large numbers of constraints.

Another purpose of this chapter is to consider expected utility in terms of relative returns and compute the exact properties of the optimal alpha, tracking error, and Sharpe ratio. These results are of interest in their own right, and allow us to assess Michaud's contribution and the extent to which the various criticisms can be deemed to be valid. Our results, whilst being highly simplified, since we do not impose the myriad of constraints that institutional portfolios typically obey, nevertheless exhibit certain key characteristics that shed light on investment issues. Furthermore, we are able to extend the problem to consider the same case with absolute, not relative, weights. This allows us to derive some new results for this problem. We present the mathematical framework in Section 2. In Section 3, we derive exact results for the relative problem, and consider the absolute problem. In Section 4, we consider some numerical calculations. Section 5 considers the case of restrictions applied to portfolios and also addresses the realistic case of inequality constraints. We discuss some computational results that will significantly speed up frontier simulations in Section 6. Our conclusions follow in Section 7.

10.2 Section 2

In this section, we discuss the role of portfolio simulation and some of the criticisms of portfolio optimization. Portfolio optimization has been criticized for being excessively sensitive to errors in the forecasts of expected returns. This leads to the optimizer choosing implausible portfolios and is a consequence of the difficulties in forecasting expected returns. Furthermore, these MV optimal portfolios lack the diversification deemed desirable by institutional investors, see Green and Hollifield (1992). A number of solutions to this problem have emerged. In some contexts, Bayesian prices on the expected returns are used to control the sample variability of the means, see, for example, Satchell and Scowcroft (2003). Practitioners often employ large numbers of constraints on the portfolio weights to control the optimizer, and we shall refer to this as the practitioner's solution. This solution has been given some support in the context of MV optimization by Jagannathan and Ma (2002, 2003).

Michaud (1998) has advocated simulating the optimization. The advantage of this is that we get some sense of the variability of the solution; however, we need to understand what the averaging in the simulation will lead to.

To motivate our analysis, we consider how Michaud (1998) carries out his resampling methodology. Quoting from Michaud (op cit, pages 17, 19, and 37):

1. "Monte Carlo simulate 18 years of monthly returns based on data in Tables 2.3 and 2.4…
2. Compute optimized input parameters from the simulated return data.

3. Compute efficient frontier portfolios...
4. Repeat steps 1–3 500 times...
5. ...Observe the variability in the efficient frontier estimation."

The assumption behind the Monte Carlo simulation of returns can vary. It can be based on historical returns and involve resampling, or it may involve using means, variance, and covariance and simulating via multivariate normality as Michaud details above; his Tables 2.3 and 2.4 contain first and second sample moments. The strength of the method comes from the law of large numbers. If we take enough replications, our sample statistics will converge to their expected values where expectation is based on the assumed population distribution. If the statistic happens to be biased, then it will converge to its expectation, which will equal the "true" value plus the bias. As we will show, the simulated average frontier is biased. This implies that the mean simulated efficient frontier will differ from the "population" efficient frontier based on the information in Step 1 by the degree of finite sample bias. Whilst this should be small for $T = 216$ monthly observations, there are lots of portfolio calculations that will be based on much shorter time periods due to the usual reasons: regime shifts, institutional change, and time-varying parameters. Furthermore, we conjecture, and subsequently show, that it is not T alone that determines bias but T and N (the number of stocks) cojointly. If N is large, even for large T, then biases can be very large indeed.

It is worth noting that the emphasis of the above approach is in terms of the MV efficient frontier analysis rather than expected utility. But as we shall show next, maximizing quadratic utility gives you a solution that is expressed solely in terms of efficient-set mathematics; the only additional information is the risk aversion coefficient (λ); as we change λ, we move along the MV frontier in any case.

Jobson (1991) derives a number of key results in this area for the conventional minimum variance frontier, and we shall refer to these results when appropriate. Stein (2002) has also derived some of our results. In a recent related paper, Okhrin and Schmid (2006) consider the standard quadratic utility maximization subject to adding up constraints. Their focus, unlike ours, is the calculation of the distributional properties of the optimal portfolio weights. Our concern, on the other hand, is the distributional properties of portfolio summary measures such as the portfolio mean return (α), the tracking error (TE), and the information or Sharpe ratio (IR). Some of our results could be deduced from those of Okhrin and Schmid (2006); in what follows, we will indicate where this occurs.

Consider the active weights $\underset{\sim}{\omega}$ and the *known* benchmark weights $\underset{\sim}{b}$, both $(N \times 1)$ vectors and both sum to 1, i.e., $\omega'i = b'i = 1$ Let μ and Ω be the $(N \times 1)$ mean vector and covariance matrix of the N asset returns, where the letter i denotes an $(N \times 1)$ vector of ones.

Our investor chooses to maximize U, where $U = \mu'(\omega - b) - \lambda/2(\omega - b)'\Omega(\omega - b)$; note that there is also a constraint $(\omega - b)'i = 0$. This is a classical MV problem equivalent, as we demonstrate, to computing the optimal frontier. It is straightforward to see that as λ ranges from 0 to ∞, we move down the frontier from the maximum expected return portfolio to the global minimum variance portfolio. This framework is widely used in finance, see Sharpe (1981), Grinold and Kahn (1999), and Scherer (2002).

Our first-order condition is:

$$\frac{\partial U}{\partial(\omega - b)} = \mu - \lambda\Omega(\hat{\omega} - b) + \hat{\theta} i = 0 \quad \text{or} \quad \hat{\omega} = b + \tfrac{1}{\lambda}\Omega^{-1}(\mu + \hat{\theta} i)$$

Using $i'(\omega - b) = 0$, we see that $\frac{1}{\lambda}(\beta + \hat{\theta}\gamma) = 0$, where $\beta = i'\Omega^{-1}\mu$, $\gamma = i'\Omega^{-1}i$ and we set $a = \mu'\Omega^{-1}\mu$. Thus, $\hat{\omega} = b + \frac{1}{\lambda}(\Omega^{-1}\mu - \frac{\beta}{\gamma}\Omega^{-1}i)$ and hence active returns α can be computed as:

$$\begin{aligned}\alpha &= \mu'(\hat{\omega} - b) = \tfrac{1}{\lambda}\left(a - \frac{\beta^2}{\gamma}\right)\\ &= \frac{1}{\lambda}\left(\frac{a\gamma - \beta^2}{\gamma}\right)\end{aligned} \tag{10.1}$$

Other terms of interest can be calculated. For example, we have:

$$\begin{aligned}\hat{\sigma}^2 &= \frac{1}{\lambda^2}\left(\Omega^{-1}\mu - \frac{\beta}{\gamma}\Omega^{-1}i\right)'\Omega\left(\Omega^{-1}\mu - \frac{\beta}{\gamma}\Omega^{-1}i\right)\\ &= \frac{1}{\lambda^2}\left(a - \frac{\beta^2}{\gamma} + \frac{\beta^2}{\gamma}\right)\\ &= \frac{1}{\lambda^2}\left(a - \frac{\beta^2}{\gamma}\right)\end{aligned} \tag{10.2}$$

and we will focus on $\hat{\sigma}$ the tracking error or standard deviation of relative returns. Finally,

$$\begin{aligned}E(U) &= \frac{1}{\lambda}\left(\frac{a\gamma - \beta^2}{\gamma}\right) - \frac{\lambda}{2}\left(\frac{1}{\lambda^2}\left(\frac{a\gamma - \beta^2}{\gamma}\right)\right)\\ &= \frac{1}{2\lambda}\left(\frac{a\gamma - \beta^2}{\gamma}\right)\end{aligned} \tag{10.3}$$

It is straightforward to compute the information ratio defined as $\hat{\alpha}/\hat{\sigma}$. Notice that in this problem all terms depend essentially on a single term $(a\gamma - \beta^2/\gamma)$ or functions of it.

10.3 Remark 1

A related formulation of the above problem is the following: min1/2 $(\omega - b)'\Omega(\omega - b)$ subject to $(\omega - b)'i = 0$ and $(\omega - b)'\mu = \pi$. Here, the Lagrangian is given by:

$$L = \tfrac{1}{2}(\omega - b)'\Omega(\omega - b) - \theta_1(\omega - b)'i - \theta_2((\omega - b)'\mu - \pi)$$

resulting in the first-order conditions:

$$\begin{aligned}\frac{\partial L}{\partial \omega} &= \Omega(\omega - b) - \theta_1 i - \theta_2\mu = 0\\ \frac{\partial L}{\partial \theta_1} &= (\omega - b)'i = 0\\ \frac{\partial L}{\partial \theta_2} &= (\omega - b)'\mu - \pi = 0\end{aligned}$$

Solving, we have $\omega - b = \theta_1\Omega^{-1}i + \theta_2\Omega^{-1}\mu$ with $\theta_1 = (\beta\pi/a\gamma - \beta^2)$ and $\theta_2 = (\gamma\pi/a\gamma - \beta^2)$.

Thus,

$$\begin{aligned}\hat{\omega} &= b + \frac{\pi\gamma}{a\gamma - \beta^2}\left(\Omega^{-1}\mu - \frac{\beta}{\gamma}\Omega^{-1}i\right)\\ \hat{\omega} &= b + \pi\hat{\hat{\omega}}\end{aligned}$$

and consequently,

$$\begin{aligned}\hat{\sigma}^2 &= \pi^2\hat{\hat{\omega}}'\Omega\hat{\hat{\omega}}\\ \hat{\sigma}^2 &= \frac{\pi^2\gamma}{a\gamma - \beta^2}\end{aligned}$$

Comparing with Equation (10.2), we see immediately that $\pi = 1/\lambda(a - \beta^2/\gamma)$.

This second problem is simply the computation of the minimum variance frontier. It differs from the earlier version in that it explicitly specifies π, the expected rate of return, rather than λ, the risk aversion coefficient.

10.4 Section 3: Finite sample properties of estimators of alpha and tracking error

To compute the biases that we need, it is necessary to calculate various expectations. Consider:

$$Q = (\mu, i)'\Omega^{-1}(\mu, i) = \begin{pmatrix} a & \beta \\ \beta & \gamma \end{pmatrix} \tag{10.4}$$

It is well known that, under normality, the sample mean $\hat{\mu} \sim N(\mu, 1/T\Omega)$, and $\hat{\mu}$ and $\hat{\Omega}$ are independent, where $\hat{\Omega}$ is the Maximum Likelihood estimator of the covariance matrix. First, by Theorem 3.2.11 of Muirhead (1982), conditional on $\hat{\mu}$, $\hat{Q}^{-1} = ((\hat{\mu}, i)'\hat{\Omega}^{-1}(\hat{\mu}, i))^{-1}$ has a central Wishart: $W_2(T - N + 1, 1/T\bar{Q}^{-1})$, where $\bar{Q} = (\hat{\mu}, i)'\Omega^{-1}(\hat{\mu}, i)$. The statistic of interest is given by $\hat{h} = \hat{\gamma}/\hat{a}\hat{\gamma} - \hat{\beta}^2$ and is the first principal element of $\hat{Q}^{-1}$. Formally, we have $\hat{h} = (1,0)\hat{Q}^{-1}(1,0)'$ and again from Muirhead (1982) Theorem 3.2.5, we have $\hat{h} \mid \hat{\mu} \sim W_1(T - N + 1, 1/T(1,0)\bar{Q}^{-1}(1,0)')$ and thus, letting $\phi = 1/T(1,0)\bar{Q}^{-1}(1,0)'$, we have

$$\frac{\hat{h}}{\phi} \mid \phi \sim \chi^2_{(\nu)}, \quad \text{where} \quad \nu = T - N + 1$$

and consequently, this result holds unconditionally.

Next we examine ϕ, noting immediately that $T\phi$ is the first principal element in $\bar{Q}^{-1}$. That is,

$$\phi = \frac{1}{T}\frac{i'\Omega^{-1}i}{(\hat{\mu}'\Omega^{-1}\hat{\mu})(i'\Omega^{-1}i) - (\hat{\mu}'\Omega^{-1}i)(i'\Omega^{-1}\hat{\mu})} \tag{10.5}$$

Now, $\hat{\mu} \sim N(\mu, 1/T\Omega)$ and thus letting $\omega = \sqrt{T}\Omega^{-1/2}\hat{\mu}$, we have $\omega \sim N(\sqrt{T}\Omega^{-1/2}\mu, I_N)$. Further, letting $c = \Omega^{-1/2}i$, we have that:

$$\phi = \frac{1}{T}\left(\frac{Tc'c}{\omega'\omega c'c - \omega' cc'\omega}\right) = \frac{1}{\omega'(I - c(c'c)^{-1}c')\omega} = \frac{1}{\omega'\bar{P}_c\omega} \tag{10.6}$$

and it follows immediately that $\omega'\bar{P}_c\omega \sim \chi^2_{(N-1,\lambda^*)}$, where $\lambda^* = T\mu\Omega^{-1/2}\bar{P}_c\Omega^{-1/2}\mu = T/h$ with $h = \gamma/a\gamma - \beta^2$. Therefore, $\phi^{-1} \sim \chi^2_{(N-1,T/h)}$ and thus the distribution of $\hat{h}$ will be given by the following ratio:

$$\hat{h} \sim \frac{\chi^2_{(v)}}{\chi^2_{(N-1,T/h)}} \tag{10.7}$$

where the two χ^2 variables in Equation (10.7) are independent. This result also appears in Stein (2002). Thus, $pdf(\hat{h})$ can be easily found using results related to noncentral F distributions. In this regard, we have from Johnson and Kotz (1972, p. 191) that the $pdf((1/h) = g)$ is, noting B( ) and ${}_1F_1$( ) to be Beta and confluent hypergeometric functions respectively,

$$pdf(g) = \frac{e^{-T/2h} g^{\frac{N-1}{2}-1}}{B\left(\frac{N-1}{2}, \frac{\nu}{2}\right)(1+g)^{\frac{N-1+\nu}{2}}} {}_1F_1\left(\frac{N-1+\nu}{2}, \frac{N-1}{2}, \frac{T}{2h}\frac{g}{(1+g)}\right)$$

Using this result and simple transformations, one can readily derive the pdf density functions for the quantities of interest, viz, $\hat{\alpha} = (1/\lambda\hat{h})$. Tracking error $= \overline{TE} = (1/\lambda\sqrt{\hat{h}})$ and the Information Ratio $= \overline{IR} = (1/\sqrt{\hat{h}})$. Thus, we have:

$$pdf(\hat{\alpha} = \omega) = \frac{\lambda e^{-T/2h} (\lambda\omega)^{\frac{N-1}{2}-1}}{B\left(\frac{N-1}{2}, \frac{\nu}{2}\right)(1+\lambda\omega)^{\frac{N-1+\nu}{2}}} {}_1F_1 \left(\frac{N-1+\nu}{2}, \frac{N-1}{2}, \frac{T}{2h}\frac{\lambda\omega}{(1+\lambda\omega)}\right), \quad \omega > 0$$

$$pdf(\overline{IR} = x) = \frac{2xe^{-T/2h} (x^2)^{\frac{N-1}{2}-1}}{B\left(\frac{N-1}{2}, \frac{\nu}{2}\right)(1+x^2)^{\frac{N-1+\nu}{2}}} {}_1F_1 \left(\frac{N-1+\nu}{2}, \frac{N-1}{2}, \frac{T}{2h}\left(\frac{x^2}{(1+x^2)}\right)\right), \quad x > 0$$

$$pdf(\overline{TE} = y) = \frac{2\lambda^2 y e^{-T/2h} (\lambda^2 y^2)^{\frac{N-1}{2}-1}}{B\left(\frac{N-1}{2}, \frac{\nu}{2}\right)(1+\lambda^2 y^2)^{\frac{N-1+\nu}{2}}} {}_1F_1 \left(\frac{N-1+\nu}{2}, \frac{N-1}{2}, \frac{T}{2h}\left(\frac{\lambda^2 y^2}{(1+\lambda^2 y^2)}\right)\right), \quad y > 0$$

From these pdfs or via that of $\hat{h}$ or g, we can easily find moments. That is, since

$$\begin{aligned} \hat{h}^{-1} &= g = \chi^2_{(N-1,T/h)}/\chi^2_{(\nu)} \\ E(g^k) &= E[(\chi^2_{(N-1,T/h)})^k]E[\chi^2_{(\nu)})^{-k}] \end{aligned}$$

and since

$$E[(\chi^2_{(N-1,T/h)})^k] = \frac{e^{-T/2h} 2^k}{\Gamma\left(\frac{N-1}{2}\right)} \Gamma\left(\frac{N-1}{2} + k\right) {}_1F_1\left(\frac{N-1}{2} + k, \frac{N-1}{2}; {}^{T}\!/_{2h}\right)$$

and

$$E((\chi^2_{(\nu)})^{-k}) = \frac{\Gamma\left(\frac{\nu}{2} - k\right)}{2^k \Gamma(\nu/2)}, \quad \nu > 2k$$

$$E[g^k] = \frac{e^{-T/2h}\Gamma\left(\frac{N-1}{2} + k\right)\Gamma\left(\frac{\nu}{2} - k\right)}{\Gamma\left(\frac{N-1}{2}\right)\Gamma(\nu/2)} {}_1F_1\left(\frac{N-1}{2} + k, \frac{N-1}{2}; T/2h\right)$$

Therefore,

$$\begin{aligned} E[(\alpha)^k] &= \lambda^{-k} E(g^k) \\ &= \frac{e^{-T/2h}\Gamma\left(\frac{N-1}{2} + k\right)\Gamma\left(\frac{\nu}{2} - k\right)}{\lambda^k \Gamma\left(\frac{N-1}{2}\right)\Gamma(\nu/2)} {}_1F_1\left(\frac{N-1}{2} + k, \frac{N-1}{2}; T/2h\right) \end{aligned} \tag{10.8}$$

$$\begin{aligned} E((\overline{TE})^k &= \lambda^{-k} E(g^{k/2}) \\ &= \frac{e^{-T/2h}\Gamma\left(\frac{N-1}{2} + \frac{k}{2}\right)\Gamma\left(\frac{\nu}{2} - \frac{k}{2}\right)}{\lambda^k \Gamma\left(\frac{N-1}{2}\right)\Gamma(\nu/2)} {}_1F_1\left(\frac{N-1}{2} + \frac{k}{2}, \frac{N-1}{2}; T/2h\right) \end{aligned} \tag{10.9}$$

$$E((\overline{IR})^k) = E(g^{k/2}) \tag{10.10}$$

In particular, if we consider the means of the three quantities, we have:

$$\begin{aligned} E(\hat{a}) &= \frac{\Gamma\left(\frac{N-1}{2} + 1\right)\Gamma\left(\frac{\nu}{2} - 1\right)}{\lambda\Gamma\left(\frac{N-1}{2}\right)\Gamma\left(\frac{\nu}{2}\right)} e^{-T/2h} {}_1F_1\left(\frac{N-1}{2} + 1, \frac{N-1}{2}; T/2h\right) \\ &= \frac{(N-1)}{\lambda(\nu - 2)} {}_1F_1\left(-1, \frac{N-1}{2}; -T/2h\right) \\ &= \frac{N-1}{\lambda(\nu-2)}\left[1 + \frac{T}{h(N-1)}\right] \\ &= \frac{N-1}{\lambda(\nu-2)} + \frac{T}{\lambda h(\nu-2)}; \quad \nu = T - N + 1 \end{aligned}$$

Since the true $\alpha = 1/\lambda h$, we can readily develop an unbiased estimator of α via a simple transformation:

$$E\left[\frac{(\nu-2)}{T}\hat{a} - \frac{N-1}{T\lambda}\right] = \alpha$$

For other quantiles, we have:

$$E(\overline{TE}) = \frac{\Gamma\left(\frac{N-1}{2}+\frac{1}{2}\right)\Gamma\left(\frac{\nu}{2}-\frac{1}{2}\right)}{\lambda\Gamma\left(\frac{N-1}{2}\right)\Gamma(\nu/2)}\,{}_1F_1\left(-\tfrac{1}{2},\frac{N-1}{2};-T/2h\right)$$

and

$$E(\overline{IR}) = \lambda E(\overline{TE})$$

Also note that since $E(\sigma^2) = E(\overline{TE}^2) = \frac{1}{\lambda}E(\hat{\alpha})$, an unbiased estimator of σ^2 is easily derived to be:

$$\hat{\sigma}^2 = \frac{1}{\lambda}\left[\frac{\nu-2}{T}\hat{\alpha} - \frac{(N-1)}{T\lambda}\right]$$

While little progress can be made with exact expressions for the expectation of $\overline{TE}$ and $\overline{IR}$, we can get more insight by considering approximations. Jobson (1991) gives similar results in that he derives the means and variances of a, β, and γ and h^{-1} and determines their marginal distributions. Stein (2002) also derives some formulae similar to ours.

We now examine a situation in which both N, the number of stocks or assets, and T, the sample size, increase in such a way that the ratio $(N-1)/T$ remains constant. We note that α, β, and γ and hence h also depend upon N and thus asymptotics here require that the terms limit to a constant, or at least we need to make assumptions about $1/h$ as a function of N. For the moment, we shall not consider this possible influence. Thus, we now let $(N-1/T) = n$ so that $N-1 = T.n$. By letting $T\to\infty$, we can readily see the effect on the moments of large N and T. For $\hat{\alpha}$, we have from our exact result:

$$E(\hat{\alpha}) = \frac{Tn}{\lambda(T(1-n)-2)}\left[1+\frac{1}{hn}\right]$$

and therefore as $T\to\infty$, we find:

$$E(\hat{\alpha}) \to \frac{1}{\lambda}\left[\frac{n}{1-n}+\frac{1}{h(1-n)}\right]$$

The corresponding results for $\overline{IR}$ and $\overline{TE}$ are given by:

$$E(\overline{IR}) \to \left(\frac{n}{1-n} + \frac{1}{h(1-n)}\right)^{1/2}$$

and

$$E(\overline{TE}) \to \frac{1}{\lambda}\left(\frac{n}{1-n} + \frac{1}{h(1-n)}\right)^{1/2}$$

We now examine portfolio optimization without a benchmark. Here, we maximize $\omega'\mu - \lambda/2\omega'\Omega\omega$ subject to $\omega' i = 1$. The associated Lagrangian is given by:

$$W = \omega'\mu - \tfrac{\lambda}{2}\omega'\Omega\omega - \theta(\omega' i - 1) \tag{10.11}$$

with

$$\frac{\partial W}{\partial \omega} = \mu - \lambda\Omega\omega - \theta i = 0$$

implying

$$\omega = \tfrac{1}{\lambda}(\Omega^{-1}\mu - \theta\Omega^{-1}i)$$

since $i'\omega = 1$, we have immediately that

$$\theta = (i'\Omega^{-1}\mu - \lambda)/i'\Omega^{-1}i$$

i.e.,

$$\theta = (\beta - \lambda)/\gamma$$

Consequently,

$$\hat{\omega} = \tfrac{1}{\lambda}\left(\hat{\Omega}^{-1}\hat{\mu} - \left(\frac{\hat{\beta} - \lambda}{\hat{\gamma}}\right)\hat{\Omega}^{-1}i\right)$$

and thus

$$\tilde{\alpha} = \hat{\mu}'\hat{\omega} = \frac{1}{\lambda}\hat{\mu}'\hat{\Omega}^{-1}\hat{\mu} - \left(\frac{\hat{\beta}-\lambda}{\hat{\gamma}}\right)\hat{\mu}'\hat{\Omega}^{-1}i$$

i.e.,

$$\tilde{\alpha} = \frac{1}{\lambda\hat{h}} + \frac{\hat{\beta}}{\hat{\gamma}} = \hat{\alpha} + \frac{\hat{\beta}}{\hat{\gamma}}$$

Also,

$$\begin{aligned}\tilde{\sigma}^2 &= \hat{\omega}'\hat{\Omega}\hat{\omega}\\ \tilde{\sigma}^2 &= \frac{1}{\lambda^2\hat{h}} + \frac{1}{\hat{\gamma}}\end{aligned}$$

Thus, we notice immediately that the active return $\tilde{\alpha}$ and the $\overline{TE}^2$ are given by our earlier results plus an additional term. Under the normality assumption, we again examine some of the statistical properties of these new estimations. We present the results below; the proofs are straightforward extensions of our earlier results.

$$\begin{aligned}E(\tilde{\alpha}) &= E(\alpha) + E(\hat{\beta}/\hat{\gamma})\\ &= E(\hat{\alpha}) + (\hat{\beta}/\hat{\gamma}\ )\\ Var(\tilde{\alpha}) &= Var(\hat{\alpha}) + Var(\hat{\beta}/\hat{\gamma}),\ \text{since } \hat{a} \text{ and } \hat{\beta}/\hat{\gamma} \text{ are independent}\\ &= \text{var}(\hat{\alpha}) + \frac{1}{T\gamma}(E(\hat{h}^{-1}) + 1)\\ E(\tilde{\sigma}^2) &= E(\hat{\sigma}^2) + \frac{T-N+1}{T\gamma}\end{aligned}$$

10.5 Remark 2

The portfolio optimization in Equation (10.11) above is the same as that considered in Okhrin and Schmid (2006). The expression for $E(\tilde{\alpha})$ can be derived from the results in their appendix. In particular, using the expression for $E(\hat{\omega}_{EU} \mid \hat{\mu})$, given on page 248 of their paper, multiplying by $\hat{\mu}$ and then taking expectations, we will get the same result as detailed above once we allow for the difference in the definition of $\hat{\omega}_1 < 0$; their divisor is $T - 1$ rather than T.

10.6 Remark 3

A similar analysis to that outlined above could be undertaken for other portfolios. In particular, following Okhrin and Schmid (2006), we can examine the global minimum variance (GMV) and the Sharpe ratio (SR) portfolios. It is then straightforward to show that in the case of the GMV portfolio, the portfolio mean $\hat{\alpha}_{GMV}$ is an unbiased estimator of $\alpha_{GMV} = \beta/\gamma$. For the SR portfolio, moments of $\hat{\alpha}_{SR}$ do not exist since conditionally it is distributed as the inverse of a Normal random variable. The nonexistence of moments for the SR portfolio return poses potential problems for those wishing to simulate optimal portfolios.

10.7 Section 4

We now illustrate the accuracy of these approximations using two contrasting numerical examples. In both cases, we have $\lambda = 12.5$ and $h = 4$, giving true values of $\alpha = 0.02$, TE $= 0.04$, and $IR = 0.5$. These values correspond to the sorts of numbers found in institutional investment for active managers measured on an annualized basis. We now consider two cases.

(i) $T = 180$, $N = 4$, so that $n = 1/60 = 0.01667$
(ii) $T = 180$, $N = 80$, so that $n = 79/180 = 0.43889$.

The results are given in the following table.

	α	TE	IR
The true values	0.02	0.04	0.5
Case I ($T = 180$, $N = 4$)			
Exact Expected Values	0.021726	0.041410	0.517621
Approx	0.021695	0.041660	0.520756
Case II ($T = 180$, $N = 80$)			
Exact Expected Values	0.100202	0.089062	1.113272
Approx	0.098218	0.088642	1.108027

What the above results illustrate is the fact that while the estimators $\hat{\alpha}$ and $\overline{TE}$ are always biased, the bias is very small when n is small. However, for large n, the bias is extremely large, being more than four times the true value for α and greater than twice the true value for TE and IR. We also notice that in both cases, the approximation is quite accurate.

Keeping our numerical results consistent with those in Scherer (2002, p. 165), we will consider two cases.

$$\text{(i)}\quad Q = \begin{pmatrix} a & \beta \\ \beta & \gamma \end{pmatrix} = \begin{pmatrix} 0.125 & 6.5 \\ 6.5 & 675.00 \end{pmatrix}$$

giving $h = \gamma / a\gamma - \beta^2 \approx 16$ and

$$\text{(ii)}\quad Q = \begin{pmatrix} 0.3 & 6.5 \\ 6.5 & 800 \end{pmatrix}$$

with $h = 4$

In each case, by choosing different values for λ (risk aversion parameter), we can generate a wide set of values for both active returns $\alpha = \mu'\omega = \frac{1}{\lambda h}$ and tracking error $\text{TE} = \left(1/\lambda\sqrt{h}\right)$. The following table highlights this relationship.

Some authors such as Grinold and Kahn (1999) express the units associated with the active return, α, and the tracking error, TE, in terms of percent. Others, e.g., Scherer (2002), use the decimal equivalent. However, shifting the units from decimal to percent will alter λ, the risk aversion parameter, by a factor of 100. That is, the λ associated with percent units will be 100th of the value of λ associated with decimal units. Thus, the following constellations of parameter values listed in the two panels below are consistent:

	$h = 16$				$h = 4$		
λ	α	TE	IR	λ	α	TE	IR
2	0.03125	0.125	0.25	12.5	0.02	0.04	0.5
0.02	3.125	12.5	0.25	0.125	2	4	0.5

In what follows, we choose the decimal representation.

	$h = 16$			$h = 4$		
λ	α	TE	IR	α	TE	IR
2	0.03125	0.125	0.25	0.125	0.25	0.5
4	0.0156	0.0625	0.25	0.0625	0.125	0.5
6	0.0104	0.04167	0.25	0.04167	0.0833	0.5
8	0.0078	0.03125	0.25	0.03125	0.0625	0.5
12.5				0.02	0.04	0.5

We now examine the tracking error optimization and the performance, i.e., relative bias of the standard estimators for different portfolio sizes, $N = 4$ and $N = 80$ with $T = 180$ in both cases.

θ:	$h = 16, \lambda = 2$			$h = 4, \lambda = 12.5$		
	$\alpha =$ 0.03125	TE = 0.125	$IR = 0.25$	$\alpha = 0.02$	TE = 0.04	$IR = 0.5$
$N = 4E(\hat{\theta})$	0.0407	0.1378	0.2757	0.0217	0.0414	0.5176
% Rel. Bias	30.24	10.24	10.28	8.50	3.50	3.52
$N = 80E(\theta)$	0.4558	0.4747	0.9494	0.1002	0.0891	1.1133
% Rel. Bias	1358.56	279.76	279.76	401.00	122.75	122.66

Again, we see evidence of large relative biases in the large N case, pointing to quite staggering inaccuracy.

10.8 Section 5: General linear restrictions

The results of the previous sections can be readily extended to incorporate general linear restrictions on the relative weights. Here we briefly outline the results; the full derivation is available from the authors upon request. We now consider the maximization of utility subject to a set of K restrictions: $R(\omega - b) = 0$, where R is a $K \times N$ matrix. The Lagrangian and the associated first-order conditions for the relative case are as follows:

$$\begin{aligned} L &= \mu'(\omega - b) - \frac{\lambda}{2}(\omega - b)'\Omega(\omega - b) + \theta' R(\omega - b) \\ \frac{\partial L}{\partial \omega} &= \mu - \lambda\Omega(\omega - b) + R'\theta = 0 \\ \frac{\partial L}{\partial \theta} &= R(\omega - b) = 0 \end{aligned}$$

Solving, we find:

$$\omega = b + \frac{1}{\lambda}\Omega^{-1}(\mu - R'(R\Omega^{-1}R')R\Omega^{-1}\mu)$$

resulting in

$$\begin{aligned} \alpha &= \mu'(\omega - b) = \frac{1}{\lambda}\left[\mu'\Omega^{-1}\mu - \mu'\Omega^{-1}R'(R\Omega^{-1}R')^{-1}R\Omega^{-1}\mu\right] \\ &= \frac{1}{\lambda}(1,0)\hat{Q}_K^{-1}(1,0)' \end{aligned}$$

and

$$\begin{aligned} \sigma^2 &= (\omega - b)'\Omega(\omega - b) \\ &= \frac{1}{\lambda^2}(1,0)\hat{Q}_K^{-1}(1,0)' \end{aligned}$$

where

$$\hat{Q}_K = (\hat{\mu}, R')'\hat{\Omega}^{-1}(\hat{\mu}, R')$$

with

$$\hat{Q}_K^{-1} \sim W_{K+1}(T - N + K, \frac{1}{T}Q_K^{-1})$$

Following earlier results, we now define:

$$\hat{h}_K = (1,0)\ \hat{Q}_K^{-1}\ (1,0)'$$

and we have immediately, corresponding to Equation (10.7):

$$\hat{h}_K \sim \frac{\chi^2_{(T-N+K)}}{\chi^2_{(N-K,(T/h_K))}}$$

where $h_K = (1,0)'Q_K^{-1}(1,0)$ with $Q_K = (\mu,R')'\Omega^{-1}(\mu,R')$.

Thus, by a simple substitution into our earlier results, we can readily specify the exact distribution and moments of $\hat{\alpha}$, $\overline{IR}$, and $\overline{TE}$. That is, we merely replace $N - 1$ by $N - K$ and h by h_K. As intuition suggests, increasing the number of restrictions is exactly the same as reducing the number of assets. However, the noncentrality parameter h_K will change as the constraints change.

This is clear from the following. If we let:

$$\underset{K\times N}{R} = \begin{pmatrix} r_1' \\ r_2' \\ r_K' \end{pmatrix}$$ where r_1' is a $1 \times N$ vector.

Then,

$$Q_K = \begin{bmatrix} \mu'\Omega^{-1}\mu & \mu'\Omega^{-1}R' \\ R\Omega^{-1}\mu & R\Omega^{-1}R' \end{bmatrix}$$

$$= \begin{bmatrix} a & \vdots & \beta_1 & \beta_1 & & \beta_K \\ \beta & \vdots & \gamma 1 & \gamma_{12} & & \gamma_{1K} \\ \vdots & \vdots & & & & \\ \beta_K & \vdots & \gamma_K & & & \gamma_{KK} \end{bmatrix} = \begin{bmatrix} a & \beta' \\ \beta & \Gamma \end{bmatrix}$$

and h_K will be the (1,1)-th element of the inverse of the $(K + 1)$-th principal minor of Q_K.

That is,

$$h_K = (a - \beta'\Gamma^{-1}\beta)^{-1}$$

That is, for $K = 1$, $h_k = h_1 = (a - \beta_1^2 / \gamma_{11})^{-1}$ and when

$$K = 2,\ h_K = h_2 = \left[a - (\beta_1 \beta_2)\begin{bmatrix}\gamma_{11} & \gamma_{12} \\ \gamma_{12} & \gamma_{22}\end{bmatrix}^{-1}\begin{pmatrix}\beta_1 \\ \beta_2\end{pmatrix}\right]^{-1} \quad \text{etc.}$$

When the restrictions are orthogonal, in the sense that $\Gamma = R\Omega^{-1} R'$ is a diagonal matrix, Γ^{-1} will have a simple representation along with h_k. In this case,

$$h_K = \left(a - \sum_{i=1}^{K} \beta_i^2 / \gamma_{ii}\right)^{-1}$$

which illustrates quite clearly that as K increases h_k also increases. To see the effect on the estimators, consider the bias in $\hat{\alpha}$. From our results in Section 3, we have:

$$E(\hat{\alpha}) - \alpha = \frac{N - K}{\lambda(T - N + K)} + \frac{1}{\lambda h_K}\left(\frac{T}{(T - N + K)} - 1\right)$$

and thus as K increases the bias will tend to be zero. In the more general case, the same argument applies as long as h_k is bounded from below.

In the case of inequality constraints, the problem is more complex. This problem has been discussed in Jagannathan and Ma (2002), although they consider upper and lower constraints on the portfolio proportions only (see Equations (10.1–10.4), p. 6, 2002).

To convert a realistic optimization into an exact problem, we consider the Kuhn–Tucker conditions, appropriate to quadratic utility.

Our problem now becomes $\max L = \mu'\omega - \frac{\lambda}{2}\omega'\Omega\omega$ as before, but now, we consider K constraints of the form $A\omega \leq b$ and also no short sales $\omega \geq 0$.

Our Kuhn–Tucker conditions are now:

$$\begin{aligned} A\omega + v &= b \\ \lambda\Omega\omega + A'u - y &= \mu \end{aligned}$$

and $\omega \geq 0$, $u \geq 0$, $y \geq 0$ and $v \geq 0$ plus the complementary constraint $\omega'y + u'v = 0$.

Because of the concavity of the objective function and linearity of the constraints, Kuhn–Tucker conditions apply and ω will be optimal if we can find

u, y, and v such that all four vectors together satisfy the above constraints. If we wished to capture explicitly the fact that some of the inequality constraints are actually equality constraints, then further refinements are necessary.

To simplify this problem but to consider the impact of constraints, we shall consider our calculations when $\omega_1 \geq 0$, but otherwise the problem is as in Equation (10.11).

Suppose that we constrained ω_1 to $\tilde{\omega}_1$ such that $\tilde{\omega}_1 \geq 0$. If the original $\hat{\omega}_1 \geq 0$, $\tilde{\omega}_1 = \hat{\omega}_1$, if $\hat{\omega}_1 < 0$, $\tilde{\omega}_1 = 0$. The distribution of $\tilde{\pi}$ and $\tilde{\sigma}^2$ will be as before if $\hat{\omega}_1 \geq 0$, but with N reduced by 1 and all parameters approximately adjusted. As we increase the number of constraints to K, say, such that $\omega \geq 0$, we get 2^K regions corresponding to all cases where constraints bite or not. In each of these regions, the distribution may differ.

Sharpe (1970), Best (2000), Best and Grauer (1991), and no doubt many others mention that a description of the constrained frontier consists essentially of solving for the corner portfolios, see, for example, Sharpe (1970, p. 66); these being the set of efficient portfolios where the set of active constraints change. Ordering these portfolios by expected return, and considering any two adjacent portfolios, fund separation will apply to all the funds in between, i.e., they can be treated as linear combinations of the two adjacent corner portfolios plus other portfolios based on the constraints that bite. Unfortunately, in the context of our problem, the stochastic nature of the means and covariances implies that the corner portfolios become stochastic, and this gives rise to different numbers of constraints holding and consequent mixtures of distributions for alpha and tracking error.

Similar issues arise if we consider a mean–variance frontier subject to inequality constraints, and the frontier will consist of different quadratic segments, between ranked corner solutions, the curvature of which are determined by the number and nature of binding constraints.

10.9 Section 6

We now consider the standard mean–variance optimal portfolio problem and use our earlier exact results to develop an efficient simulation algorithm for the frontier and 95% confidence intervals. Our confidence interval is based on the upper and lower 2.5% quantiles. Let ω be an $(N \times 1)$ vector of portfolio weights. Returns are assumed as $N(\mu, \Omega)$, i is an $(N \times 1)$ vector of ones, and π is a given level of returns.

Here, the problem is to minimize $\omega'\Omega\omega$ subject to: $\mu'\omega = \pi$ and $i'\omega = 1$, Thus, the Lagrangian is given by:

$$L = \tfrac{1}{2}\omega'\Omega\omega - \theta_1(\mu'\omega - \pi) - \theta_2(i'\omega - 1)$$

with

$$\frac{\partial L}{\partial \omega} = \Omega\omega - \theta_1\mu - \theta_2 i = 0$$
$$\frac{\partial L}{\partial \theta_1} = \mu'\omega - \pi = 0$$
$$\frac{\partial L}{\partial \theta_2} = i'\omega - 1 = 0$$

Consequently,

$$\omega = \Omega^{-1}(\mu, i)\begin{bmatrix}\theta_1\\ \theta_2\end{bmatrix}$$

where

$$\begin{bmatrix}\mu'\Omega^{-1}\mu & \mu'\Omega^{-1}i\\ i'\Omega^{-1}\mu & i'\Omega^{-1}i\end{bmatrix}\begin{bmatrix}\theta_1\\ \theta_2\end{bmatrix} = \begin{bmatrix}\pi\\ 1\end{bmatrix}$$

That is,

$$\hat{\omega} = \hat{\Omega}^{-1}(\hat{\mu}, i)\hat{Q}^{-1}\begin{bmatrix}\pi\\ 1\end{bmatrix}$$

and therefore,

$$\sigma^2 = \hat{\omega}'\hat{\Omega}\hat{\omega} = (\pi\, 1)'\hat{Q}^{-1}\begin{bmatrix}\pi\\ 1\end{bmatrix}$$
$$= p'\hat{Q}^{-1}p, \text{ where } p = \begin{bmatrix}\pi\\ 1\end{bmatrix}$$

From our earlier results, $p'\hat{Q}^{-1}p \mid \hat{\mu} \sim W_1(T - N + 1, \frac{1}{T}p\bar{Q}^{-1}p)$, where $W_1(\cdot)$ is a Wishart of dimension 1. Here, $\bar{Q} = (\hat{\mu}, i)'\Omega^{-1}(\hat{\mu}, i)$, and therefore, $p'\hat{Q}^{-1}p \sim \chi^2_{(T-N+1)} \cdot \Psi$, where χ^2_m is a chi-squared with m degree of freedom. with

$$\Psi = \frac{1}{T}p'\bar{Q}^{-1}p$$

Next, letting

$$\bar{Q} = \begin{bmatrix} \bar{a} & \bar{b} \\ \bar{b} & \bar{c} \end{bmatrix}$$

we have:

$$\Psi = \tfrac{1}{T}(\bar{a} - \bar{b}^2 / \bar{c})^{-1}(\bar{b} / \bar{c} - \pi)^2 + \frac{1}{T\bar{c}}$$

and via a transformation, $u = \sqrt{T}\Omega^{-\frac{1}{2}}\hat{\mu}$ and $\alpha = \Omega^{-\frac{1}{2}}i$, we get:

$$T(\bar{a} - \bar{b}^2 / \bar{c}) \sim \chi'^2_{(N-1, T(a-b^2/c))}$$

and

$$\bar{b}/\bar{c} \sim N(b/c, 1/Tc)$$

Therefore,

$$\sigma^2 = p'\hat{Q}^{-1}p = k_1\left[(k_2 - \pi)^2 / k_3 + \frac{1}{T\bar{c}}\right]$$

where the three random variables

$$k_1 \sim \chi^2_{(T-k+1)}, k_2 \sim N(b/c, 1/T\bar{c})$$

and $k_3 \sim \chi'^2_{(N-1, T(a-b^2/c))}$ are mutually independent.

We now have a straightforward method to generate, via simulation, the mean–variance frontier and confidence intervals, which does not require having to simulate the full portfolio.

For specified $\hat{a}$, $\hat{b}$, and $\hat{c}$ along with T and N, and taking 5,000 replications as an example:

1. Generate 5,000 observations on the three independent random variables, k_1, k_2, and k_3 i.e., k_{1j}, k_{2j}, k_{3j}, $j = 1, \ldots, 5{,}000$.

2. Select 201 values of π centered on $\hat{b}/\hat{c}$, i.e., choose an interval $\hat{b}/\hat{c} \pm 3(\hat{b}/\hat{c})$ with 100 points above $\hat{b}/\hat{c}$ and 100 below $\hat{b}/\hat{c}$.
3. Now for each value of π_ℓ, $\ell = 1,\cdots,201$, we can use

$$\sigma_j = \sqrt{k_{1j}\left(((k_{2j} - \pi_\ell)^2 / k_{3j}) + (1/T\hat{c})\right)}$$

to generate 5,000 values of σ_j.
4. From the 5,000 values of σ_j, for each π_ℓ we can calculate the mean and the 2.5 and 97.5 percentile, i.e., $\bar{\sigma}_\ell, \sigma_\ell^L$, and $\bar{\sigma}_\ell^u$ respectively.
5. Now plot the pairs of points $(\bar{\sigma}_\ell, \pi_\ell)$, $(\sigma_\ell^L, \pi_\ell)$, and $(\sigma_\ell^u, \pi_\ell)$ for $\ell = 1,\ldots,201$ giving the average mean–variance frontier and the 95% confidence limits.

From the discussion in Section 5, it is clear that a similar algorithm to that above can be used to generate the frontier in a situation with general linear restrictions. However, a simple algorithm for inequality constraints based on our approach does not seem easily attainable. We do not present results on computational efficiency gains but it is clear that for large N and T, they should be considerable.

10.10 Section 7: Conclusion

This chapter has explored the link between Monte Carlo averages and population moments to assess the merits of Monte Carlo–based optimization methods that compute average frontiers. To this end, we have collected together a number of results on exact properties of portfolio measures, some of which already exist in the literature. We extend these results to include relative and absolute return utility and relative and absolute mean–variance frontiers.

We compute biases for the optimal portfolios, alpha, volatility, and the information ratio. We detect significant biases for the case when the number of assets increases with the sample size, a case of great practical relevance. We further show that when the problem is constrained, these biases are reduced. This sheds some light on the practitioner approach to mean–variance optimization of imposing large numbers of constraints. Not only does this control the optimization, but, if the constraints are valid, it reduces the bias as well. Finally, averaging simulated optimization can be seen as a satisfactory procedure if N is small relative to T, or if N is large and K is large relative to T, or if the average optimal portfolio or its moments are bias-corrected. We have not investigated the impact of N, K, and T on the width of the simulated confidence intervals for the key parameters nor have we considered how we might extend our analysis to the Kuhn–Tucker problems discussed in Section 5; these remain topics for future research.

Acknowledgment

We wish to thank Harry Markowitz for his comments and encouragement. We also thank Phelim Boyle, participants of the 8th Annual Financial

Econometrics Conference, Waterloo, 2006, the participants of the Numerical Methods in Finance conference in Dublin, 2006, and the participants of the New Zealand Econometric Conference Dunedin, 2006, especially Isao Ishida. The first author acknowledges financial support from NSERC Canada.

References

Best, M. J. (2000). An algorithm for the solution of the parametric quadratic programming problem. In H. Fischer, B. Riedmuller, & S. Schaffler (Eds.), *Applied mathematics and parallel computing—Festschrift for Klaus Ritter*. Heidelburg: Physica-Verlag.

Best, M. J., & Grauer, R. B. (1991). On the sensitivity of mean–variance–efficient portfolios: Some analytical and computational results. *The Review of Financial Studies*, *4*, 315–342.

Bodnar, T., & Schmid, W. (2006). *Mean variance portfolio analysis under parameter uncertainty*. Frankfurt: Mimeo Department of Statistics, European University Viadrina.

Britten-Jones, M. (1999). The sampling error in estimates of mean–variance efficient portfolio weights. *Journal of Finance*, *54*, 655–671.

Frahm, G. (2007). *Linear statistical inference for global and local minimum variance portfolios*. Discussion Paper. Department of Economic and Social Statistics, University of Cologne.

Green, R. C., & Hollifield, B. (1992). When will mean–variance efficient portfolios be well diversified? *Journal of Finance*, *47*, 1785–1809.

Grinold, R. C., & Kahn, R. N. (1999). *Active portfolio management: A quantitative approach for producing superior returns and selecting superior returns and controlling risk*. New York: McGraw Hill Inc., New York.

Harvey, C., Leichty, J. C., Leichty, M. W., & Muller, P. (2004). *Portfolio selection with higher moments*. Working Paper.

Hillier, G., & Satchell, S. (2003). *Some exact results for efficient portfolios with given returns. New advances in portfolio construction and implementation*. London: Butterworth-Heinemann.

Jagannathan, R. & Ma, T. (2002). *Risk reduction in large portfolios: Why imposing the wrong constraint helps*. NBER Working Paper 8922.

Jagannathan, R., & Ma, T. (2003). Risk reduction in large portfolios: Why imposing the wrong constraint helps. *Journal of Finance*, *58*, 1651–1683.

Jobson, J. D. (1991). Confidence region for the mean–variance efficient set: An alternative approach to estimation risk. *Review of Quantitative Finance and Accounting*, *1*, 235–237.

Jobson, J. D., & Korkie, R. (1989). A performance interpretation of multivariate tests of asset set intersection, spanning, and mean–variance efficiency. *Journal of Financial and Quantitative Analysis*, *24*, 185–204.

Johnson, N. L., & Kotz, S. C. (1972). *Distributions in statistics: Continuous multivariate distributions*. New York: John Wiley & Sons.

Michaud, R. (1998). *Efficient asst management*. Cambridge: Harvard Business School Press.

Muirhead, R. J. (1982). *Aspects of multivariate statistical theory*. New York: John Wiley & Sons.

Okhrin, Y., & Schmid, W. (2006). Distributional properties of portfolio weights. *Journal of Econometrics*, *134*, 235–256.

Satchell, S. E., & Scowcroft, A. (2003). *New advances in portfolio construction and implementation*. London: Butterworth and Heinemann.

Scherer, B. (2002). *Portfolio construction and risk budgeting*. Risk Books.

Sharpe, W. F. (1970). *Portfolio theory and capital markets*. New York: McGraw-Hill.

Sharpe, W. F. (1981). Decentralised investment management. *Journal of Finance*, *36*, 217–234.

Stein, J. (2002). *Portfolio theory and market fluctuations* MSc Dissertation in Mathematical Finance. Oxford University.

11 Heuristic portfolio optimization: Bayesian updating with the Johnson family of distributions

Richard Louth

Executive Summary

This chapter seeks to advance the methodological basis for the use of non-Gaussian alternatives to traditional mean–variance analysis for large-dimension portfolio optimization problems. Through the application of a threshold acceptance algorithm and an appropriately chosen distribution from the Johnson family, we show how portfolio weights for an otherwise computationally intensive asset allocation problem can be obtained in a quick and efficient manner. The inherent flexibility and generality of this approach is further illustrated by two practical extensions. First, we introduce the idea of data reweighting as a simple and, most importantly, computationally tractable method for improving the robustness of our optimization algorithm. And second, we demonstrate how return forecasts ("alphas") can be seamlessly incorporated into the estimation procedure via Bayesian updating. In this case, we utilize an important property of the Johnson family to build a model of the joint dependence between returns and their forecasts using the class of meta-elliptical distributions.

11.1 Introduction

Contrary to overwhelming empirical evidence, the most popular commercial portfolio optimizers are still based on Gaussian fundamentals. For large-dimension problems in particular, the lure of analytical tractability and computational convenience afforded by Gaussianity appears too tempting to resist. Yet, the recent financial turmoil has reignited the debate surrounding the adequacy of the Gaussian paradigm and highlighted the potential dangers associated with Gaussian assumptions in a non-Gaussian world. Moreover, the relentless nature of Moore's Law means that computational barriers to alternative approaches are continually being eroded as new algorithms are developed and the power of parallel computing is realized.[1]

[1] In 1965, Intel cofounder Gordon Moore correctly predicted that the number of transistors that can be placed on an integrated circuit would double approximately every 2 years (Moore, 1965).

Doi:10.1016/B978-0-12-374952-9.00011-7.

The purpose of this chapter is to introduce one such alternative and to illustrate more generally how solutions to otherwise intractable optimization problems can be obtained with modest amounts of computational power via an appropriate choice of local search algorithm and parametric density specification. With regards to the former, a deterministic threshold acceptance algorithm is used to generate transitions between portfolio weights while simultaneously avoiding local solutions (Dueck & Winker, 1992). For the latter, one of the Johnson family of distributions is fitted to the time series of portfolio returns (Johnson, 1949). Johnson distributions are uniquely determined by the first four moments, which can be chosen mutually independently and so they are well suited to modeling the complex nature of asset returns.

The combination of search algorithm and density estimation procedure is then used to maximize a hitherto impractical disappointment-averse expected utility function for portfolios comprising large numbers of assets. The results of extensive simulation and out-of-sample testing reveal that the algorithm performs well under a variety of circumstances and constitutes a viable alternative to the traditional mean–variance approach to asset allocation. Since out-of-sample performance is a critical aspect of this success, we also describe two simple and most importantly, computationally efficient extensions to the baseline algorithm that further improve the robustness of the final allocation.

The first of these extensions describes how data reweighting techniques can be used to give more weight to the observations that we believe are most informative about the current state of the underlying data-generating process, whereas the second illustrates in a more direct manner how asset return forecasts can be used to guide the search process. In the case of the latter, the posterior density of portfolio returns is derived from the *meta-Gaussian* model of Kelly and Krzysztofowicz (1995).

The rest of this chapter is structured as follows. Section 11.2 surveys the literature pertaining to portfolio optimization in large dimensions, with particular emphasis on heuristic algorithms. The theoretical properties of Johnson family of distributions are reviewed in Section 11.3 along with an explanation of the estimation procedure. Section 11.4 describes the utility maximization problem and illustrates how the density estimation and threshold acceptance algorithms are combined to obtain the optimal portfolio allocation. The related ideas of data reweighting and Bayesian updating are then introduced in Sections 11.5 and 11.6 as two possible extensions to the baseline framework. Finally, the techniques are applied to a portfolio of equities from the FTSE 100 list of companies in Section 11.7. Section 11.8 concludes.

11.2 A brief history of portfolio optimization

In his seminal work, Markowitz (1952) described how an investor should allocate his or her wealth when asset returns are normally distributed. Within this framework, he showed that the complex problem of portfolio optimization reduces to the much simpler task of minimizing the variance of portfolio returns for a given mean return, or vice versa. That this approach still underpins many of the asset allocation decisions carried out within the finance industry today is a testament to the ease with which the optimal portfolio can be constructed. Even in its many modern incarnations, the mean–variance optimal portfolio is often no more than the solution to a simple quadratic programming problem.

Yet, even in the 1960s, researchers were already beginning to gather evidence against the Gaussian hypothesis.[2] In this case, only under quadratic utility does the mean–variance approach retain the property of directly maximizing expected utility.[3] For all other utility functions, mean–variance optimization can at best be interpreted as a second-order approximation of expected utility maximization (Rubinstein, 1973; Kraus & Litzenberger, 1976). Asset allocation in this context is much more complicated and can quickly become intractable, particularly in large dimensions.

Arguably, the most natural solution to this problem is to consider higher-order approximations of expected utility. As Samuelson (1970) and Kraus and Litzenberger (1976) demonstrate, we can obtain an expression for expected utility that depends linearly on the higher moments of the portfolio return by replacing the utility function with its infinite-order Taylor expansion. Unfortunately however, for utility functions such as the power variety, the series expansion only converges under very restriction conditions (Lhabitant, 1998). And even for well-behaved functional forms, there is little consensus over the appropriate point to truncate the expansion (cf. Bansal, Hsieh, & Viswanathan, 1993). What's more, this choice is made even more difficult because of the fact that the inclusion of additional moments does not necessarily improve the quality of the approximation (Brockett & Garven, 1998; Berényi, 2001). Nevertheless, the inherent flexibility of this approach and the obvious parallels with the original mean–variance framework mean that it is still widely used in practice. Recent contributions include studies by Brandt, Goyal, Santa-Clara, and Stroud (2005), de Athayde and Flores (2004), Harvey, Liechty, Liechty, and Müller (2004), and Jondeau and Rockinger (2006). All of these studies use up to fourth-order expansions in order to capture preferences over skewness and kurtosis.

The alternative to using the Taylor expansion is to continue maximizing expected utility directly. In most applications however, this requires a cumbersome numerical integration that severely limits the number of assets that can be included. For instance, in a recent study that is very much related to our approach, Duxbury (2008) outlines a utility maximization problem for Johnson-distributed asset returns and power utility. However, his choice of

[2] There is strong empirical evidence that asset return distributions are characterized by leptokurtosis and negative skewness. See Mandelbrot (1963) and Fama (1965) for early evidence.

[3] Quadratic utility has numerous unappealing properties. In particular, it implies increasing absolute risk aversion and satiation, i.e., utility that is not everywhere monotonically increasing in wealth.

method of moments estimation and brute-force optimization algorithm mean that the methods can only be applied to a maximum of five assets.[4] It appears that the combination of non-Gaussianity, large numbers of assets, and the direct maximization of expected utility necessitates an alterative approach to optimization.

The alternative approach we consider is known as the "heuristic" approach to optimization. Although there are many types of optimization algorithm that can be classified as "heuristic," and a full description of each of them is beyond the scope of this chapter, we nevertheless note that it is the class of "local search" methods that have been most fruitfully applied to portfolio optimization.[5] Within this class, there are two important subclasses: first, there are "trajectory methods" that focus on a single solution at any point in time (e.g., threshold acceptance, simulated annealing, and tabu search), and second, there are "population methods" that update sets of potential solutions simultaneously (e.g., genetic algorithms, differential evolution methods, and particle swarm optimization).

Although each local search procedure is often very different from another, they are all based on the same underlying principle. That is, we start from an initial solution that may, or may not, be randomly generated and continue the search process by comparing the current solution with potential solutions in the same neighborhood. The decision to update the current solution is governed by the "quality" of the potential solution. If the decision is "positive," the current solution is replaced by the neighboring one and is then used as the new starting point for subsequent iterations of the algorithm. The whole process is continued until a termination criterion is satisfied.

Owing to the complex and computationally burdensome nature of many of the problems encountered in finance, these algorithms are at the heart of a rapidly growing literature. For instance, the threshold acceptance algorithm has been used by Gilli and Këllezi (2002) for index tracking problems with highly nonlinear constraints, and by Gilli, Këllezi, and Hysi (2006) in the construction of optimal portfolios under different downside risk measures. In addition, Maringer (2007) demonstrates how differential evolution methods can be used for utility maximization with loss aversion.[6] Whereas the tabu search was first applied to utility maximization by Glover, Mulvey, and Hoyland (1996).

With such a variety of algorithms on offer, it is no surprise that choosing between them is a difficult task. And this is made even more complicated by the fact that each one of them depends on multiple "tuning" parameters, which, if not chosen correctly, can greatly inhibit convergence to the optimal

[4] For further examples, we refer the reader to Ramchand and Susmel (1998) or Ang and Bekaert (2002) who consider regime-switching return distributions. In both cases, the maximum number of assets is less than four.

[5] See Winker (2001), Osman and Laporte (1996), and Gilli and Winker (2004) for a comprehensive overview of optimization heuristics and a detailed discussion of their various forms.

[6] See Maringer and Parpas (2009) for further information on the differential evolution algorithm.

solution.[7] In fact, it is our belief that the choice of algorithm is secondary to the method of choosing the "tuning" parameters that motivates our final decision to use the threshold acceptance algorithm. That is, we feel that the threshold accepting (TA) algorithm has the highest level of parsimony among local search algorithms, especially when we apply the data-driven method of tuning the threshold sequence described by Gilli *et al.* (2006). However, before we describe the specifics of this algorithm in greater detail, we first of all review the properties of the Johnson family of distributions, which underlie our expected utility calculations.

11.3 The Johnson family

The Johnson family is composed of four flexible distributions which include the normal, lognormal, bounded, and unbounded (Johnson, 1949). Johnson distributions are uniquely determined by their first four moments which can be chosen mutually independently, and so they are well suited to modeling the empirical distribution of asset returns. This flexibility also means that Johnson distributions can capture the fourth-order moment patterns of many of the popular parametric models used in financial econometrics such as the Skewed Student's-*t*, Pearson Type IV, and Normal Inverse Gaussian distributions to name but a few.

Despite this flexibility, however, the use of Johnson distributions in financial applications has been severely limited. This is especially surprising given the speed and computational ease with which the parameters can be estimated. Notable exceptions include studies by Perez (2004), Jobst (2005), Yan (2005), and Duxbury (2008). Specifically, Perez (2004) uses the Johnson system as a tool to analyze and model the nonnormal behavior of hedge fund indices, while Jobst (2005) uses the Johnson density for prewhitening before his GARCH analysis of spread dynamics within European asset-backed securities markets. Yan (2005) also uses Johnson distributions in a GARCH context, but as an alternative to the more familiar Gaussian and Student's-*t* error distributions. And finally, as we have already discussed, Duxbury (2008) uses the system of Johnson distributions for asset allocation purposes in the spirit of the approach we adopt in this chapter.

11.3.1 Basic properties

In his seminal work, Johnson (1949) described a family of three probability distributions based on various transformations to normality. The first, denoted S_L, is the familiar two- or three-parameter lognormal distribution; the second,

[7] For instance, although Chang, Meade, Beasley, and Sharaiha (2000) found that simulated annealing and the genetic algorithm outperformed tabu search in a large-dimension asset allocation problem, the general view among practitioners is that this result was due to an inappropriate choice of tuning parameters rather than the poor performance of the algorithm itself.

S_B, is a bounded distribution that has been called the four-parameter lognormal distribution; and finally, S_U, is an unbounded distribution based on an inverse hyperbolic sine transform.[8] Each of the three distributions in the Johnson family employs a transformation of the original variable to yield a standard Gaussian variate, i.e.,

$$Z = \gamma + \eta g\left(\frac{X-\varepsilon}{\lambda}\right) \tag{11.1}$$

where Z is a standard normal random variable, γ and η are shape parameters, λ is a scale parameter, ε is a location parameter, and $g(.)$ is one of the following:

$$g(x) = \begin{cases} x & \text{Normal Family: } S_N \\ \ln(x) & \text{Lognormal Family: } S_L \\ \ln\left(\dfrac{x}{1-x}\right) & \text{Bounded Family: } S_B \\ \ln(x+\sqrt{x^2+1}) & \text{Unbounded Family: } S_U \end{cases}$$

Furthermore, without loss of generality, we assume that $\eta > 0$ and $\lambda > 0$ and we observe the standard convention that $\lambda \equiv 1$ and $\varepsilon \equiv 0$ for S_N and $\lambda \equiv 1$ for S_L.

From Equation (11.1), it follows that the probability density function (pdf) for X is given by:

$$f(x) = \frac{\eta}{\lambda\sqrt{2\pi}} g'\left(\frac{X-\varepsilon}{\lambda}\right) \exp\left[-\frac{1}{2}\left(\gamma + \eta g\left(\frac{X-\varepsilon}{\lambda}\right)\right)^2\right] \tag{11.2}$$

$\forall\, x \in \Omega$, where $g'(.)$ is the first derivative of the function g(.), given by:

$$g'(x) = \begin{cases} 1 & \text{Normal Family: } S_N \\ x^{-1} & \text{Lognormal Family: } S_L \\ [(1-x)x]^{-1} & \text{Bounded Family: } S_B \\ (x^2+1)^{-1/2} & \text{Unbounded Family: } S_U \end{cases}$$

and the support of the distribution is:

$$\Omega = \begin{cases} (-\infty, +\infty) & \text{Normal Family: } S_N \\ [\varepsilon, +\infty] & \text{Lognormal Family: } S_L \\ [\varepsilon, \varepsilon+\lambda] & \text{Bounded Family: } S_B \\ (-\infty, +\infty) & \text{Unbounded Family: } S_U \end{cases}$$

[8] The three-parameter lognormal distribution differs from the more common two-parameter version in that the random variable is defined on the interval $[\alpha, \infty)$ rather than $[0, \infty)$.

The corresponding cumulative distribution function (cdf) is derived by integrating Equation (11.2) over the relevant domain of integration:

$$F(x) = \Phi\left(\gamma + \eta g\left(\frac{X - \varepsilon}{\lambda}\right)\right) \tag{11.3}$$

where $\Phi(z) = (2\pi)^{-1/2}\int_{-\infty}^{z}\exp(-w^2/2)dw$ denotes the cdf of a standard Gaussian distribution.

As we have already alluded to, the system of Johnson densities constitutes a very flexible and comprehensive way of capturing the four-moment patterns of financial time series. In fact, Johnson (1949) showed that the system can capture all feasible combinations of skewness and kurtosis (see Figure 11.1). In essence, the Johnson system divides the skewness–kurtosis plane into three regions by defining two curves and the point where kurtosis = 3 and skewness = 0 (corresponding to the normal distribution). The first region is the area below the line defined by the relationship kurtosis ≤ skewness^2 + 1, which is an impossibility for any probability distribution. The second region is between this line and the curve that defines the lognormal distribution; this region is covered by the S_B distribution. The remaining region above the lognormal curve is captured by the S_U distribution. Thus, when viewed as a complete system, the Johnson family is perfect for modeling the complex distributional shapes exhibited by asset returns.

While the properties of the normal and lognormal distributions are reasonably well known, readers may not be familiar with the basic properties of the

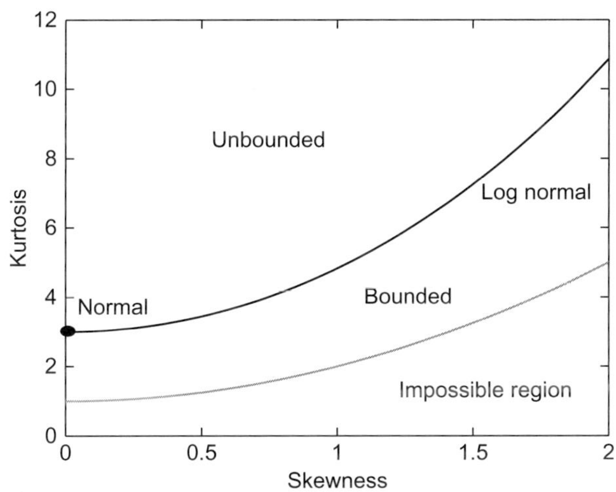

Figure 11.1 Skewness and kurtosis of the Johnson system of distributions.

S_B and S_U distributions. Although a full treatment is beyond the scope of this chapter, we nevertheless note a number of their more important features.[9] First, both are symmetric about their respective means if and only if $\gamma = 0$ and they exhibit negative (positive) skewness if $\gamma < 0$ ($\gamma > 0$). Moreover, for a fixed γ, kurtosis is increasing in η. Second, the S_U distribution is unimodal for all parameter values, whereas the S_B distribution may be either unimodal or bimodal depending on the choice of γ and η. Finally, we note that for both S_U and S_B, the density $f(x)$ and all of its derivatives tend to zero as x tends to extreme values in its support Ω. This means that the density is a perfectly smooth, infinitely differentiable, function of x for all real values of x.

11.3.2 Density estimation

Aside from providing an insight into the flexibility of the Johnson system, Figure 11.1 has a more practical application in that it is often the initial stage for many of the popular estimation methodologies used to fit Johnson densities (Johnson, 1949). For instance, Hill, Hill, and Holder (1976) use estimates of the sample skewness and kurtosis to identify the appropriate functional form of the Johnson density before implementing their method of moments procedure. Unfortunately however, this approach is infeasible for our portfolio optimization problem because the parameter estimates can only be obtained via computationally demanding recursive formulae.

With this in mind, we choose an alternative estimation procedure based on the method of quantile estimation introduced by Slifker and Shapiro (1980); see also Wheeler (1980). This method is not only easier to implement than the method of moments estimator, but simulation studies have also shown that it produces superior estimates of the parameters for the S_B and S_U distributions; see Wheeler (1980, p. 727).[10] The theoretical justification for this approach is that each member of the Johnson system can be uniquely identified by the distances between the tails and in the central portions of the distribution, much in the same way that they can also be indentified by their skewness–kurtosis values. However, in this case, computationally convenient formulae for each of the parameters are readily available and so estimation is much quicker than alternative approaches.

Mathematically, the Slifker and Shapiro (1980) algorithm begins by fixing a number $z > 0$ which implicitly defines three intervals of equivalent distance between the values $-3z$, $-z$, z, and $3z$. After choosing z, the next step is to find the cumulative probabilities of the normal distribution corresponding to each of the selected z values. For example, if $3z = 1.645$, then $\Phi(3z) = 0.95$. After obtaining the probabilities for each of the four points, $-3z$, $-z$, z, and

[9] We refer the reader to Stuart and Ord (1994) or Hahn and Shapiro (1967) for a more comprehensive treatment.

[10] The algorithms proposed by Slifker and Shapiro (1980) and Wheeler (1980) are almost identical, aside from the fact that the former uses four quantile points whereas the latter uses five.

$3z$, the corresponding quantile of the data must be estimated. This is usually done by solving the equation $(i - 0.5)/n = \Phi(.)$ for the value of i, where $\Phi(.)$ corresponds to one of the four selected points. Usually, the value of i obtained from the above equation will not be an integer, and therefore it is necessary to interpolate. The four data values obtained correspond to x_{-3z}, x_{-z}, x_z, and x_{3z}, which are no longer equally spaced.

It is these values of x that allow us to distinguish between the functional forms of the Johnson distribution. For instance, Slifker and Shapiro (1980) proved that, for a bounded symmetrical Johnson distribution, the distances between each of the outer and inner points would be smaller than the distance between the two inner points, and that the converse would be true for the unbounded case. Consequently, by letting:

$$
\begin{aligned}
m &= x_{3z} - x_z \\
n &= x_{-z} - x_{-3z} \\
p &= x_z - x_{-z}
\end{aligned}
$$

it follows that each of the four types of Johnson density are described by the following inequalities:

$$
\begin{aligned}
&\frac{mn}{p} > 1 && \text{for any } S_U \text{ distribution} \\
&\frac{mn}{p} < 1 && \text{for any } S_B \text{ distribution} \\
&\frac{mn}{p} = 1 \text{ and } \frac{m}{p} \neq 1 && \text{for any } S_L \text{ distribution} \\
&\frac{mn}{p} = 1 \text{ and } \frac{m}{p} = 1 && \text{for any } S_N \text{ distribution}
\end{aligned}
$$

The proof of this proposition can be found in Slifker and Shapiro (1980, pp. 243–246). After the functional form of the distribution has been determined, the parameters of the corresponding distribution are then derived by solving a system of equations. Expressions for the parameters are described in Appendix 11.9.1.[11]

It remains to determine the choice of $z > 0$. Slifker and Shapiro (1980) suggest that this value should be dependent on the size of the data set, although they conclude that a value around $z = 0.5$ is reasonable in most cases.[12] Our

[11] It should be noted that since the x's are continuous random variables, the probability that $mn/p^2 = 1$ is virtually zero and so it is necessary to define a tolerance level about unity.

[12] The value is usually less than 1 because a choice of $z > 1$ would make it very difficult to estimate the quantiles corresponding to $\pm 3z$ with any degree of precision.

experience is that a value between $z \in [0.65, 0.75]$ is preferred for financial applications. This choice not only reflects the large amounts of financial data available to practitioners, but also the greater emphasis on capturing the tail properties of return distributions. That is, we are taking advantage of the fact that when fitting the distribution, the Johnson approximating distribution will fit the data exactly at the chosen quantiles.

To the extent that the quality of the parameter estimates are sensitive to the choice of z, we also define an optimal value, z^*, which is based on a grid-search procedure across a range of z values. For each z value in the grid, we obtain parameter estimates using the quantile estimator and then apply Equation (11.1) to yield an approximately Gaussian series (Jobst, 2005). The Lilliefors test for normality is then used to select the optimal value of z such that the transformed data are the closest to Gaussian (Lilliefors, 1967). A comparison of our optimal z^* with various fixed values is provided in Appendix 11.9.3. In the majority of cases, we find that a fixed value of $z \in [0.65, 0.75]$ and our optimal z^* consistently deliver the lowest levels of relative bias. However, the appeal of using z^* is illustrated by the lower values of the relative root mean square error (RMSE).

We conclude this section by investigating the finite sample properties of the Slifker and Shapiro (1980) estimator and its performance under misspecification. In the latter case, we estimate the parameters of the Johnson distribution using data from an alternative parametric density. If the quantile estimator performs well, then our estimated distribution should capture important features of the true underlying density and the true value of expected utility. Our simulation results are reported in Appendix 11.9.4 for three alternative densities; they are the Normal Inverse Gaussian (Barndorff-Nielsen, 1997; Eriksson, Ghysels, & Wang, forthcoming), Pearson Type IV (Heinrich, 2004), and Skew Student's-*t* (Jondeau & Rockinger, 2003). In accordance with our discussion in Section 11.3.1, the results indicate that the method of quantiles can capture the four-moment patterns of the three alternative densities with a high degree of accuracy. For our purposes, however, the more important finding is that this is also true for the estimates of expected utility. Thus, within a portfolio optimization framework where the primary goal is to estimate expected utility, the quantile estimator appears robust to potential misspecification.

11.3.3 Simulating Johnson random variates

To obtain draws from the Johnson system of distributions, we simply reverse the operation described by Equation (11.1). That is, we generate a standard normal variate $Z \sim N(0, 1)$ and then apply the inverse translation:

$$X = \varepsilon + \lambda g^{-1}\left(\frac{Z - \gamma}{\eta}\right) \tag{11.4}$$

where for all real z we define the inverse translation function:

$$g^{-1}(z) = \begin{cases} z & \text{Normal Family: } S_N \\ \exp(z) & \text{Lognormal Family: } S_L \\ [(1 + \exp(-z))^{-1} & \text{Bounded Family: } S_B \\ (\exp(z) - \exp(-z))/2 & \text{Unbounded Family: } S_U \end{cases}$$

Repeated application of this procedure will yield a vector of draws from the appropriate Johnson density.

11.4 The portfolio optimization algorithm

In this section, we describe the investor's asset allocation problem and how the use of the threshold acceptance algorithm can be used to obtain an optimal set of portfolio weights.

11.4.1 The maximization problem

Choosing the appropriate objective function that our hypothetical investor wants to maximize is arguably the most important part of the entire portfolio optimization problem (Brandt, 2009). Although many different objective functions have been suggested, the academic literature has mainly focused on the class of hyperbolic absolute risk aversion (HARA) utility functions.[13] Within the HARA class, power or constant relative risk aversion (CRRA) utility is by far the most popular. The reason is that portfolio choices expressed as a percentage of wealth are independent of wealth, which facilitates normalizations that improve the tractability and transparency of the optimization procedure.

Despite this analytical elegance, the HARA class of utility functions have the unfortunate implication that downside losses and upside gains are treated symmetrically. This is contrary to a growing body of experimental evidence that suggests that decision makers are distinctly more sensitive to downside losses than upside gains (Tversky & Kahneman, 1991). Even in Markowitz's seminal treatise on portfolio optimization, he advocated the use of downside semivariance instead of variance as the appropriate measure of risk. It was only the intractability of semivariance that prevented him from pursuing this approach further (Granger, 2009).

In the academic literature, behavioral prospect theory has been the most common method of incorporating loss aversion into utility maximization prob-

[13] The utility function has also be replaced by numerous risk measures such as those described by Artzner, Delbaen, Eber, and Heath (1999). In this case, the utility maximization problem becomes a risk minimization problem.

lems.[14] For practical purposes, however, we prefer the more parsimonious disappointment aversion (DA) framework described by Gul (1991). In essence, DA preferences are a one-parameter extension of the expected utility framework and have the characteristic that favorable outcomes, i.e., outcomes above a certain target level, are given a lower weight relative to unfavorable outcomes. Since DA preferences are derived by relaxing the much lambasted independence axiom from standard decision theory, they allow us to retain many of the insights offered by expected utility theory and also to make comparisons between existing empirical works that use standard preference settings.

We therefore consider an investor who chooses a set of portfolio weights, w^*, in order to maximize the expected value of his or her disappointment-averse CRRA utility function:

$$U(W_{t+1}) = \frac{1}{K}\left(\int_{-\infty}^{W_{TA}} \tilde{U}(W_{T+1})dF(W_{t+1}) + A\int_{W_{TA}}^{\infty} \tilde{U}(W_{T+1})dF(W_{t+1})\right) \quad (11.5)$$

where $\tilde{U}(.)$ is the CRRA utility function,

$$\tilde{U}(W_{t+1}) = \begin{cases} \dfrac{W_{t+1}^{1-\alpha}}{1-\alpha} & \text{for } \alpha > 1 \\ \ln(W_{t+1}) & \text{for } \alpha = 1 \end{cases} \quad (11.6)$$

W_{t+1} is (normalized) end of period wealth, W_{TA} is the target level of wealth, $0 \leq A \leq 1$ is the coefficient of disappointment aversion, α is the coefficient of relative risk aversion,[15] and K is a scalar given by:

$$K = P(W_{t+1} \leq W_{TA}) + AP(W_{t+1} > W_{TA}) \quad (11.7)$$

Although this is a nonexpected utility function, CRRA preferences are a special case for $A = 1$. When $0 \leq A < 1$, individuals are averse to disappointment, i.e., outcomes below the target are weighted more heavily than the outcomes above the target. The relative reweighting of expected utility is illustrated in Figure 11.2 for $A = 0.5$ and $A = 1$.

[14] See Benartzi and Thaler (1995), Aït-Sahalia and Brandt (2001), Berkelaar, Kouwenberg, and Post (2004), and Gomes (2005) for recent applications of this approach.

[15] Larger values of this parameter signify increasingly risk-averse behavior. A value of $\alpha = 1$ corresponds to the maximization of logarithmic utility, which was popularized by Edward Thorp in the early 1960s for the game of Blackjack and subsequently applied to portfolio construction later that decade (Thorp, 1960; Thorp & Kassouf, 1967).

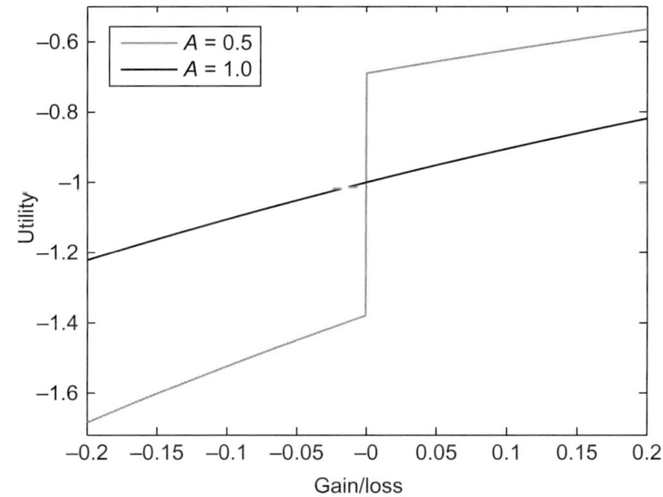

Figure 11.2 Utility from gains and losses.

In the case of $\alpha > 1$ and $A = 1$ for instance, the investor solves the following maximization problem:

$$\mathbf{w}^* = \arg\max \mathcal{U} \equiv E[U(1 + \mathbf{w}'\mathbf{r}_{t+1})]$$
$$\propto \int_{\Omega(\mathbf{w}'\mathbf{r}_{t+1})}^{\mathrm{w}} \frac{(1 + \mathbf{w}'\mathbf{r}_{t+1})^{1-\alpha}}{1-\alpha} f(\mathbf{w}'\mathbf{r}_{t-1})\, d(\mathbf{w}'\mathbf{r}_{t-1})$$

where $\mathbf{r}_{t+1}$ is a vector of returns from N assets, $\mathbf{w}$ is the corresponding vector of portfolio weights, and $f(\mathbf{w}'\mathbf{r}_{t+1})$ is the portfolio distribution that we estimate using the Johnson density algorithm outlined in the previous section.

We also introduce two types of constraint on the portfolio weights. First, the constraints $\mathbf{Bw} \leq \mathbf{b}$ restrict the percentage allocations in each sector or industry and define upper and lower bounds on the weights for each asset, whereas the gross exposure constraint $||\mathbf{w}||_1 \leq c$ prevents extreme positions.[16] In particular, $c = 1$ corresponds to the case of no short selling, while $c = \infty$ means that there are no constraint on short sales. In this study, we shall impose the constraints $w_i \geq 0$ and $\Sigma w_i = 1$, which correspond to $\mathbf{B} = -I_N$, $\mathbf{b} = 0$, and $c = 1$ in the above notation.

[16] $||\mathbf{w}||_1 = ||w_1|| + \ldots + ||w_N||$ is the L_1 norm of the weight vector. This constraint can also be interpreted as the transaction cost incurred by implementing the portfolio.

11.4.2 The threshold acceptance algorithm

The complex nature of the maximization problem described in the previous section necessitates the use of a heuristic optimization algorithm. In low dimensions, one could feasibly consider a grid-search procedure instead; however, this method becomes increasingly intractable for even a small numbers of assets (cf. Duxbury, 2008). In this section, we therefore describe the threshold acceptance optimization algorithm, which we use to derive the optimum set of portfolio weights.

TA was introduced by Dueck and Scheuer (1990) as a deterministic alternative to simulated annealing. It is a refined local search procedure that escapes local minima by accepting solutions that are not worse by more than a given threshold. The algorithm is deterministic in the sense that we fix a number of iterations, $N_{\text{Iterations}}$, and explore the neighborhood with a fixed number of steps during each iteration, N_{Steps}. The threshold is then decreased successively and reaches the value of zero in the last iteration. This procedure is then repeated N_{Restarts} times with different starting values in order to fully explore the parameter space. The optimal solution corresponds to the maximum out of all restarts. In total, the number of function evaluations is of the order O ($N_{\text{Restarts}} \times N_{\text{Iter}} \times N_{\text{Steps}}$). The pseudocode for the three constituent blocks of the TA algorithm: (1) optimization routine; (2) definition of a neighbor; and (3) choice of threshold sequence, is described in Appendix 11.9.2.

Although instructive, the pseudocode should be accompanied by two further comments. First, in the spirit of Althöfer and Koschnick (1991), who proved the convergence of the algorithm given an appropriate threshold sequence, we follow Gilli *et al.* (2006) in retrieving the threshold sequence from the empirical distribution of N_{Steps} distances between the value of the objective function for successive neighbors. And second, to generate initial solutions to the asset allocation problem, we generate weights from the beta distribution, Beta(α, β), until the constraint $\Sigma w_i = 1$ is violated. The parameters α and β describe the shape of the distribution and thus control the sparsity of the weight vector. For instance, a choice of $\alpha = \beta = 1$ corresponds to a uniform distribution that will deliver a very sparse solution to the large-dimension problem.[17] These weights are then randomly assigned to the assets in the portfolio. If there are fewer weights than assets, then the remaining weights are set to zero, and if the converse is true then we drop the appropriate number of weights and renormalize.

We conclude this section with a formal assessment of the TA algorithm. As we have already mentioned, for low-dimensional asset allocation problems, a grid-search (GS) procedure is feasible and so we select this as our benchmark algorithm. Our comparison is based on comparing the absolute difference in the optimal portfolio weights using between 6 and 18 months of data between

[17] The mean of a Beta (α, β) distribution is given by $\alpha/(\alpha + \beta)$ and so the average number of nonzero weights is given by $\|\mathbf{w}\|_0 = (\alpha + \beta)/\alpha$.

2 January 2003 and 23 December 2008 from $M = \{2, 3\}$ randomly chosen FTSE 100 stocks:

$$L(\mathbf{w}_{\mathrm{TA}}) = \frac{\|\mathbf{w}_{\mathrm{TA}} - \mathbf{w}_{\mathrm{GS}}\|_1}{\mathrm{Dim}(\mathbf{w}_{\mathrm{TA}})} 1[\mathcal{U}(\mathbf{w}_{\mathrm{TA}}) > \mathcal{U}(\mathbf{w}_{\mathrm{GS}})] \tag{11.8}$$

where $\mathrm{Dim}(\mathbf{w}_{\mathrm{TA}})$ is the dimension of the portfolio weight vector, $1[.]$ is an indicator function, and $\mathcal{U}$ is the expected utility function.[18] The results from $N = 250$ simulations are reported in Table 11.1. Even with a limited number of steps, iterations, and restarts, it appears that the TA algorithm produces similar results to a global grid-search procedure.

11.5 Data reweighting

Dealing with an unknown and changing data-generating process is undoubtedly the greatest challenge that one faces when trying to model asset returns, especially when "success" is measured in terms of out-of-sample performance.

Table 11.1 Threshold acceptance versus grid search

N_{Steps}	2 Assets		3 Assets	
	$N_{\mathrm{Restarts}} = 10$	$N_{\mathrm{Restarts}} = 20$	$N_{\mathrm{Restarts}} = 10$	$N_{\mathrm{Restarts}} = 20$
100	0.0070	0.0053	0.0189	0.0140
	0.0394	0.0383	0.0689	0.0626
250	0.0038	0.0022	0.0133	0.0072
	0.0207	0.0114	0.0515	0.0352
500	0.0038	0.0008	0.0046	0.0024
	0.0310	0.0037	0.0233	0.0117
750	0.0035	0.0023	0.0070	0.0024
	0.0336	0.0291	0.0502	0.0211
1000	0.0031	0.0003	0.0051	0.0009
	0.0298	0.0012	0.0313	0.0061

This table describes the mean (upper entry) and standard deviation (lower entry) of $L(\mathbf{w}_{\mathrm{TA}})$, based on $N = 250$ simulations using $N_{\mathrm{Iter}} = 3$. The grid-search procedure is based on 2 runs with 200 points in the first run and 100 in the second. We set $z = 0.65$ for convenience, and assume a coefficient of relative risk aversion equal to two, i.e., $\alpha = 2$, with no disappointment aversion, i.e., $A = 1$. All calculations are performed using an Intel Core 2 Duo 1.86GHz Processor and 2GB RAM.

[18] Our original intention was to consider $M = \{2,3,4\}$ stocks. However, including a fourth stock would entail 1.7 billion grid-point evaluations per simulation, and 425 billion evaluations in total. This highlights the infeasibility of a grid-search approach, especially when a relatively fine grid is used.

Even though one may have access to vast quantities of historic data, most practitioners would therefore caution against the use of the entire history for modeling purposes or the naïve assumption that the information content of all subsets of the data is identical. The popularity of "moving window" procedures that ignore observations beyond a certain arbitrary distance in the past, while equally weighting recent ones, is a testament to this.

Although we agree with the sentiment behind by the removal of old data, we believe that information decay is a more gradual process than implied by the binary nature of the moving window approach. To capture this idea, we introduce a vector of nondecreasing weights $\varpi_t \geq 0 \ \forall \ t$, such that $\Sigma\varpi_t = 1$ (cf. Mittnik & Paolella, 2000). By giving more weight to recent observations, it is hoped that parameter estimates will more closely reflect the "current" value of the "true" parameter. It is interesting to note that the idea of weighting recent events more heavily is embedded in the decision weight function from prospect theory. Such "recency effects" have been discussed by Kahneman and Tversky (1979), Hogarth and Einhorn (1992), and Kahneman (1995) who all argued that recently sampled outcomes receive greater weight than earlier sampled ones.

To determine the (weighted) parameter estimates, we combine the standardized weights, $\{\varpi_t\}$, with the quantile estimator described in Section 11.3.2. Specifically, we sort the data into ascending order and then continue to sum the weights of the sorted data until the appropriate quantile value has been reached (Boudoukh, Richardson, & Whitelaw, 1998). For instance, to calculate the qth quantile, we start from the lowest return and keep accumulating the weights until q is reached. Linear interpolation is used between adjacent points to achieve a more accurate estimate of the desired quantile.

Following Mittnik and Paolella (2000), we consider two weight schemes, a geometric scheme, for which $\varpi_t \propto \rho^{T-t}$, and a hyperbolic scheme, for which $\varpi_t \propto (T - t + 1)^{\rho-1}$. In both cases, a value of $\rho < 1$ ($\rho > 1$) means that the most recent observations are given relatively more (less) weight than those values far in the past, while $\rho = 1$ corresponds to standard moving window estimation. The two weight schemes are illustrated in Figure 11.3.

11.6 Alpha information

To the extent that they provide additional insights into the future evolution of the price process, asset return forecasts ("alphas") have long been an integral part of portfolio optimization.[19] The objective of this section is therefore to illustrate how such forecasts can be incorporated into the existing framework via the method of Bayesian updating. Formally, if we let X denote the portfolio return whose realization x is being forecasted, and if we let Y denote the point

[19] In spite of mixed evidence surrounding the issue of return predictability (Goyal & Welch, 2008; Campbell & Thompson, 2008; Boudoukh, Richardson, & Whitelaw, 2008), financial forecasts are still a crucial input into the asset allocation process.

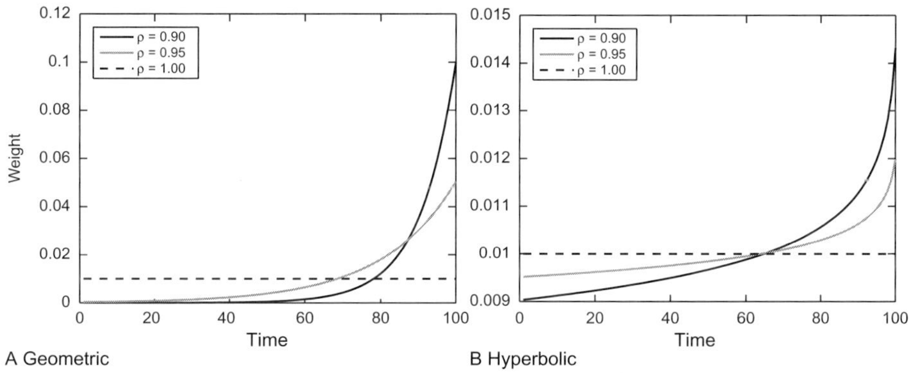

Figure 11.3 Weight schemes: (a) geometric and (b) hyperbolic.

forecast whose realization y constitutes a point prediction of X, then the purpose of this section is to derive the posterior distribution of X conditional on the realized point prediction $Y = y$.

At the heart of this problem is the need to model the joint dependence between realized returns, X, and their forecasts, Y, given that their marginal distributions are members of the Johnson family. Fortunately, for our purposes, there is now a burgeoning literature on the topic of constructing multivariate distributions from specified marginals.[20] In theory, therefore, after appropriately transforming the data, well-established results from copula theory can be used to model the dependence between returns and their forecasts. The only caveat is that the choice of copula needs to be computationally tractable.

Unlike traditional copula theory however, we choose to transform our Johnson marginals into standard Gaussian distributions using Equation (11.1), instead of uniform distributions, so that dependence can be modeled in the computationally convenient bivariate Gaussian framework. This approach leads to the well-known *meta-Gaussian* model of Kelly and Krzysztofowicz (1995), Kelly and Krzysztofowicz (1997).[21] In these papers, the bivariate *meta-Gaussian* distribution, H, and density, h, of (X, Y) take the form:

$$H(x,y) = P(X \leq x, Y \leq y) = B(Z_X(x), Z_Y(y) \mid \varphi) \tag{11.9}$$

$$h(x,y) = \frac{g(x)\,k(y)}{\sqrt{1-\varphi^2}} \exp\left[-\frac{\varphi}{2(1-\varphi^2)}\left(\varphi Z_X(x)^2 - 2Z_X(x)Z_Y(y) + \varphi Z_Y(y)^2\right)\right] \tag{11.10}$$

[20] See Sklar (1959) for an early reference, and Joe (1997) or Nelsen (2006) for a survey of the recent literature.

[21] These results are easily extendable to the larger class of elliptical distributions. This leads to the *meta-elliptical* model of Fang, Fang, and Kotz (2002).

where B denotes the bivariate standard Gaussian distribution, φ is the associated Pearson correlation coefficient, g and k are the marginal densities of X and Y, respectively, and Z_X and Z_Y are standard Gaussian transforms of (X, Y):

$$\begin{aligned} V &= Z_X(X) \\ W &= Z_Y(Y) \end{aligned}$$

where $Z(.)$ is given by Equation (11.1). Applying these transforms to our sample data yields the joint sample of (transformed) realized returns and forecasts, $\{(v, w)\} = \{(Z_X(x), Z_Y(y))\}$, and an estimate of the correlation coefficient, $\hat{\varphi} = Cor(v, w)$.

Given this structure, we can straightforwardly derive the conditional distribution of realized returns, X, given a point forecast, $Y = y$. Since,

$$\begin{pmatrix} V \\ W \end{pmatrix} \sim N\left(\begin{bmatrix} 0 \\ 0 \end{bmatrix}, \begin{bmatrix} 1 & \varphi \\ \varphi & 1 \end{bmatrix}\right)$$

then $V \mid (W = w) \sim N(\varphi w, 1 - \varphi^2)$, and so upon transforming the variates into the original space, the distribution of X, conditional on $Y = y$ is given by:

$$H(x|y) = \Phi\left(\frac{Z_X(x) - \varphi Z_Y(y)}{\sqrt{1 - \varphi^2}}\right) \tag{11.11}$$

where $\Phi(.)$ is the standard Gaussian cdf. Moreover, using Equation (11.10), the corresponding density function is given by:

$$h(x|y) = \frac{h(x, y)}{k(y)} = \frac{1}{\sqrt{1 - \varphi^2}} \exp\left[\frac{1}{2} Z_X(x)^2 - \frac{1}{2}\left[\Phi^{-1}(H(x|y))\right]^2\right] g(x) \tag{11.12}$$

Thus, via the process of Bayesian updating, we have updated the original portfolio return density, $g(x)$, using the forecast $Y = y$ to yield a posterior density, $h(x|y)$, which can be used for the purposes of expected utility maximization as described in Section 11.4.1.

To make these ideas concrete, and to illustrate the notion of Bayesian updating more generally, we conclude this section with a simple example. Suppose that the prior return distribution, $g(x)$, is described by an unbounded Johnson density, S_U, whereas the marginal forecast density, $k(y)$, is of the bounded variety, S_B, and exhibits a much lower variance than the density of realized

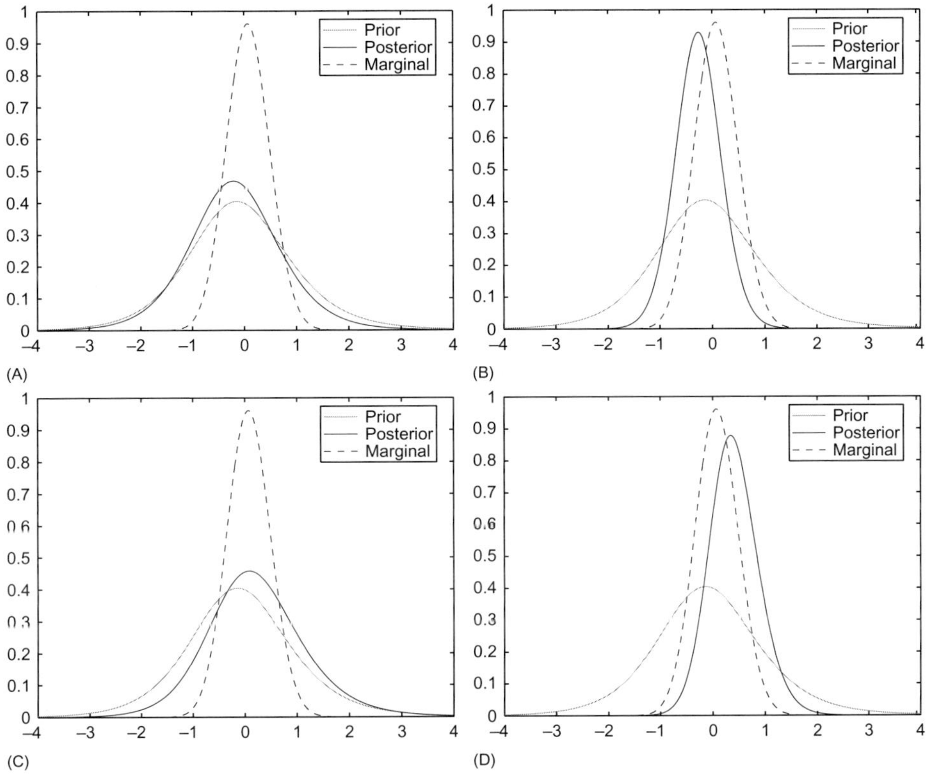

Figure 11.4 Prior, marginal, and posterior density estimates: (A) $\varphi = 0.5$ and $y = 0$; (B) $\varphi = 0.9$ and $y = 0$; (C) $\varphi = 0.5$ and $y = 0.3$; and (D) $\varphi = 0.9$ and $y = 0.3$.

returns.[22] Figure 11.4 illustrates how the posterior density reflects a compromise between the prior and marginal densities for different values of the correlation coefficient, $\varphi \in \{0.5, 0.9\}$, and current forecast, $Y = y \in \{0, 0.3\}$. Although highly stylized, the figure illustrates the fundamental idea that the the higher the correlation between returns and their forecasts, the closer the posterior density resembles the forecast distribution and the more the location is influenced by the current forecast, $Y = y$.

11.7 Empirical application

After outlining the baseline utility maximization algorithm and two extensions, we now focus our attention on practical implementation and the choice of "tuning" parameters. Wherever possible, we endeavor to describe data-driven

[22] This is likely to occur naturally in practical applications because of the inequality $\mathrm{Var}(X) \geq \mathrm{Var}(E(X\Omega))$, where Ω is the information set used to forecast X.

rules for each of these choices and so, for illustrative purposes, we apply our ideas to a real-world dataset involving the FTSE 100 list of companies.[23] Our sample begins on 2 January 2003 and ends on 23 December 2008, and therefore includes the recent period of financial turmoil linked to the credit crunch.

11.7.1 The decay factor, ρ

In order to operationalize the idea of data reweighting, it remains to choose an optimal value of ρ. Although cross-validation is arguably the most natural method of estimating ρ^* (Stone, 1974; Fan & Yao, 2000), the relative abundance of FTSE 100 data in both the time and cross-sectional dimensions leads us to consider an alternative strategy.

We start by creating an asset universe that comprises 6 months of daily FTSE 100 data from $U_1 \sim \text{Uniform}[3, 10]$ randomly selected equities between 2 January 2003 and 23 December 2008. The optimization algorithm is then applied to the data and the optimal portfolio weights are obtained for various values of $\rho \in (0, 1]$. One-month out-of-sample expected utility is recorded and the entire process is repeated $N = 250$ times for different combinations of equities and time periods in order to deliver an optimal ρ that is robust under a variety of circumstances. For both geometric and hyperbolic weight schemes, we find that the optimal value of ρ is strictly less than unity, which confirms our earlier suspicions surrounding the superior information content of the most recent data (see Figure 11.5 for the hyperbolic case).

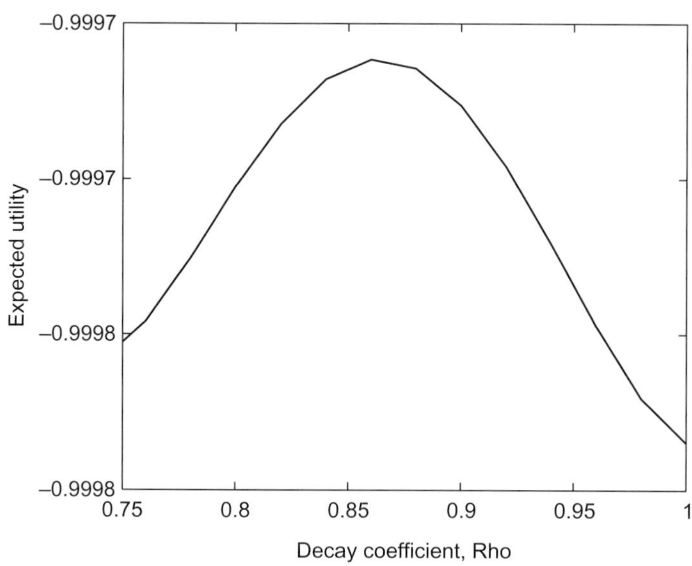

Figure 11.5 Optimal hyperbolic decay coefficient, ρ, for 6 months of data.

[23] All data are from Yahoo! Finance, sampled at a daily frequency and measured in Pounds Sterling (£).

Although instructive, the optimal ρ implied by these figures was calibrated using 6 months of historical data, i.e., $T = 168$, and so there is no guarantee that it will still be optimal for other data lengths. In response, we derive a data-dependent formula for selecting the optimal decay factor, $\rho \equiv \rho\,(T)$, using the generalized logistic transform. Specifically, we repeat the above simulation procedure, but this time we randomly choose between 3 and 18 months of data, i.e., $T \sim$ Uniform [56, 504], for each of the $N = 250$ repetitions. The resulting $\{T_i, \rho_i^*\}_{i=1}^{N}$ are then used to estimate the parameters C_1 and C_2 in the logistic transform:

$$\rho^* = \frac{U - L}{1 + C_1 \exp(C_2(T - \bar{T}))} + L \tag{11.13}$$

where $\bar{T} = 280$ and L and U are the upper and lower asymptotes of ρ^*, respectively. After linearizing the transform:

$$\ln\left(\frac{U - L - \rho^*}{\rho^* - L}\right) = \ln(C_1) + C_2(T - \bar{T}) \tag{11.14}$$

we set $L = 0.5$ and $U = 1$ and then apply standard least-squares estimation to yield $\hat{C}_1 = 0.2612$ and $\hat{C}_2 = 0.0094$. Figure 11.6 provides a graphical illustration of the resulting rule for choosing the decay factor.

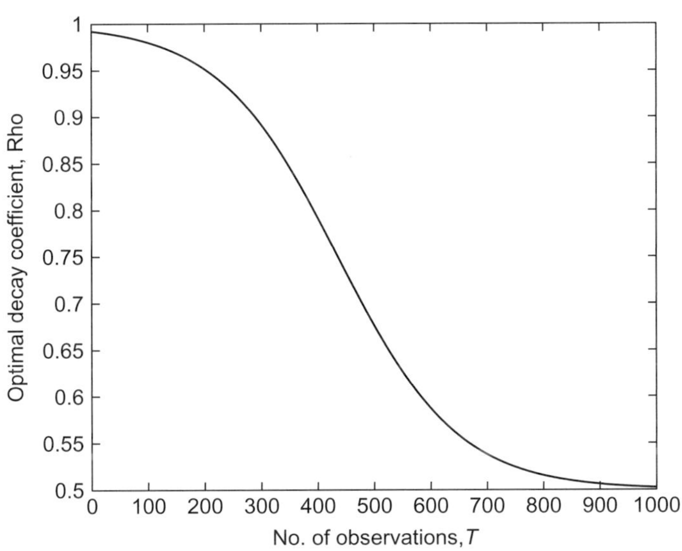

Figure 11.6 Optimal hyperbolic decay coefficient, ρ.

11.7.2 The coefficient of disappointment aversion, A

Up to this point, we have always set the coefficient of disappointment aversion equal to unity, i.e., $A = 1$. This was partly for computational convenience but mainly because we did not have much intuition about what value is reasonable in the context of portfolio optimization. In order to remedy this problem, we illustrate how a suitable value can be chosen by relating the idea of disappointment aversion to the concept of maximum drawdown (MDD). By doing so, a preferred value of A can be chosen on the basis of the maximum drawdown that one is willing to concede over a 1-month holding period.

We define a drawdown, DD_N, as the price difference resulting from a continuous fall of a security over the next N days:

$$DD_N = \sum_{t=1}^{\tau} r_t \tag{11.15}$$

where $\tau \equiv N \wedge (\inf\{t \geq 1 : r_t \geq 0\} - 1)$ and $DD_N = 0$ if $\tau = 0$. Analogously, we can define the N-day maximum drawdown, MDD_N, as the maximum of the $j = 1, \ldots, D_N$ drawdowns occurring over the next N days:

$$MDD_N = \max_{j \leq D_N} DD_{N,j} \tag{11.16}$$

where $D_N \in [0, [(N + 1)/2]]$ denotes the number of drawdowns in N days.[24]

On the basis of these definitions, we calculate 1-month out-of-sample MDD for various values of $A \in [0, 1]$ using $N = 250$ randomly generated asset universes based on U_1 ~ Uniform [3, 18] months of data from U_2 ~ Uniform [3, 10] equities. The mean MDD, described in Figure 11.7 for $\alpha = 2$, reveals a monotonically decreasing relationship between disappointment aversion and MDD, and can therefore be used as a guide for choosing A on the basis of a given level of drawdowns. For instance, if a hypothetical investor is willing to tolerate an MDD of 5% in the proceeding month, then the coefficient of disappointment aversion should be set at approximately $A = 0.8$.

11.7.3 The importance of non-Gaussianity

As we stated in the introduction, the primary purpose of this chapter is to advance the use of non-Gaussian alternatives to traditional mean–variance analysis for asset allocation purposes. In this section, we therefore provide a formal comparison between our Johnson-based TA algorithm and Markowitz's mean–variance approach to optimization. At the heart of this comparison is a fundamental question about the accuracy and reliability of mean–variance as an approximation.

[24] The maximum number of draw downs is bounded from above by $[(N + 1)/2]$, where $[z]$ denotes the integer part of z.

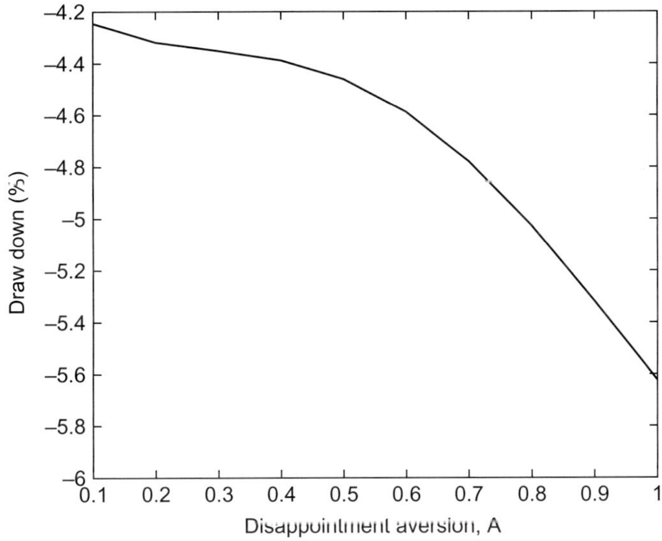

Figure 11.7 Maximum monthly drawdown for various levels of disappointment aversion.

For the class of elliptical distributions, which includes the normal, Student's-*t*, and Laplace, Chamberlain (1983) has shown that mean–variance approximation of expected utility is exact for all utility functions. In real-world applications, however, portfolios often contain derivatives and other exotic products and so the elliptical assumption becomes less and less plausible and questions surrounding the accuracy of mean–variance resurface. Indeed, casual inspection of empirical return distributions suggests that even equity returns themselves may not be elliptically distributed.

In a nonelliptically distributed world, mean–variance does not guarantee a sufficiently good approximation to expected utility and therefore direct maximization may be preferred. To explore this idea in more detail, we randomly select $U_1 \sim$ Uniform [6, 18] months of data from $U_2 \sim$ Uniform [15, 30] equities and then compare the 1-month out-of-sample performance of the TA algorithm against Markowitz's MV approach. The global MV weights are the solution to the following quadratic programming problem:

$$w^*_{MV} = \arg\min w'\Sigma w \tag{11.17}$$

where $w_i \in [0, 1]$, $i = 1, \ldots, N$, and $\sum w_i = 1$. This process is then repeated $N = 250$ times for different combinations of equities and time periods. A range of out-of-sample performance metrics are reported in Table 11.2 for four different levels of risk aversion, $\alpha \in \{1, 2, 5, 10\}$. These include the Sharpe ratio, R_p/σ_p, the Sortino ratio, R_p/σ_p^D, the Upside potential ratio,

Table 11.2 Mean–variance versus threshold acceptance

	$\alpha = 1$	$\alpha = 2$	$\alpha = 5$	$\alpha = 10$	MV
Rp	1.7583	1.7086	2.1237	2.5064	−1.7540
σ_p	5.2908	5.1813	4.8791	4.4295	4.7384
Skewness	−0.0440	−0.0613	−0.0402	−0.0257	−0.0916
Kurtosis	3.3176	3.3019	3.2090	3.3476	3.0328
Sharpe	1.7650	1.7496	1.8451	1.8935	1.5767
Sortino	4.8249	4.6845	4.8729	5.1897	4.1756
Upside	0.8104	0.8053	0.8152	0.8214	0.8014
Calmar	0.9294	0.9036	0.9769	1.0636	0.8222
max w_i	0.3574	0.3295	0.3090	0.2912	0.0919
No_p	0.6321	0.6436	0.6632	0.6855	1.0000

This table compares the (annualized) out-of-sample performance of the mean–variance approach with our Johnson-based threshold acceptance algorithm. The algorithm is allowed 15 minutes of computational power for each of the $N = 250$ simulations, which corresponds to setting $N_{\text{Restarts}} = 50$, $N_{\text{Iter}} = 4$, and $N_{\text{Steps}} = 3000$. We set $z = 0.65$ for convenience, and assume no disappointment aversion, i.e., $A = 1$. All calculations are performed using an Intel Core 2 Duo 1.86 GHz Processor and 2GB RAM.

$\sum_{t=1}^{28} \max[r_t, 0]/(28\sigma_p^D)$, and the Calmar ratio, $R_p/|MDD_p|$.[25] To compare diversification, we also report the maximum portfolio weight, max w_i, and the proportion of assets that are included in the optimal portfolio, $\text{No}_p = 1 - \|\mathbf{w}\|_0/U_2$.

According to the four performance metrics, direct maximization of the expected utility appears superior to the mean–variance criterion for all levels of risk aversion. What's more, the "No_p" statistics indicate that this outperformance is achieved with only 63–69% of the assets in the universe, which means that including transactions costs into our analysis will only strengthen this result. In contrast, our threshold acceptance portfolios appear less diversified than the corresponding mean–variance portfolios owing to their greater emphasis on one asset. However, we can easily nullify this issue by defining upper bounds for the weights, $\mathbf{w} \leq \mathbf{b}$, or by imposing a minimum number of assets to be included in the final portfolio $\|\mathbf{w}\|_0 \geq d$. Alternatively, we could introduce a direct preference for diversification by considering alternative utility functions of the form $U(\mathbf{w}'\mathbf{r}) + \lambda f(\mathbf{w})$, where $f(\cdot)$ is a measure of diversification and λ describes the strength of these preferences. While nonlinearities and augmented utility functions are trivial extensions to our baseline algorithm, they lead to many additional complications within the mean–variance framework.

[25] We define $1 + R_p = \prod_{t=1}^{28}(1 + r_t)$ as the cumulative portfolio return, $\sigma_p = \sqrt{\sum_{t=1}^{28} r_t^2/28}$ as the standard deviation, and $\sigma_p^D = \sqrt{\sum_{t=1}^{28} \min[r_t, 0]^2/28}$ as the downside standard deviation.

11.8 Conclusion

The obvious appeal of the mean–variance paradigm is that it captures the two fundamental aspects of portfolio choice—diversification and the trade-off between risk and reward—in an analytically tractable and easily extendable framework. Mean–variance portfolios are, however, optimal only when asset returns are elliptically distributed or when investors have quadratic utility functions. In all other cases, there is no guarantee that the optimal portfolio will be chosen. Yet, this approach still continues to underlie the overwhelming majority of the asset allocation decisions made within the finance industry. It seems that practitioners believe, rightly or wrongly, that the benefits gained from considering alternative approaches with more realistic assumptions do not outweigh their associated costs.

The purpose of this chapter has been to challenge this long-held belief and in doing so, to illustrate how solutions to otherwise intractable large-dimension optimization problems can be obtained with modest amounts of computational power via the combination of threshold acceptance search algorithm and Johnson distribution specification. Together they allow us to adopt a brute-force approach to expected utility maximization, which does not rely upon Taylor approximation methods or restrictive functional forms. Most importantly, the framework is sufficiently flexible to allow for a wide range of distributional shapes, covering the entire skewness–kurtosis plane no less, a plethora of potential objective functions, all of which may exhibit discontinuity, nonlinearity, etc., and all manner of portfolio constraints.

Since performance measurements are often based on out-of-sample metrics, we also introduced the ideas of data reweighting and Bayesian updating via alpha information as two simple and, most importantly, computationally efficient extensions of the baseline algorithm for improving the robustness. We then applied these techniques to a real-world dataset based on the FTSE 100 list of companies where we found evidence pertaining to superior performance of the TA algorithm against its MV counterpart. Moreover, we also showed how the results of estimation could be used to derive simple rules for the determination of the various "tuning" parameters that are a critical input into the optimization process.

In summary, we have sought to contribute to the ongoing discussion between practitioners and academics in order to advance the methodological basis for the use of non-Gaussian alternatives to traditional mean–variance analysis for large-dimension portfolio optimization problems. Especially if technological progress continues at the same pace witnessed in recent decades, the computational convenience of mean–variance analysis will become less important as practitioners shift their preferences toward algorithms that are based on more realistic, albeit computationally more burdensome, assumptions. To this end, we hope that our exposition has illustrated the complexity of the problems that can now be solved and will stimulate future developments in this area.

Appendix

Parameters of the Johnson density

Slifker and Shapiro (1980, pp. 243–246) report the following parameter estimates for each of the four Johnson distributions using their method of quantiles estimator. Interested readers should consult the original article for a full derivation of these formulae.

Parameters estimates of the S_U distribution

$$\eta = \frac{2z}{\cosh^{-1}\left[\frac{1}{2}\left(\frac{m}{p}+\frac{n}{p}\right)\right]} \quad (\eta > 0), \quad \gamma = \eta \sinh^{-1}\left[\frac{\frac{n}{p}-\frac{m}{p}}{2\left(\frac{mn}{p^2}-1\right)^{1/2}}\right]$$

$$\lambda = \frac{2p\left(\frac{mn}{p^2}-1\right)^{1/2}}{\left(\frac{m}{p}+\frac{n}{p}-2\right)\left(\frac{m}{p}+\frac{n}{p}+2\right)^{1/2}} \quad (\lambda > 0)$$

$$\varepsilon = \frac{x_z + x_{-z}}{2} + \frac{p\left(\frac{n}{p}-\frac{m}{p}\right)}{2\left(\frac{m}{p}+\frac{n}{p}-2\right)}$$

Parameters estimates of the S_B distribution

$$\eta = \frac{z}{\cosh^{-1}\left(\frac{1}{2}\left[\left(1+\frac{p}{m}\right)\left(1+\frac{p}{n}\right)\right]^{1/2}\right)} \quad (\eta > 0)$$

$$\gamma = \eta \sinh^{-1}\left[\frac{\left(\frac{p}{n}-\frac{p}{m}\right)\left[\left(1+\frac{p}{m}\right)\left(1+\frac{p}{n}\right)-4\right]^{1/2}}{2\left(\frac{p^2}{mn}-1\right)}\right]$$

$$\lambda = \frac{p\left[\left(\left(1+\frac{p}{m}\right)\left(1+\frac{p}{n}\right)-2\right)^{2}-4\right]^{1/2}}{\left(\frac{p^{2}}{mn}-1\right)} \quad (\lambda > 0)$$

$$\varepsilon = \frac{x_{z}+x_{-z}}{2} - \frac{\lambda}{2} + \frac{p\left(\frac{p}{n}-\frac{p}{m}\right)}{2\left(\frac{p^{2}}{mn}-1\right)}$$

Parameters estimates of the S_L distribution

$$\eta = \frac{2z}{\ln\left(\frac{m}{p}\right)}, \quad \gamma = \eta \ln\left[\frac{\frac{m}{p}-1}{p\left(\frac{m}{p}\right)^{1/2}}\right], \quad \lambda = 1, \quad \varepsilon = \frac{x_{z}+x_{-z}}{2} - \frac{p\left(\frac{m}{p}+1\right)}{2\left(\frac{m}{p}-1\right)}$$

Parameters estimates of the S_N distribution

$$\eta = \frac{2z}{m}, \quad \gamma = -\eta\frac{x_{z}+x_{-z}}{2}, \quad \lambda = 1, \quad \varepsilon = 0$$

Threshold acceptance pseudocode

The utility maximization algorithm comprises three constituent blocks: (1) the optimization routine; (2) the definition of a neighbor; and (3) choice of threshold sequence. The pseudocode for each of them is described below.

Algorithm 1: The optimization routine

1. Compute the threshold sequence, $\tau = \{\tau_I\}$, in accordance with Algorithm 3.
2. For $R = 1, \ldots, N_{\text{Restarts}}$.
3. Randomly generate a current solution $\mathbf{w}^{c}$ by drawing random weights from a beta distribution, Beta(α, β), such that $\|\mathbf{w}^{c}\|_{1} = 1$. The α and β control the sparsity of the weight vector. We set $\alpha = 1.5$ and $\beta = (N-1)\alpha$, where N is the number of assets in the universe.

4. For $I = 1,\ldots, N_{\text{Iterations}}$.
5. For $S = 1,\ldots, N_{\text{Steps}}$.
6. Generate a neighbor $\mathbf{w}^{\text{n}} \in \mathcal{N}(\mathbf{w}^{\text{c}})$ in accordance with Algorithm 2 and compute $\Delta = \mathcal{U}(\mathbf{w}^{\text{c}}) - \mathcal{U}(\mathbf{w}^{\text{n}})$.
7. If $\Delta < \tau_I$ then redefine $w^c = w^n$.
8. Record $\mathcal{U}(\mathbf{w}_{\mathbf{R}}^*) = \mathcal{U}(\mathbf{w}^{\mathbf{c}})$ and $\mathbf{w}_{\mathbf{R}}^* = \mathbf{w}^{\mathbf{c}}$.
9. The optimal solution, $\mathbf{w}^*$, is given by $\mathbf{w}^* = \max_R(\mathbf{w}_{\mathbf{R}}^*)_{R=1}^{N_{\text{Restarts}}}$.

Algorithm 2: The definition of a neighbor

1. Randomly select an asset with positive weight, i, and one other asset, $j \neq i$.
2. Draw a uniform random number, $U = $ Uniform $(0, w_{max})$, where w_{max} is the maximum increment allowed.
3. If $w_i^c - U \geq 0$ and $w_j^c + U \leq 1$ then $w_i^n = w_i^c - U$ and $w_j^n = w_j^c + U$,
   Else if $w_i^c - U \leq 0$ and $w_j^c + U \leq 1$ then $w_i^n = 0$ and $w_j^n = w_j^c + w_i^c$,
   Else if $w_i^c - U \geq 0$ and $w_j^c + U \geq 1$ then $w_i^n = w_i^c + w_j^c - 1$ and $w_j^n = 1$,
   Else if $w_i^c - U \leq 0$ and $w_j^c + U \geq 1$ then repeat Steps 2 and 3.

Algorithm 3: The threshold sequence

1. Set an $N_{\text{Iterations}}$-dimensional vector of percentage rejections that will be used in each stage of the iteration process. For instance, the vector $\xi = (0.75, 0.5, 0.25, 0)$ will reject 25% of the most distant neighbors in the first iteration, 50% in the second, etc.
2. Randomly choose $\mathbf{w}^{\text{c}}$.
3. For $S = 1, \ldots, N_{\text{Steps}}$.
4. Compute $\mathbf{w}^{\text{n}} \in \mathcal{N}(\mathbf{w}^{\text{c}})$ in accordance with Algorithm 3 and compute $\Delta_S = \mid \mathcal{U}(\mathbf{w}^{\text{c}}) = \mathcal{U}(\mathbf{w}^{\text{n}}) \mid$.
5. Compute the empirical distribution, F_Δ of Δ_S, $S = 1, \ldots, N_{\text{Steps}}$.
6. Compute the threshold sequence $\tau_I = F_\Delta^{-1}(\xi_I)$ for $I = 1, \ldots, N_{\text{Iterations}}$.

The optimality of z^*

The purpose of this section is to investigate the optimality of z^*. To this end, we start by uniformly generating parameters from an unbounded Johnson density, S_U, with supports given by $\varepsilon \in [-0.1, 0.1]$, $\lambda \in [0.1, 0.5]$, $\gamma \in [0, 0.2]$, and $\eta \in [1, 2]$. The random parameters are then used to simulate a vector of length $N \in \{100, 250, 500\}$ from the density in accordance with Equation (11.4). Finally, the quantile estimator, described in Section 11.3.2, is applied to the simulated series and the parameter estimates are compared to their true values. This procedure is repeated 10,000 times in order to obtain an estimate of the relative bias and RMSE for each parameter and length of data:

$$\text{Bias} = 100E\left[\frac{\hat{\varepsilon} - \varepsilon_0}{\varepsilon_0}\right]$$

$$\text{RMSE} = 100\sqrt{\frac{E\left[(\hat{\varepsilon} - \varepsilon_0)^2\right]}{\varepsilon_0^2}}$$

where $\hat{\varepsilon}$ is estimator and ε_0 is the true parameter value. The results are presented in Table 11.3.[26]

In the majority of cases, we find that a fixed value of $z = 0.7$ and our optimal z^* deliver the lowest levels of relative bias. However, the appeal of using z^* when compared to a fixed value is highlighted by the lower values of the relative RMSE. Allowing z to be chosen by the data leads to a lower variability in the parameter estimates because of the enhanced ability to match the data generating process.

The quantile estimator under misspecification

In this section, we explore the limitations of the method of quantiles, described in Section 11.3.2, when the true density is not one of the Johnson family. To compare the performance of the estimation procedure, we choose three popular densities that have been used to model financial time series. These are the Normal Inverse Gaussian (Barndorff-Nielsen, 1997; Eriksson *et al.*, forthcoming), Pearson Type IV (Heinrich, 2004), and Skew Student's-*t* (Jondeau & Rockinger, 2003). In order to facilitate comparisons between the results, we select their parameters such that the theoretical moments correspond to an annualized mean return of 10%, a standard deviation of 20%, skewness of −0.5, and kurtosis of 5. We then compute relative bias and RMSE estimates for the first four moments of the estimated Johnson density as well as expected utility. For completeness, we also report the mean absolute error, $L_1(f)$, the mean squared error, $L_2(f)$, and the Kullback–Leibler distance, $KL(f)$:

$$L_1(f) = \int |f(x) - \hat{f}(x)| dx \tag{11.18}$$

$$L_2(f) = \int (f(x) - \hat{f}(x))^2 dx \tag{11.19}$$

$$KL(f) = \int_{-\infty}^{\infty} f(x) \log \frac{f(x)}{\hat{f}(x)} dx \tag{11.20}$$

[26] Note: We find almost identical results using the other members of the Johnson family.

Table 11.3 Choice of z: bias and RMSE

Parameter	N	$z = 0.5$	$z = 0.6$	$z = 0.7$	$z = 0.8$	$z = 0.9$	$z = z^*$
	100	14.98	3.29	−5.41	−6.32	−7.91	3.00
		311.14	241.92	218.69	225.31	377.29	173.77
	250	5.13	−3.13	2.22	0.31	2.03	−0.88
		136.55	116.00	113.52	128.81	156.60	88.66
	500	0.40	0.78	2.22	−0.53	3.29	1.72
		80.07	71.68	72.95	85.29	111.90	58.63
λ	100	31.31	14.64	5.16	0.15	35.62	3.10
		96.59	61.43	36.18	31.15	74.24	33.70
	250	8.74	4.48	1.73	0.08	−1.39	1.22
		39.82	25.43	20.49	18.70	19.73	19.09
	500	4.42	1.47	0.85	0.29	0.00	0.32
		23.69	16.14	14.03	13.26	14.07	12.99
γ	100	−36.22	−6.34	4.72	−0.66	10.42	−69.84
		434.60	428.61	437.53	457.36	733.37	381.55
	250	−8.07	3.68	8.97	5.91	1.82	−25.11
		271.76	255.29	258.10	277.38	345.05	203.42
	500	4.77	3.58	5.47	3.54	4.34	−13.09
		176.27	159.42	167.90	189.21	234.98	142.34
η	100	−1.73	2.28	4.11	2.47	25.80	3.56
		28.48	26.88	24.52	23.22	42.38	24.16
	250	3.14	4.04	2.81	1.73	1.21	1.96
		23.77	19.60	16.36	14.80	15.67	14.65
	500	3.77	1.76	1.39	1.16	1.43	0.79
		18.71	12.95	11.07	10.48	11.35	9.92

This table describes the relative bias (upper entry) and RMSE (lower entry) estimates (%) for the parameters of the unbounded Johnson density, S_U, based on 10,000 replications for each time period and choice of z. The choice of z^* is based on a 20-point grid-search procedure with support [0.4, 1] and the Lilliefors (1967) test for normality.

where f is the true density, $\hat{f}$ is the estimated Johnson density, and log denotes the base 2 logarithm. The results are reported in Table 11.4.

In accordance with our discussion in Section 11.3.1, the results indicate that the method of quantiles can capture the four-moment patterns of the three alternative densities—NIG, Pearson Type IV, and Skew-T—with a high degree of accuracy, especially for $N \geq 250$. For our purposes, however, the more pertinent finding is that the relative bias and RMSE estimates of expected utility are small. In each case, the estimated relative bias is less than 0.25% with a relative RMSE of no more than 5%. To interpret these magnitudes, we calculate the certainty equivalent return differential, CED, along the lines of Chopra and Ziemba (1993):

$$\text{CED} = 100 \times \frac{U^{-1}(\mathcal{U}_0) - U^{-1}(\hat{\mathcal{U}})}{U^{-1}(\mathcal{U}_0)} \tag{11.21}$$

Table 11.4 Performance of the method of quantiles in the case of density misspecification

	True: NIG			True: Pearson Type IV		
	$N = 100$	$N = 250$	$N = 500$	$N = 100$	$N = 250$	$N = 500$
Mean	0.64(21.83)	0.15(13.74)	0.07(9.85)	0.66(21.59)	0.69(13.82)	0.34(9.74)
Std	1.86(13.49)	0.69(7.33)	0.46(5.01)	0.62(11.29)	−0.17(6.67)	−0.39(4.60)
Skewness	−9.03(143.60)	1.71(81.11)	3.91(54.70)	−23.86(121.88)	−17.04(70.19)	−15.71(48.94)
Kurtosis	45.32(159.83)	18.38(51.10)	11.64(30.42)	16.88(94.76)	−2.57(33.82)	−8.34(21.25)
EU	0.22(4.59)	0.04(1.53)	0.02(1.08)	0.06(3.08)	−0.07(1.50)	−0.06(1.05)
L_1	0.1493	0.0943	0.0669	0.1477	0.0931	0.0650
L_2	0.0386	0.0149	0.0074	0.0372	0.0143	0.0068
KL	0.0353	0.0149	0.0076	0.0335	0.0143	0.0071
	True: Skew-T			**True: Johnson Unbounded**		
	$N = 100$	$N = 250$	$N = 500$	$N = 100$	$N = 250$	$N = 500$
Mean	−1.28(21.87)	−1.24(13.93)	−0.53(10.12)	1.10(21.62)	0.32(13.72)	0.21(9.76)
Std	1.13(11.35)	0.05(6.85)	0.16(4.79)	0.95(11.37)	0.14(6.52)	−0.05(4.58)
Skewness	−5.33(119.62)	0.42(72.11)	3.62(49.95)	−16.06(135.36)	−8.84(76.93)	−5.36(52.64)
Kurtosis	15.17(77.20)	−0.76(34.22)	−4.26(22.57)	26.67(99.72)	6.13(36.46)	0.66(21.52)
EU	0.26(2.51)	0.14(1.53)	0.06(1.11)	0.02(2.47)	−0.01(1.49)	−0.02(1.06)
L_1	0.1472	0.0963	0.0691	0.1472	0.0926	0.0660
L_2	0.0367	0.0154	0.0078	0.0368	0.0140	0.0070
KL	0.0333	0.0147	0.0077	0.0331	0.0142	0.0073

This table describes the relative bias (RMSE) estimates of the first four moments and expected utility, as well as the three distance criteria using 10,000 replications for each time period, $N \in \{100, 250, 500\}$. We set $z = 0.65$ for convenience, and assume a coefficient of relative risk aversion equal to two, i.e., $\alpha = 2$, with no disappointment aversion, i.e., $A = 1$. The parameter values for the four distributions, using the notation given in each of the references, are: $\alpha = 7.06$, $\beta = -1.58$, $\delta = 0.26$, and $\mu = 0.16$ (NIG); $a = 0.47$, $\lambda = 0.22$, $\nu = 1.80$, and $m = 4.46$ (Pearson Type IV); $\mu = 0.1$, $\sigma = 0.2$, $\eta = 7.49$, and $\lambda = -0.18$ (Skew Student's-t); and $\gamma = 0.54$, $\eta = 2.12$, $\varepsilon = 0.20$, and $\lambda = 0.36$ (Unbounded Johnson, S_U).

Table 11.5 Certainty equivalent return differential, CED

No. Obs	NIG	Pearson	Skew-T	S_U
100	0.2175	0.0554	0.2577	0.0198
250	0.0436	0.0746	0.1411	0.0103
500	0.0220	0.0649	0.0613	0.0196

This table describes the percentage certainty equivalent return differential for our expected utility calculations using Equation (11.21).

where $\mathcal{U}_0$ is the value of expected utility under the true density, $\hat{\mathcal{U}}$ is our estimate of expected utility, and U is the familiar power utility function. We see that for large numbers of observations, the bias in our estimates of expected utility translates into a certainty equivalent return differential of around six basis points for the Pearson and Skew-T distributions and around two basis points for the NIG and S_U distributions (see Table 11.5). Thus, within a portfolio optimization framework where the primary goal is to estimate expected utility, the quantile estimator appears robust to potential misspecification.

References

Aït-Sahalia, Y., & Brandt, M. (2001). Variable selection for portfolio choice. *Journal of Finance*, *56*, 1297–1351.

Althöfer, I., & Koschnick, K. U. (1991). On the convergence of "threshold accepting." *Applied Mathematics and Optimization*, *24*, 183–195.

Ang, A., & Bekaert, G. (2002). International asset allocation with regime shifts. *Review of Financial Studies*, *15*, 1137–1387.

Artzner, P., Delbaen, F., Eber, J. M., & Heath, D. (1999). Coherent measures of risk. *Mathematical Finance*, *9*, 203–228.

Bansal, R., Hsieh, D., & Viswanathan, S. (1993). A new approach to international arbitrage pricing. *Journal of Finance*, *48*, 1719–1747.

Barndorff-Nielsen, O. E. (1997). Normal inverse Gaussian distributions and stochastic volatility modelling. *Scandinavian Journal of Statistics*, *24*, 1–13.

Benartzi, S., & Thaler, R. (1995). Myopic loss aversion and the equity premium puzzle. *Quarterly Journal of Economics*, *110*, 73–92.

Berényi, Z. (2001). Accounting for illiquidity and non-normality of returns in the performance assessment. Working Paper, University of Munich.

Berkelaar, A. B., Kouwenberg, R., & Post, T. (2004). Optimal portfolio choice under loss aversion. *Review of Economics and Statistics*, *86*(4), 973–987.

Boudoukh, J., Richardson, M., & Whitelaw, R. (1998). The best of both worlds: A hybrid approach to calculating value at risk. *Risk*, *11*(5), 64–67.

Boudoukh, J., Richardson, M., & Whitelaw, R. (2008). The myth of long-horizon predictability. *The Review of Financial Studies*, *24*(4), 1577–1605.

Brandt, M. W. (2009). Portfolio choice problems. In Y. A. Sahalia & L. P. Hansen (Eds.), *Handbook of Financial Econometrics* (pp. 429–448). Amsterdam: North-Holland.

Brandt, M. W., Goyal, A., Santa-Clara, P., & Stroud, J. R. (2005). A simulation approach to dynamic portfolio choice with an application to learning about return predictability. *Review of Financial Studies*, *18*, 831–873.

Brockett, P. L., & Garven, J. R. (1998). A reexamination of the relationship between preferences and moment orderings by rational risk averse investors. *Geneva Papers on Risk and Insurance Theory*, *23*, 37–127.

Campbell, J. Y., & Thompson, S. B. (2008). Predicting excess stock returns out of sample: Can anything beat the historical average? *The Review of Financial Studies*, *24*(4), 31–1509.

Chamberlain, G. (1983). A characterization of the distributions that imply mean–variance utility functions. *Journal of Economic Theory*, *29*, 185–201.

Chang, T. J., Meade, N., Beasley, J. E., & Sharaiha, Y. M. (2000). Heuristics for cardinality constrained portfolio optimisation. *Computers and Operations Research*, *27*, 1271–1302.

Chopra, V., & Ziemba, W. (1993). The effect of errors in means, variances, and covariances on optimal portfolio choice. *Journal of Portfolio Management*, *19*(2), 6–12.

de Athayde, G. M., & Flores, R. G. (2004). Finding a maximum skewness portfolio: A general solution to three-moments portfolio choice. *Journal of Economic Dynamics and Control*, *28*, 1335–1352.

Dueck, G., & Scheuer, T. (1990). Threshold accepting: A general purpose algorithm appearing superior to simulated annealing. *Journal of Computational Physics*, *90*, 161–175.

Dueck, G., & Winker, P. (1992). New concepts and algorithms for portfolio choice. *Applied Stochastic Models and Data Analysis*, *8*, 159–178.

Duxbury, C. (2008). An investigation into the impact of skewness and kurtosis on asset allocation for investors with power utility behaviour. Masters Thesis, University of London.

Eriksson, A., Ghysels, E., & Wang, F. (2009). The normal inverse Gaussian distribution and the pricing of derivatives. *The Journal of Derivatives*, 16(3), 23–37.

Fama, E. (1965). The behavior of stock market prices. *Journal of Business*, *38*, 34–105.

Fan, J., & Yao, Q. (2000). *Nonlinear time series. Nonparametric and parametric methods*. New York: Springer Series in Statistics.

Fang, H. B., Fang, K. T., & Kotz, S. (2002). The meta-elliptical distributions with given marginals. *Journal of Multivariate Analysis*, *82*, 1–16.

Gilli, M., & Këllezi, E. (2002). *Financial engineering, e-commerce, and supply chain*. Kluwer Applied Optimization Series.

Gilli, M., & Winker, P. (2004). A review of heuristic optimization methods in econometrics. Research Paper, Swiss Finance Institute.

Gilli, M., Këllezi, E., & Hysi, H. (2006). A data-driven optimization heuristic for downside risk minimization. *The Journal of Risk*, *8*(3), 1–18.

Glover, F., Mulvey, J. M., & Hoyland, K. (1996). Meta-heuristics: Theory and applications. Kluwer Applied Optimization Series.

Gomes, F. J. (2005). Portfolio choice and trading volume with loss averse investors. *Journal of Business*, *78*, 675–706.

Goyal, A., & Welch, I. (2008). A comprehensive look at the empirical performance of equity premium prediction. *The Review of Financial Studies*, *24*(4), 1455–1508.

Granger, C. W. J. (2009). The Methodology and Practice of Econometrics: A Festschrift in honour of David F Hendry. In J. Castle & N. Shephard (Eds.), *In Praise of Pragmatics in Econometrics* (pp. 429–448). Oxford University Press.

Gul, F. (1991). A theory of disappointment aversion. *Econometrica*, *59*, 667–686.

Hahn, G. J., & Shapiro, S. S. (1967). *Statistical models in engineering*. New York: John Wiley & Sons.

Harvey, C. R., Liechty, J. C., Liechty, M. W., & Müller, P. (2004). Portfolio selection with higher moments. Working Paper, Duke University.

Heinrich, J. (2004). A guide to the Pearson Type IV distribution. Working Paper, University of Pennsylvania.

Hill, I. D., Hill, R., & Holder, R. L. (1976). Fitting Johnson curves by moments. *Applied Statistics*, *25*(2), 180–189.

Hogarth, R. M., & Einhorn, H. J. (1992). Order effects in belief updating: The belief-adjustment model. *Cognitive Psychology*, *24*, 1–55.

Jobst, A. (2005). European securitization: A GARCH model of CDO, MBS and Pfandbrief spreads. *Journal of Structured Finance*, *12*(1), 55–80.

Joe, H. (1997). Multivariate models and dependence concepts number 73. In *Monographs in statistics and probability*. London: Chapman and Hall.

Johnson, N. L. (1949). Systems of frequency curves generated by methods of translation. *Biometrika*, *36*, 149–176.

Jondeau, E., & Rockinger, M. (2003). Conditional volatility, skewness and kurtosis: Existence, persistence, and comovements. *Journal of Economic Dynamics and Control*, *27*(10), 1699–1737.

Jondeau, E., & Rockinger, M. (2006). Optimal portfolio allocation under higher moments. *European Financial Management*, *12*, 1.

Kahneman, D. (1995). Varieties of counterfactual thinking. In N. J. Roese & J. M. Olson (Eds.), *What might have been: The social psychology of counterfactual thinking* (pp. 375–396). New Jersey: Hillsdale.

Kahneman, D., & Tversky, A. (1979). Prospect theory: An analysis of decision under risk. *Econometrica*, *47*, 263–291.

Kelly, K. S., & Krzysztofowicz, R. (1995). Bayesian revision of an arbitrary prior density. In *Proceedings of the section on Bayesian statistical science* (pp. 50–53). American Statistical Association.

Kelly, K. S., & Krzysztofowicz, R. (1997). A bivariate meta-Gaussian density for use in hydrology. *Stochastic Hydrology and Hydraulics*, *11*(1), 17–31.

Kraus, A., & Litzenberger, R. H. (1976). Skewness preference and the valuation of risk assets. *Journal of Finance*, *31*, 1085–1100.

Lhabitant, F. S. (1998). On the (ab)use of Taylor series approximations for portfolio selection, portfolio performance and risk management. Working Paper, University of Lausanne.

Lilliefors, H. W. (1967). On the Kolmogorov–Smirnov test for normality with mean and variance unknown. *Journal of the American Statistical Association*, *62*, 399–402.

Mandelbrot, B. (1963). The variation of certain speculative prices. *Journal of Business*, *35*, 394–419.

Maringer, D. (2007). Risk preferences and loss aversion in portfolio optimization. Working Paper, University of Essex.

Maringer, D., & Parpas, P. (2009). Global optimization of higher order moments in portfolio selection. *Journal of Global Optimization*, *43*, 219–230.

Markowitz, H. M. (1952). Portfolio selection. *Journal of Finance*, *7*, 77–91.

Mittnik, S., & Paolella, M. S. (2000). Conditional density and value-at-risk prediction of Asian currency exchange rates. *Journal of Forecasting*, *19*, 313–333.

Moore, G. (1965). Cramming more components onto integrated circuits. *Electronics Magazine*, *38*, 8.

Nelsen, R. B. (2006). *An introduction to copulas.* (2nd ed.). New York: Springer.

Osman, I. H., & Laporte, G. (1996). Metaheuristics: A bibliography. *Annals of Operations Research*, *63*, 513–623.

Perez, P. G. (2004). An approach to the non-normal behavior of hedge fund indices using Johnson distributions. Working Paper, Instituto Tecnologico Autonomo de Mexico.

Ramchand, L., & Susmel, R. (1998). Volatility and cross correlation in global equity markets. *Journal of Empirical Finance*, *5*(4), 397–416.

Rubinstein, M. E. (1973). The fundamental theorem of parameter-preference security valuation. *Journal of Financial and Quantitative Analysis*, *8*, 61–69.

Samuelson, P. (1970). The fundamental approximation theorem of portfolio analysis in terms of means, variances and higher moments. *Review of Economic Studies*, *37*, 42–537.

Sklar, A. (1959). Fonctions de répartition à *n* dimensions et leurs marges. *Publications de l'Institut Statistique de l'Université de Paris*, *8*, 229–231.

Slifker, J. F., & Shapiro, S. S. (1980). The Johnson system: Selection and parameter estimation. *Technometrics*, *22*(2), 239–246.

Stone, M. (1974). Crossvalidatory choice and assessment of statistical prediction. *Journal of the Royal Statistical Society*, *36*, 33–111.

Stuart, A., & Ord, J. K. (1994). *Kendall's advanced theory of statistics, Volume 1: Distribution theory*. London: Edward Arnold.

Thorp, E. O. (1960). Fortune's formula: The game of blackjack. *Notices of the American Mathematical Society*, 935–936.

Thorp, E. O., & Kassouf, S. T. (1967). *Beat the market: A scientific stock market system.* New York: Random House.

Tversky, A., & Kahneman, D. (1991). Loss Aversion in Riskless Choice: A Reference-Dependent Model. *Quarterly Journal of Economics*, *106*(4), 1039–1061.

Wheeler, R. E. (1980). Quantile estimators of Johnson curve parameters. *Biometrika*, *67*(3), 725–728.

Winker, P. (2001). *Optimization heuristics in econometrics: Applications of threshold accepting*. Chichester: Wiley.

Yan, J. (2005). Asymmetry, fat-tail, and autoregressive conditional density in financial return data with systems of frequency curves. Working Paper, University of Iowa.

12 More than you ever wanted to know about conditional value at risk optimization

Bernd Scherer

Executive Summary

We split our critique on conditional value at risk (CVaR) into implementation and concept issues. While implementation issues can be overcome at the cost of sophisticated statistical procedures that are not yet widely available, they pose a strong objection against current naive use of CVaR. Estimation error sensitivity amplified by approximation error and difficulties in modeling fast updating scenario matrices for nonnormal multivariate return distributions will stop many practitioners from applying CVaR. More limiting in our view, however, is the inability of CVaR to integrate well into the way investors think about risk. Averaging across small and extremely large losses, i.e., giving them the same weight, does not reflect rising risk aversion against extreme losses, which is probably the most agreeable part of expected utility theory.

12.1 Introduction: Risk measures and their axiomatic foundations

The world of investments offers a puzzling variety of alternative risk measures. It almost seems as if we could invent risk measures freely. However, it needs to be stressed that we cannot. Risk measures must be supported by axioms (statements that are generally believed to be true). These axioms can then be used to build or criticize risk measures and, of course, a risk measure is only as convincing as the axioms it is built on. Discussing risk measures without reference to their axioms is a pointless exercise.

We distinguish between axioms on random variables (actuarial or statistical axioms) and axioms on behavior (economic axioms). *Economic axioms* require a risk measure or more generally an objective function to be consistent with decision making under uncertainty formulated by Neumann and Morgenstern (1944), i.e., expected utility optimization. This is very familiar to economists and builds the backbone of mean–variance optimization by Markowitz (1952).

Doi:10.1016/B978-0-12-374952-9.00012-9.

He showed that investor preferences (expected utility) can be approximated locally by mean and variance of risky investments alone. This approximation works well as long as returns are not "too nonnormal" (certainly violated with nonlinear payoffs) or "too volatile" (or equivalently with a "too long" time horizon), which would make a local approximation increasingly bad behaved for not so infinitesimal (larger) returns as pointed out in Markowitz and Levy (1979). The computational convenience offered by Markowitz inspired countless applications. However, academics and practitioners alike soon started to ignore the fact that mean–variance investing was designed as an approximation to maximizing expected utility under the above conditions. Instead, they expressed their disapproval of variance as a measure of risk and invented (rather than derived from first principles) new statistical risk measures.[1] I call them statistical risk measures as they have been selected to capture particular features of a return distribution and not whether they more closely resemble expected utility maximization. Apart from a few dispersion measures (e.g., mean–absolute deviation rather than variance), a large variety of downside risk measures have been engineered. These measures in particular had no relation to plausible utility functions. Reconciliation with an expected utility optimization framework has therefore been *ex post*. Academia invented utility functions that would support a given measure to provide a theoretic fig leave. However, on close inspection these new utility functions are at odds with investors' preferences. An example for this or more provocatively a low point in this development was the "invention" of value at risk (*VaR*). It is defined as the maximum loss that will not be exceeded with a specified confidence (usually 95%) at the end of a given time period (usually a year).[2] Investors are assumed not to worry about the extent of losses exceeding *VaR*, which in fact makes them risk-seeking in the tail of a return distribution. We are very unlikely to meet this type of investors in reality. In any case, there was little evidence that mean and *VaR* would approximate utility better than mean and variance for realistic problems. Given the rise in statistical rather than economic risk measures, Artzner, Delbaen, Eber, and Heath (1997) developed a set of *statistical axioms* designed to address the shortcomings of statistical risk measures and, in particular, value at risk. For a risk measure to be coherent (pass all axioms), it must be monotonous (larger losses means larger risks), positive homogeneous (increasing positions by a multiple must increase risk by the same multiple[3]), invariant to translations (adding cash does not reduce risk), and subadditive

[1] Observing that returns are not really normally distributed, many authors addressed the shortcomings of variance without relating their measures to expected utility maximization. Portfolio selection then wrongly became a field for statistics, engineering, and operations research. While expected utility maximization got pushed into the background, it nevertheless was always there as the only way to correctly solve the problem, showing von Neumann's original genius.

[2] We use the 95% confidence limit throughout this Chapter.

[3] The credit crisis in 2008 has revealed that this axiom is likely to be the weakest, given that liquidity might dry up for very large positions.

(portfolio risk is lower than the sum of stand-alone risks). The last axiom is the most intuitive as diversification should decrease risk. However, this need not be the case with *VaR*. Suppose we add two marginal distributions with both long left tails (i.e., extreme losses) together. Looking at each marginal distribution in isolation, these extreme events will not be picked up by value of risk owing to its ignorance on what is looming in the tail of a distribution. Extreme losses are simply not frequent enough to identify themselves at the 95% confidence level. However, when combined, the losses diversify into the joint distribution and so the *VaR* of the combined portfolio will rise. In other words, *VaR* will provide the wrong diversification advice. Given that *VaR* so blatantly fails when distributions are highly skewed (it works fine if distributions are elliptical, but then it adds no additional informational value to variance), one should not call it a risk measure in the first place. This led many to the adoption of conditional value at risk (*CVaR*), which offers not only computational advantages (it can be optimized with a linear program, while *VaR* can only be dealt with heuristically due to its nonconvex nature as we have seen in the diversification example). Contrary to *VaR*, *CVaR* is a coherent risk measure, i.e., it passes ARTZNER's statistical axioms. It does so by looking into the tail of a distribution, i.e., it averages across all losses that exceed a 95% confidence bound. While this looks like good news at first glance, the remainder of this text will confront the reader with implementation challenges and conceptional problems for *CVaR*.

Implementation problems center around the fact that *CVaR* is extremely sensitive to estimation error (more than other risk measures) and approximation error (this issue is unique to all scenario optimization problems and not existent in mean–variance investing) to name only two. While implementation issues can be overcome (at the cost of sophisticated statistical procedures that are not yet widely available), they pose a strong objection against current naive use of *CVaR*. *Conceptional problems* are in our view more limiting. What we point out here is the inability of *CVaR* to integrate well into the way investors think about risk. Averaging across small and extremely large losses, i.e., giving them the same weight in your risk calculation does not reflect rising risk aversion against extreme losses, which is probably the most agreeable part of economic decision theory.

12.2 A simple algorithm for *CVaR* optimization

We start with a brief description around the technicalities of *CVaR* optimization. Let us first define an auxiliary variable, e_s, for each of $s = 1,\ldots, S$ scenarios.

$$e_s = \max\left[0, VaR - \sum_{i=1}^{n} w_i r_{is}\right] \tag{12.1}$$

It measures the excess loss of a portfolio consisting of $i = 1,\ldots, n$ securities with respective weights w_i that pay off r_{is} in scenario s. There are no restrictions on the distribution of scenarios. We can, for example, include nonlinear positions as they would typically arise from call or put options.[4] This is a major advantage relative to mean–variance optimization that cannot deal with these instruments. Suppose $VaR = -20\%$ and $\sum_{i=1}^{n} w_i r_{is} = -25\%$. We then get $e_s = \max[0, -20\% - -25\%] = 5\%$, i.e., a 5% excess. In practice, we will need many scenarios to approximate a continuous portfolio distribution from a grid of discrete scenarios. We will come back to this point when we discuss approximation error in *CVaR* optimization. *CVaR* will always be larger than *VaR* (because it represents the average of losses larger than value at risk) and, hence, we can write it as:

$$CVaR = VaR - \frac{1}{\alpha}\left(\frac{1}{S}\sum_{s=1}^{S} e_s\right) \tag{12.2}$$

where $\frac{1}{S}\sum_{s=1}^{S} e_s$ denotes the average loss across *all* scenarios and α the probability of a loss larger than *VaR* (i.e., 5% for a 95% confidence level), which leverages this average up to adjust for the fact that we are looking for a loss conditional on being in the tail. A complete optimization problem looks like the following linear program, where $\bar{\mu}_i$ denotes the average return for asset i and μ^* stands for the targeted portfolio return.

$$\max_{VaR, w_i, e_s} \; VaR - \frac{1}{\alpha}\left(\frac{1}{S}\sum_{s=1}^{S} e_s\right) \tag{12.3}$$

$$\sum_{i=1}^{n} w_i \bar{\mu}_i \geq \mu^* \tag{12.4}$$

$$e_s \geq VaR - \sum_{i=1}^{n} w_i r_{is} \tag{12.5}$$

$$e_s \geq 0 \tag{12.6}$$

$$w_i \geq 0 \tag{12.7}$$

[4] Option positions need to be path independent as *CVaR* optimization still remains a one-period model. We can, for example, not include the payoff from a look-back call on a stock index as end of period option payoffs do not relate to end of period index payoffs.

Optimization software like NUOPT™ for S-PLUS™ can deal with these problems efficiently. See Scherer and Martin (2005) for a complete set of code for various kinds of problems including transaction costs, cardinality constraints, etc. A closer look at Equations (12.3)–(12.7) will reveal the mechanics. Given that excesses are constrained to be positive via Equation (12.6), any loss larger than value at risk will have a negative impact on the objective (12.3). Excesses can be kept small by choosing w_i in Equation (12.5) to prevent e_s from becoming large. However, if portfolio returns were all positive for a given set of weights, how can we prevent e_s from becoming negative? The optimizer can now increase *VaR* in order to improve Equation (12.3) and relax Equation (12.5). Given a linear objective function with linear inequality constraints, we have a well-developed technology at hand to solve this type of problems. This also applies to wider problems like the inclusion of integer variables in order to model cardinality constraints (max number of assets) or lot sizes. Risk budgets for *CVaR* are easily calculated by taking the numerical derivative:

$$\frac{dCVaR}{dw_i} = \frac{CVaR(w_i+\Delta)-CVaR(w_i-\Delta)}{2\Delta} \tag{12.8}$$

for a given scenario matrix, where Δ represents an infinitesimal change in the weight of the *i*th asset. For large enough scenario matrices, this function will be sufficiently smooth. Alternatively, Tasche (1999) has shown that we can estimate Equation (12.8) as expected value from a large set of scenarios (drawn from bootstrapping or Monte Carlo simulation):

$$\frac{dCVaR}{dw_i} = -E(r_i|r_p \le VaR_p) = \frac{1}{\alpha}\left(\frac{1}{S}\sum_{s=1}^{S} r_{is}\cdot I_s\right) \tag{12.9}$$

where $I_s = 1$ for $r_p \le VaR_p$ and zero otherwise. Again, i denotes the respective asset and VaR_p stands for portfolio *VaR*. We simply calculate the expected performance contribution from asset i given that portfolio return falls below portfolio *VaR*.[5]

The practical usefulness of *CVaR* for portfolio construction relies on our ability to construct a predictive scenario matrix, i.e., a rectangular array of dimension $S \times n$ that contains returns r_{is} for $i = 1,\ldots,n$ assets (columns) across $s = 1,\ldots,S$ scenarios (rows). Scenarios need to be representative (reflect expectations about future returns), approximate (approximate the distribution closely), parsimonious (offer a relatively limited set of scenarios to remain computationally feasible), and finally arbitrage free (if the number of scenarios is less than the number of assets involved, our hypothetical economy is not

[5] Note that for *VaR* we would use $\frac{dVaR}{dw_i} = -E(r_i|r_p = VaR_p)$. However, given that the equality sign will never be true for continuous distributions, we could use the kernel estimator suggested by Tasche (2007). Tasche also provides R/S-Plus code on his web page, so we don't need to describe the procedure here.

complete and arbitrage possibilities might exist). Virtually, all implementation problems relate to the scenario matrix and its generation.

12.3 Downside risk measures

12.3.1 Do we need downside risk measures?

Mean–variance optimization and mean–*CVaR* optimization yield identical results if returns are normally distributed. The efficient set (i.e., the set of optimal portfolios) is identical as the return distribution would be completely determined by the first two moments. In a normally distributed world, *CVaR* optimization has nothing to exploit to its advantage. Downside risk measures need distributional asymmetry (positive or negative skewness) for their justification. While we know the world is not normally distributed most of the time, we need to ask ourselves the following question to assess the merit of deviating from mean–variance optimization: *Are the deviations from symmetry statistically significant and do these deviations display an exploitable pattern, i.e., are they stable?*

Throughout this chapter, we will use monthly total return data for 10 US sector indices (Oil, Basic Materials, Industrials, Health Care, Consumer Goods, Consumer Services, Telecom, Utilities, Financials, Technology) for the last 20 years (from March 1989 to March 2009) to illustrate all our results. In order to assess the stability of risk estimates, we split the sample into two 10-year periods and estimate the excess skewness for each series and time period. We then run a regression of the form:

$$skew_{t+1} = \beta_0 + \beta_1 skew_t + \varepsilon_{t+1} \tag{12.10}$$

If deviations from normality were stable, we would expect $\beta_0 = 0$ and $\beta_1 = 1$ together with a high R^2. For the data set above, we estimate $\beta_0 = 0.00003$, $\beta_1 = 2.187$ with an R^2 of 68%. The slope coefficient β_1 is highly significant with a t-value of 4.2. Data and regression fit are displayed in Figure 12.1.

Deviations from symmetry seem to be reasonably stable and given both subperiods have shown a negative skew, we would expect *CVaR* optimization to arrive at different solutions than mean–variance investing. Of course, these results cannot be generalized and a different investment universe might exhibit different properties.

12.3.2 How much momentum investing is in a downside risk measure?

Let us first distinguish between dispersion measures (volatility, mean–absolute deviation) and downside risk measures (*CVaR*, minimum regret, etc.).[6] While dispersion measures look at the whole distribution (they measure risk as dispersion

[6] See Scherer (2007) for a review of these measures.

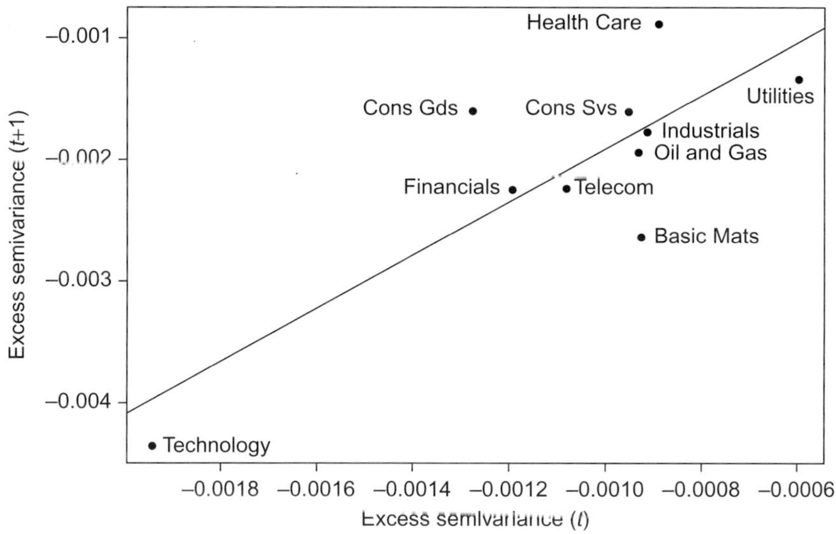

Figure 12.1 Are deviations from symmetry stable? In order to assess the stability of estimates for nonnormality, we split the sample into two 10-year periods and estimate the excess skew for each series and time period. The solid line represents fitted values from Equation (12.10).

around mean returns), downside risk measures focus exclusively on losses (often defined as shortfalls below a given target return). This sounds in line with a more layman interpretation of risk (risk only arises from unexpected losses not from unexpected profits), but it carries some underappreciated side effects. First, this view ignores the fact that any asset that goes up a lot might also go down a lot. We might not have seen losses in a particular asset for a long time (Japanese equity market between 1975 and 1992), but that does not mean we are not exposed (Japanese equities fell about 70% in the following 3 years). In other words, what can go up can go down. *CVaR* will not realize that large upside deviations from average growth are a sign of volatility and should not be ignored in risk considerations. Second (and very much related to the point above), *CVaR* will create a bias toward momentum investing.[7] Assets that show positive momentum will not only exhibit positive returns, but also lower risks as a sequence of positive returns will reduce the measured downside risk. What is hoped to be a more appropriate risk measure will tend to load up on a momentum factor. Backtests on realized return data claiming the superiority of downside risk measures relative to dispersion measures should be very mindful to this and might give a misleading picture.

[7] While it is true that momentum will also cause autocorrelation in returns and therefore lead to an underestimate in volatility, this effect is much less pronounced.

12.3.3 Will downside risk measures lead to "under-diversification?"

How can we measure diversification? The simple (and maybe naïve) answer is by measuring volatility. Suppose we perform the following thought experiment. Portfolio A holds 100 assets and exhibits 15% of risk, while portfolio B holds 10 assets and also exhibits 15% of risk. Which one is riskier? If you believe 100% in your risk model, both portfolios must look equally risky to you. The alternative view is that investors are not sure about their risk models and they face the risk of extreme returns in any single asset, e.g., second or even first moments might not even exist.[8] Under this scenario, holding a large number of small positions is preferable.[9] This is why diversification (or better dispersion in portfolio weights) plays an important role in practical portfolio construction and why diversification measures (weight dispersion, concentration measures, etc.) contain some information not captured in a risk measure. In other words, in a world where extreme moves are possible, investors are likely to underestimate the required extent of diversification. Recently, the field of robust optimization started to put regularity constraints on weights to enforce diversification.[10] We will now look into the dispersion optimal *CVaR* portfolios weights. Are they more or less diversified than mean–variance portfolios? To make portfolios "comparable," we minimize risk subject to a return requirement of 78.5 basis points per month, which is the average of maximum and minimum sector return. In other words, our portfolio should be placed in the "middle" of the efficient frontier. The resulting optimal portfolios can be found in Exhibit 2 and Exhibit 3. We see that *CVaR* portfolios in this example are much more concentrated than mean–variance portfolios. Consumer goods, for example, exhibit moderate volatility risk but high tail risk. Given that we have little reason to believe that tail risk can be reliably estimated (see the next section on estimation risk), we feel uncomfortable with such an "under-diversified" portfolio, in particular in the presence of what might turn out as extreme tail events. One should read the above statement carefully. We cannot generalize that *CVaR* optimal portfolios will *always* be more concentrated than mean–variance portfolios. Counterexamples are easy enough to engineer (Figures 12.2 and 12.3).

It is well known that assets with positive (co)skewness and small (co)kurtosis are favored by *CVaR* optimization. The "researcher" can pick his universe and time horizon to arrive at the desired result. However, in our experience, under-diversified *CVaR* portfolios can be observed much more often than not. The reason for this should be obvious as it is equally well

[8] As the number of observations increases, a single observation will dominate all other observations. For a Cauchy distribution variance will become infinite as the number of observations grows.

[9] If returns exhibit "wild randomness" as used by Taleb (2005), investors are better off with very diversified portfolios even at the expense of these portfolios looking mean–variance inefficient due to "over-diversification."

[10] See Uppal, DeMiguel, Garlappi, and Nogales (2008).

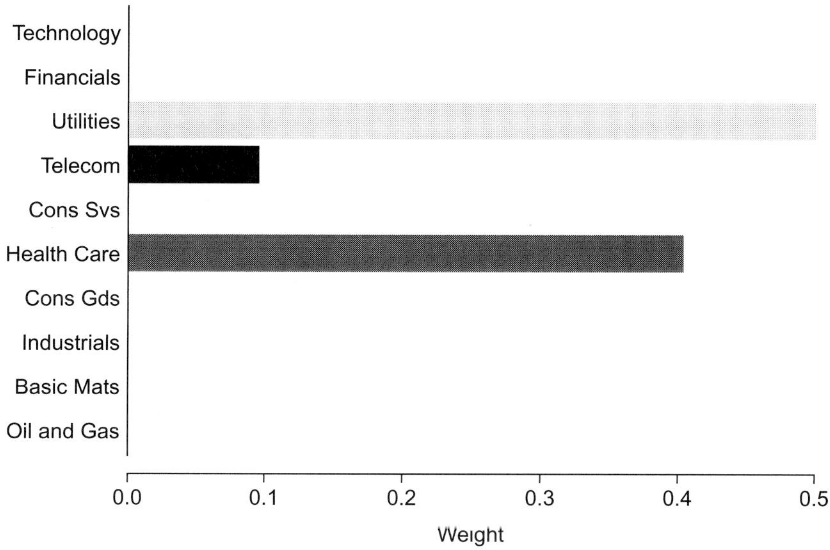

Figure 12.2 Optimal *CVaR* portfolio weights. Minimize *CVaR* subject to an excess return target of 78.5 bps per month. Scenarios are taken from the unconditional sector returns for the period March 1989–March 2009.

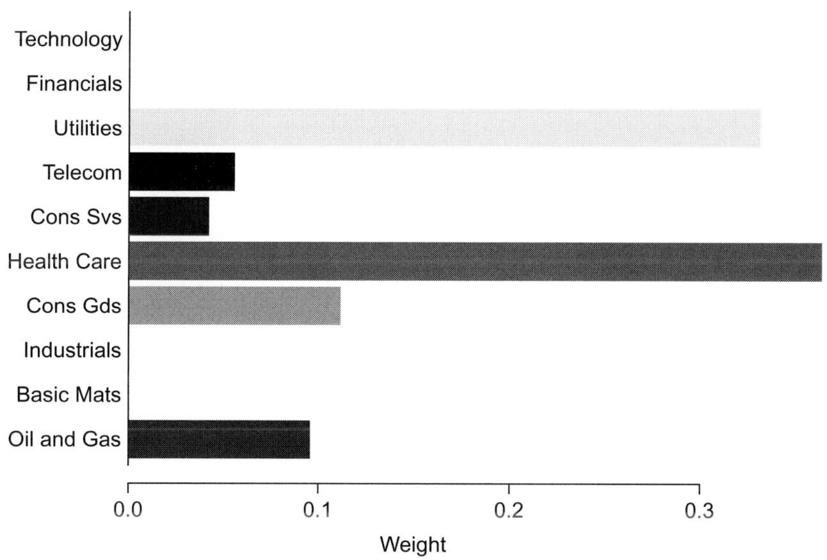

Figure 12.3 Optimal mean–variance portfolio weights. Minimize *CVaR* subject to a required excess target of 78.5 bps per month. Variance estimates are taken from the unconditional sector returns for the period March 1989–March 2009.

known that estimation error in conjunction with long-only constraints causes extreme, i.e., concentrated mean–variance portfolios. Given that *CVaR* is much more sensitive to estimation error, we conjecture without proof that *CVaR* optimal portfolios will also tend to be more concentrated for very much the same reasons. There are not too many assets with positive skewness and small kurtosis.

12.4 Scenario generation I: The impact of estimation and approximation error

12.4.1 Estimation error

Estimation error is a serious concern in portfolio construction. If we have little confidence that our inputs used to describe the randomness of future returns are accurate, we will also have little confidence in the normative nature of optimal portfolios. Any portfolio optimization process will spot high return, low risk, low common correlation opportunities and try to leverage on them. These are precisely the estimates that will be the most error laden, as high returns with little risk and low correlation are not an equilibrium proposition but "free lunches." This is the economic basis of the "error maximization" argument and it is one of the most serious objections to portfolio optimization of practitioners and academics alike. We will not attempt to review the vast literature on how to best deal with estimation error ranging from Bayesian statistics to robust statistics and robust portfolio optimization. However, we need to make the point that not all risk measures are equally sensitive to estimation error.[11] How does the estimation error in *CVaR* compare to alternative risk measures (Figure 12.4)?

We employ a simple bootstrapping exercise, where we take the returns for an arbitrary sector (here oil) sample downside and dispersion risk measures 1,000 times and plot the distribution of percentage deviations from the sample risk measure (which serves as the true risk measure). Exhibit 4 summarizes our results. Downside risk measures like value at risk (*VaR*), *CVaR*, and semivariance (*SV*) are many times more sensitive to estimation error than dispersion measure like mean–absolute deviation (*MAD*) or volatility (*VOL*). This should not be surprising given that dispersion measures are symmetric risk measures that use all available return information. Downside risk measures on the other hand use at best half of the information in the case of semivariance and only extreme returns for value at risk and *CVaR*. *CVaR* in particular is very sensitive to a few outliers in the tail of the distribution.[12]

[11] After all, this is the foundation of robust statistics, i.e. not all statistics have the same sensitivity to outliers.

[12] Note that we can estimate the minimum variance portfolio without having to rely on return expectations. Whatever return expectations investors have, they will not change the minimum variance portfolio. This does not apply to the minimum *CVaR* portfolio, the higher the return, the lower the conditional value at risk.

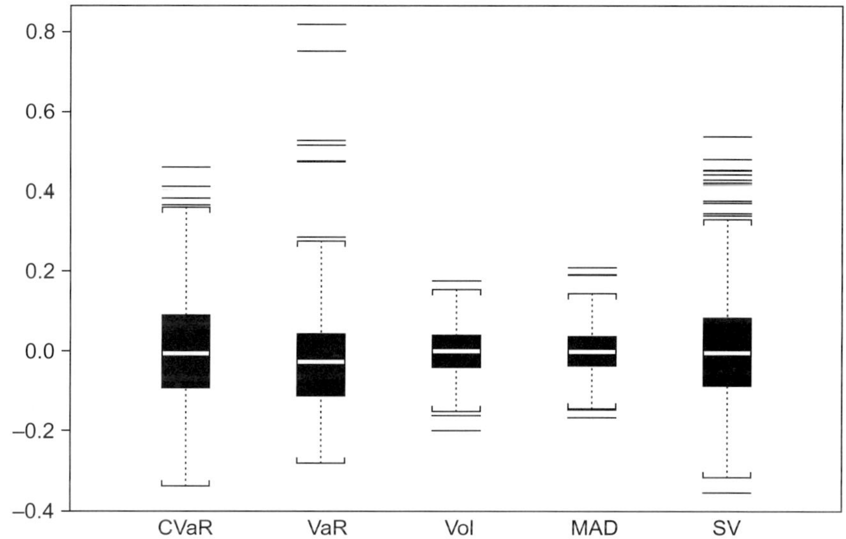

Figure 12.4 Estimation error for alternative risk measures. Original data for the Oil & Gas return series are bootstrapped 1,000 times and risk measures are calculated from each resampling. The distribution of estimated risk measures is summarized in box-plots.

12.4.2 Approximation error

We have seen in Section 12.2 that *CVaR* optimization is essentially scenario optimization, which models variability in returns by simulating (or bootstrapping if done nonparametrically) a large numbers of scenarios. Once scenarios are drawn, uncertainty is essentially removed and we optimize a deterministic problem. The reliability of its solution depends on its ability to approximate a continuous multivariate distribution from a discrete number of scenarios. The difficulty to do this increases with the required *CVaR* confidence level (the further we go into the tail) and the number of assets involved (the number of conditional tails that need to be estimated). This is why *CVaR* optimization is usually applied at an asset-allocation level rather than on a large portfolio of individual stocks. Interestingly, this has not been widely addressed in the finance literature, while it is well known in the stochastic optimization literature. We will engage in a simulation exercise to raise this point more clearly. To isolate approximation error, we use the following two-step approach. First, we estimate a variance–covariance matrix from historical data.

Second, we simulate 240, 480, 1,200, and 2,400 return draws (assuming normality) and adjust the generated scenarios to match the estimated mean return and variance with the original data for each asset. This step is repeated 1,000 times and the optimized portfolios (the return target is halfway between minimum and maximum sector return, i.e., we should always arrive in the

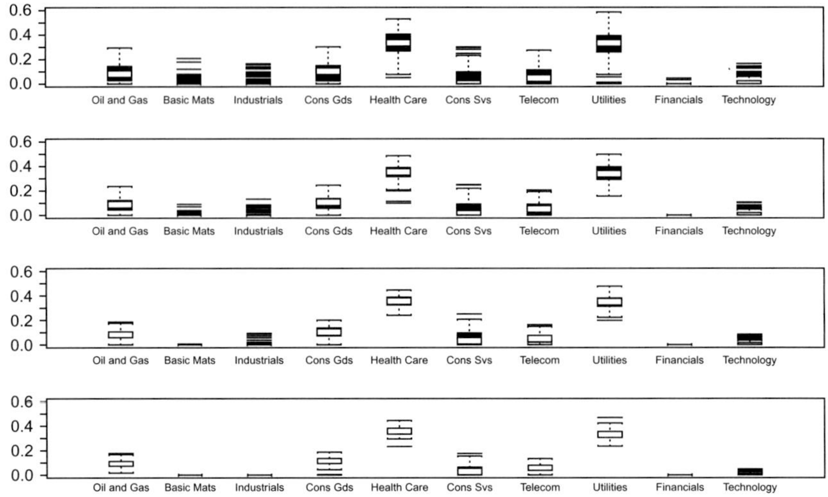

Figure 12.5 Approximation error for alternative size of scenario matrices. The variability of optimal weights is described in the above box-plots. The top panel uses 1,000 optimizations with 240 generated scenarios (to approximate a 10-asset problem), while the last panel uses 2,400 generated scenarios. The intermediate panels use 480 and 1,200 scenarios respectively. Approximation error (measured as the indicated dispersion shown in the above box-plots) falls with an increasing number of employed scenarios.

"middle" of the efficient frontier) are visualized in below box-plots provided in Figure 12.5.[13] Given that we employ a 10-asset problem only, it becomes clear that even for small problems we cannot rely on historical data. Note that even for 2,400 monthly returns (which equals 200 years of data) in the bottom panel, the estimation error is still far from negligible. Not only do we not have the luxury of such data, but even if we did we would find these data unlikely to be usable due to nonstationarity. Given the above simulation results, the minimum advice that needs to be given is to limit the application of *CVaR* optimization to problems with a small set of assets. In any case, investors should check the accuracy of their results with alternative samples for the scenario matrix to estimate the impact of approximation error. They should ask: *How many scenarios do we need to generate to get the approximation error down to a predefined level?* A *CVaR* optimization based on a nonparametric historic scenario matrix (in other words, a download of historical return series from a data provider) is likely to disappoint due to its large in-built approximation error, even if there was no estimation error at all.

[13] Note that our setup (normal scenarios) offers no advantage over mean–variance investing; we just build an extremely slow and imprecise mean–variance optimizer.

12.5 Scenario generation II: Conditional versus unconditional risk measures

We start with the most obvious way to build a scenario matrix, i.e., we build it from historical returns. Under the assumption of stationary returns (means and covariances do not change over time), each observation period (e.g., a month) corresponds to a scenario. Think of it as downloading data from a data provider and arranging them in a spreadsheet, where rows correspond to scenarios and columns to assets. With this approach, we will not only face considerable approximation error and will need to limit ourselves to a few assets, but we have essentially chosen an unconditional and nonparametric approach to scenario generation: unconditional, because we do not make our views of future risk conditional on risk factors or a particular (e.g., most recent) market environment, and nonparametric, because we do not impose any assumptions on return distributions. The nonparametric nature of the historical approach (and its limited ability to generate a sufficient number of scenarios for all but extremely small problems) can easily be dealt with.[14] We could, for example, select the best fitting distributions for each marginal asset distribution and "glue" these distributions together using a copula function of your choice. Given that we can now draw a large number of scenarios from this setup, we can make the approximation error very small. However, what remains is the unconditional nature of the employed distribution. Sudden shifts in volatility regimes cannot easily be married with nonnormal return distributions. One of the few methods used by practitioners to overcome this issue is combining GARCH models with measures of nonnormality. First, we fit a GARCH model to a single return series. This model will provide us with a series of standardized residuals that hopefully lost their autocorrelation in squared returns. However, these residuals might still exhibit serious deviations from normality. We can now use the forecasted volatilities from our GARCH model (which will be very responsive to recent market events) to scale our residuals (which contain information about nonnormality of returns) up or down. This represents a nonparametric way to deal with deviations from normality but nothing stops us to here to fit a nonnormal distribution or an extreme value model to our residuals. This allows us to simulate a large number of scenarios for each asset combining risk updates conditional on the current environment with an unconditional interpretation of nonnormality. Of course, we could deal with both issues by calculating the risk neutral probability density function from option markets (capture the conditional nature of the return distribution completely, i.e., dispersion and nonnormality). Marginal distributions can then be "tied

[14] One might be tempted to use bootstrapping multiperiod returns (monthly from weekly data) for scenario generation to construct a large number of scenarios. Given the well-known problems to maintain the original nonnormality in bootstrapped data, this does not look like a viable idea. Bootstrapping will eventually create normal returns (central limit theorem) and, hence, leave *CVaR* optimization with no advantage over mean–variance alternatives.

together" using copula functions. Owing to the sophistication required for this approach, almost all *CVaR* models used in practice work with the unconditional distribution. The year 2008 surely taught us that this is not a good idea. Finally, users of *CVaR* should also be aware that we have no established literature on building multivariate nonnormal predictive distributions.[15] This is a major disadvantage relative to the well-developed literature on Bayesian methods in combination with multivariate normal distributions, in particular, since *CVaR* is even more sensitive to estimation error than variance.

12.6 Axiomatic difficulties: Who has *CVaR* preferences anyway?

Let us first recall the definition of *CVaR*. For a confidence level of, for example, 95%, we simply average the 5% worst-case returns for a given portfolio and scenario matrix to arrive at *CVaR*. However, averaging worst-case returns (i.e., giving them equal weights) essentially assumes that an investor is risk neutral in the tail of the distribution of future wealth. *The fact that* CVaR *attaches equal weight to extreme losses is inconsistent with the most basic economic axiom used in our very first (micro) economics class: investors prefer more to less at a decreasing rate. As a corollary, they do certainly not place the same weight (disutility) on a large loss and total ruin.* Although *CVaR* is a coherent risk measure (it ticks all boxes on statistical axioms), it does fail well-accepted economic axioms we all have accepted in our first microeconomics class. We could, of course (and some have done that[16]), introduce utility functions that are linear below a particular threshold value to technically conform to expected utility maximization.

Where do we go from here? Are we stuck in a dead end? Recognizing the shortcomings of *VaR* and *CVaR*, Acerbi (2004) introduced the so-called spectral risk measures as the newest innovation in the risk manager's toolbox. Spectral risk measures attach different weights to the *i*th quantile of a distribution. They are coherent risk measures as long as the weightings the quantile receives are a nondecreasing function of the quantile. In other words, the 96% quantile must get at least the same weighting than the 95% quantile. This is in stark contrast to *VaR* where, for example, the 95% is assigned a weighting of 1, whereas all other quantiles get a weight of 0. *CVaR*, however, is coherent (attaches equal weight, i.e., nondecreasing weight to all quantiles above *VaR*). It still has the unpleasant property that investors evaluate losses larger than

[15] With the exception of multivariate mixtures of normal distributions that are difficult to estimate and even more demanding to put informative priors on (we need priors for two covariance matrices and two return vectors).

[16] Kahneman and Tversky (1979) and their Nobel prize–winning work focused on deriving utility functions from experiments. This is somehow odd, as the scientists' role is to provide normative advice and guide individuals to better decision making rather than "cultivating" their biases.

VaR with their expected value. This risk neutrality in the tail is not plausible at all. Spectral risk measures can help here. We could, for example, suggest a weighting function like:

$$\phi(p) = \frac{\lambda e^{-\lambda(1-p)}}{1-e^{-\lambda}} \tag{12.11}$$

with $\lambda > 0$, as part of the definition of the spectral risk measure:[17]

$$M_{\phi}(X) = \int_0^1 \underbrace{\phi(p)}_{\substack{weighting \\ function}} \underbrace{F_X^{-1}(p)}_{\substack{loss \\ quantile}}(p)dp \tag{12.12}$$

Higher losses (larger values of p) get larger weights and weightings increase even further if λ increases. Spectral risk measures allow us to include risk aversion in the risk measure by allowing a (subjective) weighting on quantiles. Interpreting $\lambda > 0$ as the coefficient of absolute risk aversion from an exponential utility function, we (re)introduced utility-based risk measures through the backdoor.

We calculate Equation (12.12) via numerical integration for alternative risk aversion coefficients and distributions. Table 12.1 provides an example. We see that spectral risk measures explode much faster for an increase in risk aversion if the underlying returns follow a "*t*" rather than a normal distribution. In other words, the possibility of tail events has an amplifying effect on the risk measure depending on risk aversion (weighing function). Put more bluntly, spectral risk measures offer utility optimization in disguise. This concludes a rather unproductive academic research cycle. After 50 years of research, we are back to expected utility maximization. While Markowitz (1952) tried to approximate the correct problem, many of his followers were getting further

Table 12.1 Spectral risk measures. We numerically integrate Equation (12.12) to arrive at values for our spectral risk measure with weighting function (12.11). Under a fat-tailed *t*-distribution, spectral risk measures explode much faster for an increase in risk aversion than under a normal distribution

Risk aversion λ	Normal distribution	*t*-distribution
1	0.27	3.8
5	1.08	18.26
10	1.50	34.69
25	1.95	79.69
100	2.51	274.79

[17] This is just one example based on the exponential utility function.

away from solving the original problem by solving related but yet different problems. Each of their solutions showed economic and sometimes statistical shortcomings that could not be reconciled with basic economic axioms. While Acerbi (2004) tried to fix many of theses issues, he really (willingly or unwillingly) reinstated expected utility maximization. Science works, but sometimes it works very slowly. Given these implications, it is of little surprise that spectral risk measures had yet little impact on both the theoretical as well as the applied literature. After all, investors are well advised to read the original literature of the 1950s, or its more recent reincarnations like Kritzman and Adler (2007). There is no need for a new framework. Expected utility maximization is the route to follow. The implementation problem will be identical, but the conceptional flaw of *CVaR* can be avoided. Computationally, we can always piecewise linearize a given utility function and use linear programming technology.

12.7 Conclusion

We split our critique on *CVaR* into implementation and concept issues. While implementation issues can be overcome at the cost of sophisticated statistical procedures that are not yet widely available, they pose a strong objection against current naïve use of *CVaR*. Estimation error sensitivity amplified by approximation error and difficulties in modeling fast updating scenario matrices for nonnormal multivariate return distributions will stop many practitioners from applying *CVaR*. More limiting in our view, however, is the inability of *CVaR* to integrate well into the way investors think about risk. Averaging across small and extremely large losses, i.e., giving them the same weight, does not reflect rising risk aversion against extreme losses, which is probably the most agreeable part of expected utility theory.

Acknowledgment

I thank K. Scherer for many valuable suggestions. All errors remain mine.

References

Acerbi, C. (2002). Spectral measures of risk: A coherent representation of subjective risk aversion. *Journal of Banking and Finance*, *26*, 1505–1518.

Artzner, Ph., Delbaen, F., Eber, J.-M., & Heath, D. (1997). Thinking coherently. *RISK*, *10*, 68–71.

Kahnemann, D., & Tversky, A. (1979). Prospect theory: An analysis of decision under risk. *Econometrica*, *47*(2), 263–291.

Kritzman, M., & Adler, T. (2007). Mean variance versus full scale optimisation: In and out of sample. *Journal of Asset Management*.

Markowitz, H. (1952). Portfolio selection. *Journal of Finance*, 7, 77–91.
Markowitz, H., & Levy, H. (1979). Approximating expected utility by a function of the mean and variance. *American Economic Review*, 69, 308–317.
Neumann, J., & Morgenstern, O. (1944). *Theory of games and economic behavior*. Princeton University Press.
Scherer, B. (2007). *Risk budgeting and portfolio construction* (3rd ed.). London: Riskwaters.
Scherer, B., & Martin, D. (2005). *Modern portfolio optimization with NuOPT*. New York: Springer.
Taleb, N. (2005). *The black swan*. Random House.
Tasche, D. (1999). *Risk contributions and performance measurement*. Working Paper.
Tasche, D. (2007). *Capital allocation for credit portfolios with kernel estimators*. Working Paper.
Uppal, R., DeMiguel, V., Garlappi, L., & Nogales, F. (2008). *A generalized approach to portfolio optimization: Improving performance by constraining portfolio norms*. SSRN Working Paper.

Index